DEVELOPMENTS IN AGRICULTURAL
AND MANAGED-FOREST ECOLOGY, 12

plant production and management under drought conditions

OTHER TITLES IN THIS SERIES

1. Plant Modification for More Efficient Water Use
   by J.F. Stone (Editor)
   1975 x + 320 pp.

2. Tree Ecology and Preservation
   by A. Bernatzky
   1978 viii + 358 pp.

3. Cycling of Mineral Nutrients in Agricultural Ecosystems
   by M.J. Frissel (Editor)
   1978 viii + 356 pp.

4. Interactions Between Non-Pathogenic Soil Microorganisms and Plants
   by Y.R. Dommergues and S.V. Krupa (Editors)
   1978 xii + 476 pp.

5. Ecology of Root Pathogens
   by S.V. Krupa and Y.R. Dommergues (Editors)
   1979 x + 282 pp.

6. Soil Disinfestation
   by D. Mulder (Editor)
   1979 xiv + 368 pp.

7. Management of Semi-arid Ecosystems
   by B.H. Walker (Editor)
   1979 x + 398 pp.

8. Management and Analysis of Biological Populations
   by B.S. Goh
   1980 x + 288 pp.

9. Comparison of Forest Water and Energy Exchange Models
   by S. Halldin (Editor)
   1979 xi + 258 pp.

10. Alternative Methods of Agriculture
    by R. Boeringa (Editor)
    1980 vi + 200 pp.

11. Wetlands of Bottomland Hardwood Forests
    by J.R. Clark and J. Benforado (Editors)
    1981 xviii + 402 pp.

DEVELOPMENTS IN AGRICULTURAL
AND MANAGED–FOREST ECOLOGY 12

plant production and management under drought conditions

Papers presented at the Symposium, 4–6 October 1982, held at Tulsa, OK, U.S.A.

Edited by

JOHN F. STONE
Department of Agronomy, Oklahoma State University, Stillwater, OK 74708, U.S.A.

WAYNE O. WILLIS
Agricultural Research Service, U.S. Department of Agriculture, Fort Collins, CO, U.S.A.

ELSEVIER SCIENCE PUBLISHERS B.V.
Amsterdam — Oxford — New York — Tokyo 1983

ELSEVIER SCIENCE PUBLISHERS B.V.
Molenwerf 1
P.O. Box 211, 1000 AE Amsterdam, The Netherlands

Distributors for the United States and Canada:

ELSEVIER SCIENCE PUBLISHING COMPANY, INC.
52 Vanderbilt Avenue
New York, NY, 10017

Reprinted from *Agricultural Water Management,* Vol. 7, Nos. 1–3, pp. 1–390.

Library of Congress Cataloging in Publication Data
Main entry under title:

Plant production and management under drought conditions.

(Developments in agricultural and managed-forest ecology ; 12)
Sponsored jointly by the Oklahoma State University Center for Water Research, and others.
1. Droughts--Congresses. 2. Plants, Effect of drought on--Congresses. 3. Crops and climate--Congresses. 4. Plants--Drought resistance--Congresses.
I. Stone, John F. II. Willis, Wayne O., 1928-
III. Oklahoma State University. Center for Water Research. IV. Series.
S600.7.D76P53 1983 631.5'8 83-11631
ISBN 0-444-42214-5 (U.S.)

ISBN 0-444-42214-5 (Vol. 12)
ISBN 0-444-41515-7 (Series)

Printed in The Netherlands

CONTENTS

PREFACE

This publication represents the continuing interest in crop production of a group of scientists in the Great Plains of the United States. A common agricultural factor over the Great Plains is limited water for plant production in the subhumid to semiarid environment. Crop production under drought or variable water stress conditions is a continuing problem. The Great Plains Agricultural Council serves to focus the interests of scientists and agencies towards agricultural problems in the west central portion of the United States. The Research Committee of the Great Plains Agricultural Council has several research and coordinating committees. Two of these are GPC-1, Evapotranspiration; and GPC-7, Remote Sensing. These committees were involved in the publication of a treatise entitled *Plant Modification for More Efficient Water Use*. The publication was a result of a symposium held in 1973 in Stillwater, OK, and was published as Volume 1 in the Elsevier Series on Developments in Agriculture and Managed-Forest Ecology. In the intervening decade, advances in the subject prompted GPC-1 and GPC-7 to organize a symposium which would focus on the extension of such information to crop management and production under drought conditions and to highlight research needs for future development of the management.

J.F. Stone and W.O. Willis were named co-chairmen of the symposium committee. There were five subcommittees for subject matter areas. The chairmen and members of the program advisory committees were *Weather Modification*, J. Schuelein (Chairman), B.L. Blad, J.R. Barnett, D. Eddy; *Soil Plant and Air Effects*, B.L. Blad (Chairman), E.T. Kanemasu, R.J. Reginato, P.W. Unger; *Plant Breeding and Genetics*, J.R. Welsh (Chairman), K.F. Schertz, V.A. Johnson; *Physiology of Stress*, W.R. Jordan (Chairman), C.Y. Sullivan, D.R. Krieg, R.M. Castleberry; and *Remote Sensing*, J.L. Heilman (Chairman), C.L. Wiegand, E.L. Maxwell and E.T. Kanemasu. J.R. Johnston and C.H. Schmidt were Administrative Advisers for CPC 1 and GPC 7, re spectively. The symposium was sponsored jointly by the Oklahoma State University Center for Water Research, N.N. Durham (Director); Oklahoma Water Resources Board, J.R. Barnett (Director); Oklahoma Agricultural Experiment Station, C.B. Browning (Director); and the Great Plains Agricultural Council, L. Lucas (Chairman).

The symposium papers assembled for this special publication were reviewed and edited by the section chairmen and the two symposium cochairmen. Authors were asked to consider the present and future technology for management of plant and environmental factors for plant production under drought conditions in the Great Plains. In addition, the concerns of economics and energy were to be kept in perspective. The intent was to emphasize management strategy and to present information for user orientation. This publication is dedicated to those who will find it of use and subsequently utilize the information in crop production or as a convenient point of departure for further research.

March 1983
JOHN F. STONE
Department of Agronomy, Oklahoma State University
Stillwater, OK 74078, U.S.A.

WAYNE O. WILLIS
Agricultural Research Service, U.S. Department of Agriculture
Fort Collins, CO, U.S.A.

REFERENCE

Stone, J.F. (Editor), 1975. Plant Modification for More Efficient Water Use. Developments in Agricultural and Managed-Forest Ecology, Vol. 1. Elsevier, Amsterdam/Oxford/New York, x + 320 pp. Reprinted from Agric. Meteorol., Vol. 14, No. 1/2 (1974).

Introduction to Part 1

WEATHER MODIFICATION

While the research community labors to place cloud seeding on a secure scientific foundation, some people with major weather interests at stake have confidence in the technology and are already investing in operational programs; in fact, some have been doing so for over 30 years. The need for more rain and less hail has created a demand for cloud seeding services, even in the face of serious and sometimes outspoken scientific doubt about the basis for it.

Although the current concepts of cloud seeding are much more complex and the operational requirements much more demanding than a decade ago, cloud seeding projects are perceived as good investment by many users. All indications are that operational cloud seeding will be used more and over wider areas in the future.

In the 1978 report to the U.S. Secretary of Commerce, the Weather Modification Advisory Board indicated the overriding problem for operational could seeding programs has been that, with the exception of a few cases, the predictability of the results of seeding is not to the point where operations can be conducted with confidence.

The papers that follow address a number of issues of concern to organizers of operational programs including: results of the latest round of field experiments on the augmentation of summer rainfall from convective clouds in the northern Great Plains of the United States; societal and environmental factors; and various interacting factors that should be considered as a part of an effective evaluation. Finally, the approach utilized by one state government to develop an operational cloud seeding program incorporating these features is discussed.

JAMES W. SCHUELEIN
Oklahoma Water Resources Board
Oklahoma City, OK, U.S.A.

Agricultural Water Management, 7 (1983) 3—14
Elsevier Science Publishers B.V., Amsterdam — Printed in The Netherlands

AUGMENTATION OF RAINFALL FROM SUMMER CUMULUS CLOUDS

A.S. DENNIS

Division of Atmospheric Resources Research, Bureau of Reclamation, U.S. Department of the Interior, P.O. Box 25007, Denver, CO 80225 (U.S.A.)

(Accepted 5 February 1983)

ABSTRACT

Dennis, A.S., 1983. Augmentation of rainfall from summer cumulus clouds. *Agric. Water Manage.*, 7: 3—14.

The latest round of field experiments related to augmentation of rainfall from summer cumulus clouds over the Great Plains is described briefly. HIPLEX (High Plains Cooperative Program) was designed to reduce scientific uncertainty associated with attempts to increase rainfall by cloud seeding in the High Plains. Field programs were conducted at three sites. A randomized experiment, HIPLEX-1, was set up in Montana to test a physical hypothesis linking dry ice seeding to rainfall from cloud base.

HIPLEX has confirmed the importance of ice processes to precipitation formation in cumulus clouds of the High Plains and the dominance of mesoscale convergence/divergence fields on convective cloud development. HIPLEX-1 clearly showed increases in cloud ice concentrations associated with dry ice seeding 2 and 5 min after treatment time. Some changes in the subsequent steps in the physical hypothesis linking seeding to increased rainfall at cloud base are indicated but cannot be firmly established with the present data set.

INTRODUCTION

Weather modification has been used as a weapon against drought in the Great Plains of the United States for over 30 years. Many of the projects have involved attempts to increase rainfall by seeding summer cumulus clouds. Despite frustration because of the lack of clearcut evidence of rainfall increases, the attempts have continued because of the perception that even small increases in rainfall would reduce significantly the impact of drought.

It is sometimes asked whether or not enough clouds exist during droughts to make seeding them practical. Certainly one could limit the impact of a drought by seeding the marginal conditions that exist before and after the drought and around the edges of the drought-stricken area. Furthermore, climatological data indicate that, even in the middle of a severe drought, enough clouds exist to justify attempts at seeding them (Huff and Semonin, 1975).

The conclusion just stated is based on the assumption that rainfall increases can be produced by seeding cumulus clouds. Seeding experiments designed to test that assumption have been conducted over the Great Plains and the Mississippi Basin for over 30 years (Coons and Gunn, 1951; Braham et al., 1957; Panel on Weather and Climate Modification, 1966, 1973; WMAB, 1978a, b). The experiments have increased in complexity, with each round of experiments featuring improved instrumentation, data handling, and cloud modeling work as compared to its predecessors. However, the physical hypotheses tested by the experiments have not changed much. Coons and Gunn (1951) considered speeding the coalescence of cloud droplets into raindrops by waterspray or hygroscopic agents and the initiation of the Bergeron Process in supercooled clouds by seeding them with dry ice or silver iodide. The Bergeron Process involves the growth of ice particles into precipitation embryos because of the difference in equilibrium vapor pressure over ice and over supercooled water surfaces, which causes simultaneous evaporation from supercooled droplets and deposition of water vapor on the ice particles.

This paper describes the results of recent experiments designed to test some of the same physical hypotheses.

HIPLEX

The most recently completed program on the modification of cumulus clouds in the Central United States is HIPLEX. HIPLEX was undertaken by the Bureau of Reclamation in 1973 in response to a directive from the Office of Management and Budget. The general objective of HIPLEX was to reduce scientific uncertainties associated with attempts to increase rainfall by cloud seeding in the High Plains region. HIPLEX was a cooperative venture between the Federal Government and State governments. The Canadian Government, universities, and private firms also participated. HIPLEX included economic and environmental studies in addition to work on methods for modifying cumulus clouds.

Fieldwork was conducted in Texas, Kansas and Montana. HIPLEX made extensive use of newly available equipment and techniques. Notable among them were PMS (particle measuring systems) probes on aircraft for measuring cloud and precipitation particles, computers on aircraft for real-time analysis of cloud situations, recording of radar data on tape, mesoscale networks of automated weather stations reporting via satellite relay stations, and advanced numerical cloud models for simulating seeding experiments in computers.

HIPLEX-1, a randomized experiment conducted in the Miles City, Montana area during the summers of 1979 and 1980, is of particular interest. It was designed to test the applicability of a specific physical hypothesis to isolated cumulus congestus clouds in the Miles City region. The hypothesis involved the formation of ice crystals by seeding with dry ice, growth of some ice crystals by deposition and riming into graupel, and fallout of the graupel particles (perhaps already melted into raindrops) from cloud base (Silverman, 1980). This physical hypothesis, which involves a specific version of the Bergeron Process, was adopted on the basis of several years' work which showed that Montana clouds did not have characteristics generally considered conducive to success in hygroscopic seeding (e.g., Cotton, 1982), and that dry ice was a satisfactory seeding agent for forming ice crystals in them at temperatures between -10°C and 0°C (Holroyd et al., 1978).

The experimental unit for HIPLEX-1 was a single cumulus congestus cloud. The seeding agent was dry ice, which was dropped from a jet aircraft at a rate of 0.1 kg km^{-1} on passes through the cloud near the -10°C level, or at cloud top, whichever was lower (in elevation).

Test clouds were selected principally on the basis of data collected on a pretreatment pass by a King Air instrumented aircraft operating near the -8°C level. The target clouds are classified as types A1, A2 and B (Silverman, 1980). The differences in behavior between seeded and unseeded clouds were expected to be most readily detected in type A1 clouds and hardest to detect in type B clouds. Type A1 clouds were considered unlikely to precipitate naturally because of warm tops (-6 to -12°C) and a presumed absence of ice multiplication processes. Type A2 clouds were clouds in the same temperature range but in which an ice multiplication process was considered likely to occur. Type B clouds were considered likely to precipitate naturally, with or without ice multiplication, due to their relatively cold tops (mostly -12 to -20°C). Type B clouds were selected for experimentation only when no suitable type A clouds were available. The experimental unit was limited in time to 40 min, as observations and numerical modeling results have indicated that the hypothesized physical process should take place in less than 40 min. Flight procedures were worked out to ensure proper coordination between the seeding aircraft and the King Air instrumented aircraft, which systematically worked downward from the -8°C level to monitor the precipitation process. Details are given in 'The Design of HIPLEX-1' (BuREC, 1979).

The statistical design of HIPLEX-1 included double-blind randomization, the selection of ten primary response variables (Table I) to test the steps of the physical hypothesis, and the development of new statistical methods based on MRPP (multi-response permutation procedures) for testing for possible differences between seed and no-seed cases (Dennis et al., 1980). Simulation experiments indicated 50 to 100 test cases would be required for conclusive results to be obtained.

Only 20 test cases were recorded during the 2 years, 1979 and 1980, that HIPLEX-1 was in the field. However, the experimental design permitted the drawing of some useful conclusions from the truncated data set.

The following section, which summarizes present knowledge concerning aspects of augmentation of rainfall from cumulus clouds, draws upon the results of HIPLEX in general and of HIPLEX-1 in particular.

TABLE I

HIPLEX-1 primary response variables

1. CIC 2, Cloud ice concentration, 2 min after treatment
2. CIC 5, Cloud ice concentration, 5 min after treatment
3. CCR 5, Concentration of crystals rimed, 5 min after treatment
4. PIC 8, Precipitating ice number concentration, 8 min after treatment
5. MVD 8, Mean volume diameter of precipitating ice particles, 8 min after treatment
6. AWC 8, Average liquid water concentration, 8 min after treatment
7. TFPI, Time to first precipitating ice (particles with diameters > 0.6 mm in concentrations > $0.1\ l^{-1}$)
8. TFE, Time to first radar echo (15 dBZ)
9. (a) TIPA, Time to initial precipitation at +10°C level, aircraft measurement
   (b) TIPR, Time to initial precipitation at +10°C level, SWR-75 radar (15 dBZ)
10. (a) RERA, Radar-estimated rainfall at +10°C level, Z-R relationship adjusted by hydrometeor measurements
    (b) RERC, Radar-estimated rainfall at +10°C level, using a constant Z-R relationship
    (c) AER, Aircraft-estimated rainfall at +10°C level

CURRENT VIEWS ON POTENTIAL FOR RAINFALL AUGMENTATION

Convective processes, including rainfall, are dominated by synoptic and mesoscale systems which control cloud size and organization. Observations to support this statement have been reported from Florida (Ulanski and Garstang, 1978), Texas (Tex. Dep. Water Resour., 1982) and Montana (Miller et al., 1981). In general, convergence values of $5 \times 10^{-4}\ s^{-1}$ or more (in absolute terms) over several hundred square kilometers are required to produce a major convective storm. The mesoscale convergence is associated with air rising at rates approaching $0.5\ m\,s^{-1}$ in the lower troposphere. Matthews and Silverman (1980) have used a numerical model simulation to demonstrate how mesoscale lifting increases the ability of the atmosphere to support deep, moist convection.

Most convective rainfall is produced by clusters or lines of convective cells which require mesoscale support, rather than by isolated cells. Table II, based on Jurica et al. (1981), illustrates the point for west Texas. The small amounts of rainfall contributed by isolated convective clouds is attributable to the relatively small amounts of water vapor condensed in such clouds and their low precipitation efficiencies.

It has been suspected for a long time that isolated cumulus clouds are inefficient precipitation producers because they are dissipated by mixing with ambient dry air before the cloud droplets have sufficient time to coalesce into raindrops or be collected by solid hydrometeors (Braham, 1981). Evidence on this point compiled by Canadian researchers (Schemenauer et al., 1981) during experiments in Ontario and the Northwest Territories and during the HIPLEX experiments at Miles City is shown in Table III. The set of 'Miles City' clouds referred to in Table III is not the same as the set of

TABLE II

Distribution of convective summer rainfall in west Texas by radar echo type, 1978—80

Type of echo	Frequency of occurrence (%)	Rainfall contribution (%)
Cell	3	0.1
Small cluster	34	4.3
Large cluster	41	27.5
Nested cluster	19	67.9
Unclassified	3	0.2
Total	100	100.0

After Jurica et al. (1981).

TABLE III

Average lifetimes of cumulus clouds

	First pass liquid water concentration ($g\,m^{-3}$)	Lifetime to 0.1 $g\,m^{-3}$ (min)	Number of clouds
Yellowknife, N.W.T.			
1975	0.50	>20	8
1976	0.40	12	35
Thunder Bay, Ont.			
1977	0.88	8	15
1978	1.44	6	23
Miles City, MT			
1979	0.64	12	11
1980	0.34	3	10

After Schemenauer et al. (1981).

20 HIPLEX-1 clouds. However, the Canadian results on cloud lifetimes at Miles City are supported by detailed studies of the evolution of HIPLEX-1 clouds by Cooper (1981).

Table III shows that some clouds with significant liquid water concentrations on a first aircraft pass have less than 0.1 $g\,m^{-3}$ of supercooled cloud water left 5 min later. As the time required for the Bergeron Process to produce precipitation from ice crystals is 10 to 15 min, it is not surprising that many isolated cumulus clouds fail to produce precipitation. Table III also shows interesting variations in cloud behavior with location and from one summer to the next.

HIPLEX results have tended to confirm earlier findings concerning the initiation of natural precipitation in cumulus clouds. In the southern plains, where cloud-base temperature is often +10 to +15° C, precipitation is usually initiated by coalescence; however, major storms always involve the Bergeron Process and, possibly, hail formation in their later stages as the cloud tops rise far above the 0° C isotherm (Tex. Dep. Water Resour., 1982). In the northern plains, where cloud base temperatures are typically 0 to +10° C, precipitation generally forms through the Bergeron Process (Hobbs et al., 1980). These findings agree with earlier observations.

As noted above, the design of HIPLEX-1 provided a mechanism for monitoring links in the physical hypothesis linking dry ice seeding to precipitation in supercooled cumulus clouds.

A first look at the results of the 20 test cases indicates that dry ice seeding produced additional ice crystals (Table IV). Fig. 1, taken from Cooper et al. (1981), demonstrates clearly the marked difference in CIC2 that existed between seeded and unseeded clouds.

Table V shows *P*-values obtained from the MRPP analysis of the HIPLEX-1 primary response variables tabulated to date (P. Mielke, private communication, 1982). A *P*-value is simply the probability that the difference between the seed and no-seed cases for the variable and sets of cases in question arose entirely by chance. The *P*-values confirm that dry ice seeding produced changes in the concentrations of ice crystals at 2 and 5 min after treatment time in both type A1 and type B clouds. The pooled data also show a significant difference in the concentration of rimed crystals 5 min after treatment time.

TABLE IV

Average values of HIPLEX-1 primary response variables

Response variable	Average value	
	Unseeded	Seeded
CIC 2 (l^{-1})	0.2	15.8
CIC 5 (l^{-1})	1.1	20.8
CCR 5 (l^{-1})	0.5	45.2
PIC 8 (l^{-1})	0.02	0.05
MVD 8 (mm)	0.97	1.13
AWC 8 (g m^{-3})	0.03	0.04
TFPI (s)	790[a]	650
TFE (s)	1 750[a]	1 440[a]
TIPA (s)	2 110[a]	1 930[a]
TIPR (s)	2 210[a]	2 270[a]
AER (mm km^2)	0.37	1.23
RERC (mm km^2)	0.0015	0.32

[a] Average reflects inclusion in calculations of default values for non-precipitating clouds.

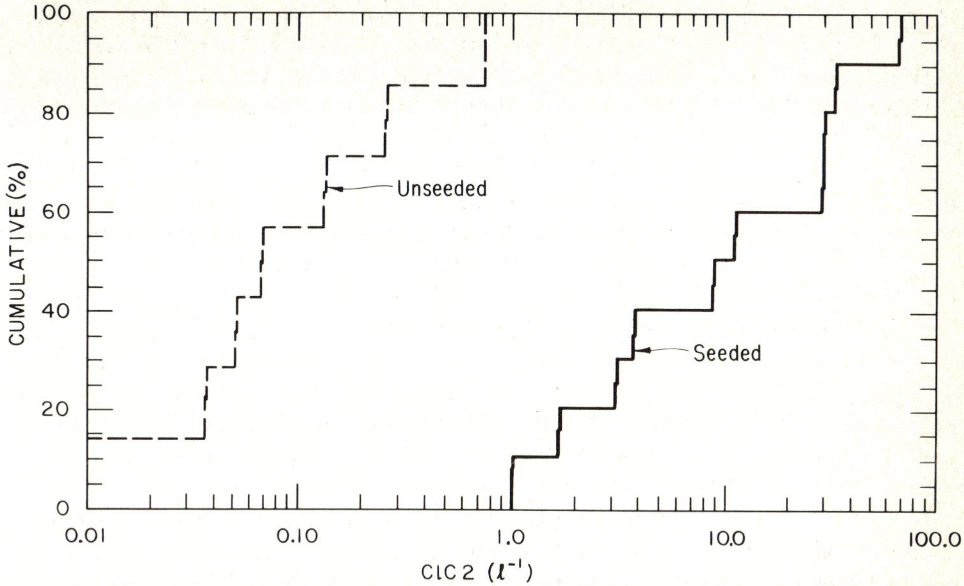

Fig. 1. Cumulative distribution of CIC 2, cloud ice concentration 2 min after treatment time, in unseeded and seeded HIPLEX-1 clouds (after Cooper et al., 1981).

TABLE V

P-values from application of MRPP to HIPLEX-1 primary response variables taken individually

Response variable	Type A1 cases (3 NS, 4 S)		Type B cases (5 NS, 8 S)		Type (A1 + B) cases (8 NS, 12 S)	
	Raw data	Ranks	Raw data	Ranks	Raw data	Ranks
CIC 2	0.057	0.029	0.039 6	0.010 9	0.009 39	0.000 34
CIC 5	0.057	0.029	0.079 3	0.063 7	0.038 2	0.011 4
CCR 5	0.257	0.143	0.065 3	0.102	0.043 8	0.017 3
PIC 8	0.143	0.229	0.428	0.523	0.289	0.460
MVD 8	0.400	0.429	0.278	0.249	0.267	0.195
AWC 8	0.343	0.314	0.922	0.832	0.405	0.467
TFPI	0.029	0.029	1.000	1.000	0.227	0.178
TFE	0.486	0.600	0.761	0.598	0.505	0.375
TIPA	0.743	0.600	0.864	0.810	0.622	0.441
TIPR	(1.000)[a]	(1.000)[a]	0.838	0.887	0.814	0.770
AER	1.000	1.000	0.110	0.196	0.578	0.529
RERC	(1.000)[a]	(1.000)[a]	0.487	0.674	0.334	0.712

[a]Parentheses indicate situations for which all the values of the response variable were equal, so that no separation between seed and no-seed cases can be determined.

Observations by Rodi (1981) and two-dimensional cloud modeling studies by Kopp et al. (1980) in support of HIPLEX have indicated that organized internal motions as well as turbulence must be invoked to explain the spread of seeding material or ice crystal plumes inside a convective cloud. Rodi (1981) found that the standard deviation of the plume width reached 1 km in about 1 000 s, in agreement with earlier findings. HIPLEX-1 seeded clouds had ice crystals well distributed throughout them by 10 min or so after treatment time. Unfortunately, while turbulence was distributing the ice crystals, it was also hastening dissipation of the smaller clouds by mixing ambient air into them.

There is some evidence that precipitating ice was more abundant in seeded than in unseeded clouds 8 min after treatment time (Fig. 2), but the difference is not statistically significant (Table V). The later links in the physical hypothesis relating seeding to rain at cloud base are not distinguishable in the HIPLEX-1 statistical results (Fig. 3, Tables IV and V). It had been expected, for example, that AWC 8, the cloud water concentration measured by the cloud physics airplane 8 min after treatment time, would show a decrease in the seeded clouds as the cloud water was taken up by the artificially produced precipitation embryos. However, no statistically significant difference was detected, apparently because mixing with dry air around the clouds, rather than depletion by growing ice crystals, dominated the evolution of the liquid water.

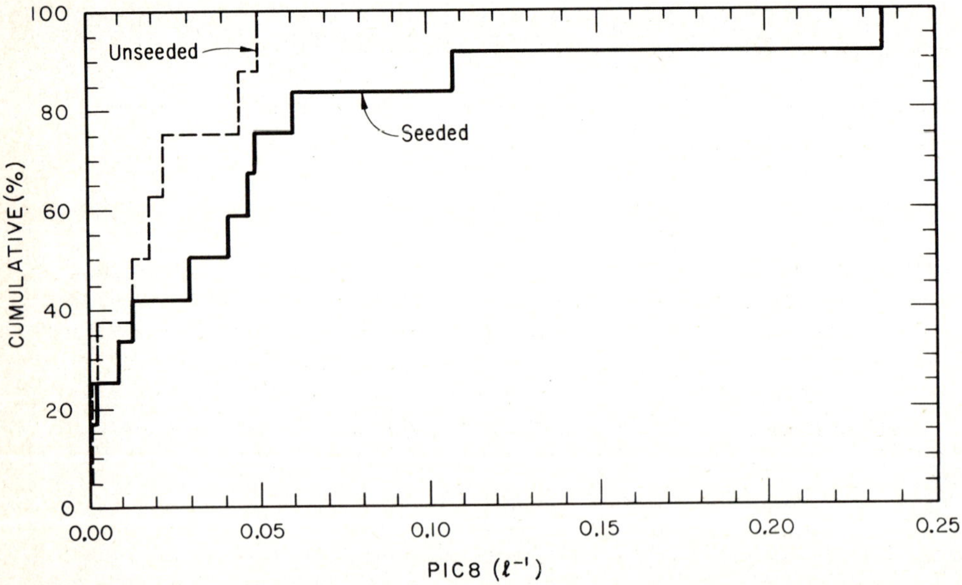

Fig. 2. Cumulative distribution of PIC 8, precipitating ice concentration 8 min after treatment time, in unseeded and seeded HIPLEX-1 clouds (after Cooper et al., 1981).

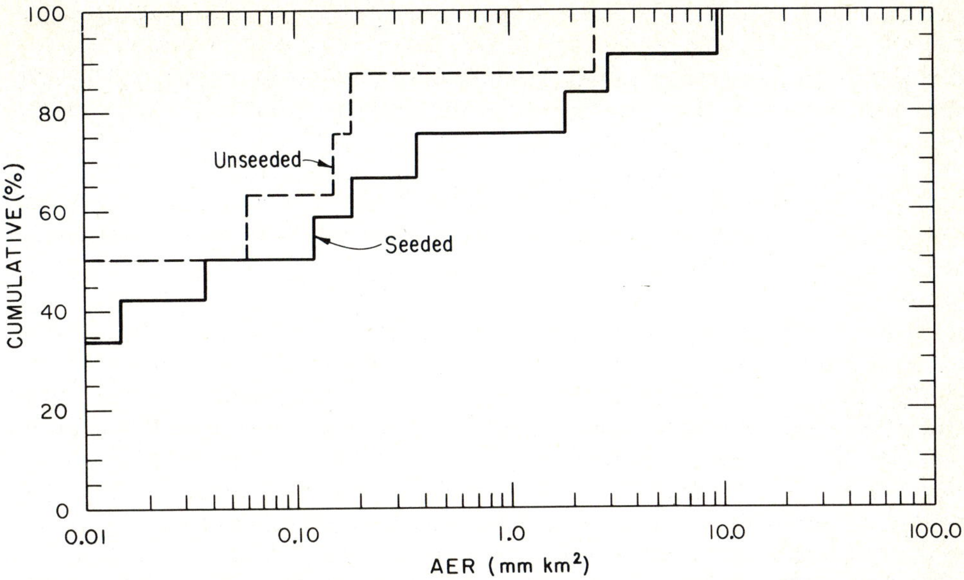

Fig. 3. Cumulative distribution of AER, aircraft estimated rainfall, in unseeded and seeded HIPLEX-1 clouds (after Cooper et al., 1981).

The appearance of natural precipitating ice at the sampling altitude in unseeded HIPLEX-1 clouds further masks differences between seed and no-seed cases. There is some evidence that the natural ice formed at higher levels and was carried down by downdrafts (Rodi, 1981). There is also evidence of differences in subsequent growth mechanisms, with aggregation of ice crystals predominating in seeded clouds and riming in unseeded clouds (A.B. Super, private communication, 1982).

FUTURE OUTLOOK

Although the HIPLEX-1 results have not demonstrated net rainfall increases due to seeding, they do support the physical hypothesis. HIPLEX-1 has shown beyond any doubt that dry ice seeding produces ice crystals in supercooled cumulus clouds and that some of the ice crystals grow to precipitation embryo size in about 5 min, as theory predicts. If the artificial embryos remained in a favorable growth environment, say in a multicelled cloud, they would grow into graupel or snowflakes. However, natural precipitation forms in large clouds in a time frame comparable to that for artificial precipitation, so the artificial embryos would face competition. Therefore, a seeding hypothesis for significant rainfall increases at the ground must be more complex than the straightforward one enunciated for HIPLEX-1. Some persons have suggested that dynamic effects related to latent heat release play a role in the total impact of seeding upon rainfall in the Great

Plains (Tex. Dep. Water Resour., 1982), but there is not enough evidence to evaluate their suggestions.

In any case, studies to identify criteria which would distinguish clouds that persist from clouds that do not would be worthwhile. We have already noted that mesoscale features control the organization and size of convective clouds and, hence, the rainfall they can produce. One encouraging finding from the Texas HIPLEX work is that the mesoscale features controlling convection can be identified. Rawinsonde networks with spacings of 200 km have been found adequate to locate mesoscale features associated with heavy convective outbreaks. The mesoscale features can be identified objectively by computer analyses using filter techniques (D. Matthews, private communication, 1982).

An ability to identify and follow mesoscale features opens up many possibilities in future cloud seeding experiments. For one thing, it would be possible to concentrate seeding experiments upon clouds in regions favoring further growth, rather than upon clouds destined to disappear a few minutes after their selection as test cases. The ability to measure mesoscale convergence would also provide an estimate of what the clouds would have done in the absence of seeding, thereby sharpening the statistical analysis.

Continued attention to numerical modeling, including simulation of seeding effects, should improve our understanding of possible physical mechanisms linking seeding to precipitation at the ground. Present indications are that the effect of seeding varies greatly among individual clouds, which is one likely reason why many statistical experiments have not been conclusive. It is not likely that two- or three-dimensional models could be run in real-time as an aid to cloud seeding operations, but it is reasonable to suppose that long-term modeling efforts will improve the physical hypotheses and thereby help in the design of future experiments.

ACKNOWLEDGEMENTS

The author wishes to express his appreciation to Dr. Bernard Silverman, Dr. Arlin Super and Mr. David Matthews of the Bureau of Reclamation for stimulating discussions and to Drs. Paul Mielke and Ken Berry of Colorado State University, for providing the *P*-values in Table V.

REFERENCES

Braham, R.R., Jr., 1981. Designing cloud seeding experiments for physical understanding. Bull. Am. Meteorol. Soc., 62: 55—62.

Braham, R.R., Jr., Battan, L.J. and Byers, H.R., 1957. Artificial nucleation of cumulus clouds. In: Cloud and Weather Modification. Meteorol. Monogr. 2(11), American Meteorological Society, Boston, MA, pp. 47—85.

BuREC, 1979. The Design of HIPLEX-1. Rep., Division of Atmospheric Resources Research, Bureau of Reclamation, U.S. Department of the Interior, Denver, CO, 271 pp.

Coons, R.D. and Gunn, R., 1951. Relation of artificial cloud-modification to the production of precipitation. In: T.F. Malone (Editor), Compendium of Meteorology. American Meteorological Society, Boston, MA. pp. 235—241.

Cooper, W.A., 1981. Characteristics of seeded and unseeded HIPLEX-1 clouds. In: Extended Abstracts, Eighth Conference on Inadvertent and Planned Weather Modification, October 5—7, 1981, Reno, NV. American Meteorological Society, Boston, MA, pp. 120—121.

Cooper, W.A., Holroyd, E.W., Middleton, J.R. and Super, A.B., 1981. Response variables of the HIPLEX-1 experiment. In: Extended Abstracts, Eighth Conference on Inadvertent and Planned Weather Modification, October 5—7, 1981, Reno, NV. American Meteorological Society, Boston, MA, pp. 118—119.

Cotton, W.R., 1982. Modification of precipitation from warm clouds — a review. Bull. Am. Meteorol. Soc., 63: 146—160.

Dennis, A.S., Brown, B.G., Smith, P.L., Jr., Sengupta, S., Mielke, P.W., Jr., Berry, K.J. and Brockwell, P.J., 1980. Statistical methods for evaluation of HIPLEX-1. In: Papers presented at the Third WMO Scientific Conference on Weather Modification, July 21—25, 1980, Clermont-Ferrand, France, Vol. 1. World Meteorological Organization, Geneva, Switzerland, pp. 189—196.

Hobbs, P.V., Politovitch, M.K. and Radke, L.F., 1980. The structure of summer convective clouds in eastern Montana. I: Natural clouds. J. Appl. Meteorol., 19: 645—663.

Holroyd, E.W., III, Super, A.B. and Silverman, B.A., 1978. The practicability of dry ice for on-top seeding of convective clouds. J. Appl. Meteorol., 17: 49—63.

Huff, F.A. and Semonin, R.G., 1975. Potential of precipitation modification in moderate to severe droughts. J. Appl. Meteorol., 14: 974—979.

Jurica, G.M., Haragan, D.R. and Leary, C.A., 1981. Investigations of summer convective cloud systems in the Texas High Plains. Tech. Rep. LP-174, Texas Tech University, Lubbock, TX, 205 pp.

Kopp, F.J., Orville, H.D., Farley, R.D., Hirsch, J.H., 1980. Numerical simulation of dry ice cloud seeding experiments in HIPLEX. In: Papers presented at the Third WMO Scientific Conference on Weather Modification, July 21—25, 1980, Clermont-Ferrand, France, Vol. 1. World Meteorological Organization, Geneva, Switzerland. pp. 197—203.

Matthews, D.A. and Silverman, B.A., 1980. Sensitivity of convective cloud growth to mesoscale lifting: A numerical analysis of mesoscale convective triggering. Mon. Weather Rev., 108: 1056—1064.

Miller, J.R., Jr., Doneaud, A.A., Priegnitz, D.L., Viswanath, L. and Engel, T., 1981. Surface mesoscale wind and moisture structures as potential storm predictors for a dry continental climate. In: Extended Abstracts, Eighth Conference on Inadvertent and Planned Weather Modification, October 5—7, 1981, Reno, NV. American Meteorological Society, Boston, MA, pp. J7—J8.

Panel on Weather and Climate Modification (G.J.F. MacDonald, Chairman), 1966. Weather and Climate Modification Problems and Prospects. Publ. 1350, National Academy of Sciences/National Research Council, Washington, DC, Vol. I, 28 pp.; Vol. II, 198 pp.

Panel on Weather and Climate Modification (T.F. Malone, Chairman), 1973. Weather and Climate Modification: Problems and Progress. National Academy of Sciences, Washington, DC, 258 pp.

Rodi, A.R., 1981. Study of the fine-scale structure of cumulus clouds. Rep. AS 131, University of Wyoming, Laramie, WY, 328 pp.

Schemenauer, R.S., Isaac, G.A., Strapp, J.W. and MacPherson, J.I., 1981. Aircraft assessment of weather modification potential. In: Extended Abstracts, Eighth Conference on Inadvertent and Planned Weather Modification, October 5—7, 1981, Reno, NV. American Meteorological Society, Boston, MA, pp. 108—109.

Silverman, B.A., 1980. HIPLEX-1: A convective rainfall enhancement experiment. In: Papers presented at the Third WMO Scientific Conference on Weather Modification, July 21—25, 1980, Clermont-Ferrand, France, Vol. 1. World Meteorological Organization, Geneva, Switzerland, pp. 165—172.

Tex. Dep. Water Resour., 1982. Texas HIPLEX summary report. Final report, Texas Department of Water Resources, Austin, TX, 176 pp.

Ulanski, S.L. and Garstang, M., 1978. The role of divergence and vorticity in the life cycle of convective rainfall. Part I: Observation and analysis. J. Atmos. Sci., 35: 1047–1062.

WMAB, 1978a. The Management of Weather Resources, Vol. I. Report to Secretary of Commerce, Weather Modification Advisory Board (H. Cleveland, Chairman). U.S. Government Printing Office, Washington, DC, 229 pp.

WMAB, 1978b. The Management of Weather Resources, Vol. II. Report to Secretary of Commerce, Weather Modification Advisory Board (H. Cleveland, Chairman). U.S. Government Printing Office, Washington, DC, 92 pp.

Agricultural Water Management, 7 (1983) 15—21
Elsevier Science Publishers B.V., Amsterdam — Printed in The Netherlands

SOCIETY'S INVOLVEMENT IN PLANNED WEATHER MODIFICATION

STANLEY A. CHANGNON, Jr.
Illinois State Water Survey, P.O. Box 5050, Station A, Champaign, IL 61820 (U.S.A.)
(Accepted 22 November 1982)

ABSTRACT

Changnon, S.A., Jr., 1983. Society's involvement in planned weather modification. *Agric. Water Manage.*, 7: 15—21.

Study of the societal and environmental aspects of planned weather modification have helped point to the issues and concerns that weather modification practitioners, state and federal officials, and supporters of weather modification programs must consider if the utilization of the precipitation modification technologies is to proceed in a meaningful way within our society. The understanding of these issues and how to address them is of particular importance in implementing and operating a state program. Social, economic, and legal impacts call for state involvement in project development, project funding, regulation and monitoring, research, and in public relations. The degree of state involvement varies according to the length of a project and the degree of certainty of the capability to enhance precipitation.

INTRODUCTION

Well-conducted precipitation modification operations are generally viewed as providing benefits in most states. Much of the motivation for precipitation enhancement has arisen out of local interests, often spawned in dry weather. The federal government has remained out of the encouragement, operational, and control roles, but has played a major role in the research and development of weather modification.

A variety of reports have dealt with the roles that local, state, and federal agencies can and should play to properly implement and then manage planned weather modification (Changnon, 1980). It seems appropriate to consider these ideas in light of the evolving use of precipitation enhancement by states as a means to increase water supplies. The socio-economic issues, environmental concerns, legal questions, and institutional arrangements related to purposeful weather change stand as potential problems that state water resource planners must consider.

SOME BASIC TRUTHS

To properly consider a state's policy options for precipitation modification, a few basic 'givens', or truths, should be kept in mind.

0378-3774/83/$03.00

First, rain modification is an environmental change beginning in the atmosphere and if successful, it is translated to man and the biosphere. Secondly, precipitation modification in most areas will produce impacts to the benefit or disbenefit of agriculture, and to a lesser extent for water supplies. In a few weather locales water for power generation is sought through weather modification.

Third, sizable weather changes, such as 50% increases in summer rainfall or 80% decreases in hail loss are unlikely events now and in the foreseeable future. The best realistic estimates for future, reliable modification technologies are 10 to 25% increases in rainfall under certain conditions, and possibly 30 to 50% reductions in hail loss.

Another interesting fact is that there is a prevailing perception that there has been a great deal of social controversy generated by weather modification; however, this is wrong. There have been a few controversies and these have been more widely publicized than widespread.

IMPACTS OF PRECIPITATION ENHANCEMENT VARY WITH TIME AND AREA AND WITH CAPABILITIES

The types of societal and environmental impacts that are derived from precipitation modification vary considerably based on changing time and space scales, as well as success of modification. For example, assessors of environmental impacts indicate that projects ranging in length from 1 to 5 years produce very different impacts, in many respects, from those projects that last 10 to 30 years. Similarly, impacts produced by precipitation modification in one county will vary greatly from those that would occur from modification of weather over half of a state.

Another 'scale' factor that related to the impacts of weather modification is the degree of success. That is, how sizable is the change in the weather being produced?

This obvious factor is also interrelated to another which deals with 'what is being changed'? Typically, projects are performed to increase rain or snow. However, if such modification of clouds and storms occurs in order to make rain, there may be effects on the production of hail and on the amount of clouds, wind-speeds, and frequency of lightning. All these changes need to be considered, whether they are done on purpose or accidentally in the modification effort. An example of how the effects of altered rain and hail can influence the economic outcome relates to statistics from northwest Kansas (Changnon et al., 1977). The 1975 value of the average net income harvested per ha with no weather change is US$10. A 50% hail decrease increases the income of $11.10 (a gain of $1.10).A 50% hail decrease plus a 10% rain increase (at the same time) would produce an income of $12.20 (a gain of $2.20). However, a 50% hail decrease also occurring with a 10% rain decrease would produce an income of $9.84, representing a loss of $0.16 per ha. Hence, the type of weather change produced will greatly affect the resulting economic impacts.

RANGE OF IMPACT

A better concept of how the impacts from planned precipitation modification projects (assuming some reasonable degree of success can be established) can vary spatially is revealed by examining some of the known impacts from small, short-duration projects, and comparing these with those from large area, long-duration projects.

Small short-term projects

Here, we talk about a project covering 250 to 5 000 km^2 typically achieving only moderate rain or hail changes; and having a relatively short duration (5 years or less). In general, research has shown that the impacts of such a project will occur only locally and generally be negligible. The producers of food (farmers) will get an increase in income and some improvement in the year-to-year stability of income. There will be a few minor effects, generally imperceptible, on local agri-businesses and monetary institutions. There wil be no other environmental impacts beyond those noted in the weather change. A social conflict could occur because of any one of several reasons, however. If scientific consensus over attainable modification remains uncertain, as it is now in many places, and there has been little community involvement in the decision process involved in the project, local interest groups may develop. Potential controversy may evolve by the end of the fourth or fifth year. Conflict potential is enhanced if decreases in rainfall are perceived, or if the area has two or more crops with very different water needs.

Large long-duration projects

Projects that cover large ($>$10 000 km^2) areas of a state and last for at least 5 years have evolved when modification capabilities have been reasonably well identified and/or state interest is very high. Such a project produces a much greater series of impacts, both in project area and beyond the project, than does the small short duration project.

Economically, such a program in a state like Illinois would lead to (or be part of) the spread of the technology's usage over several other states in the Midwest and the Great Plains (rain or hail modification). The advantages to the farmer in the modification area (which came early in the adoption period) will be largely lost in the large area project involving sizable segments of the producers. However the individual farmer's income will be more stable, but the ultimate gain of large scale usable technology will be to the US consumer. Farmers growing the same crops but in the areas without weather modification will be hurt. As the national production goes up, the prices likely will go down.

Environmentally, the impacts could be wide-ranging if the programs are addressing precipitation changes. Ecologists who have investigated weather modification believe that the small-scale, short duration projects will produce imperceptible environmental changes. However, they perceive that large-scale, long duration (greater than 5 years) projects will produce impacts that they really cannot predict. There is also uncertainty over the effects of the silver (silver iodite is often used as the cloud seeding material) on soils and microorganisms. Clearly, a measurable change in precipitation over a large area for many years changes the climate and will have some influence on the regional hydrology, and hence, on plants and certain animals. There are environmental unknowns that must be considered in big projects.

In the societal area, one can expect there to be local conflicts and ensuing legal battles in big area projects. A more definitive capability in weather modification will lead to more clear identification of losses (i.e., too much rain for some on a seeded day). Systems and institutions to compensate for valid loss claims must be developed. Mechanisms for public participation in decisions about the program must be developed, hopefully in a logical way, on the local and state scales. The region will have less outmigration along with a host of other lesser societal impacts.

Another area of impacts will relate to government activities. For example, the design, operation, and evaluation of the projects will certainly evolve as a governmental activity, or be under governmental control with private business performing the task. The scale of field operations will shift to large project areas each covering 25 000 to 40 000 km^2. The operations will likely be a very complex mixture of aircraft, radars, and other facilities. Major activities will develop in the operational, monitoring and evaluation area with a governmental system likely to evolve that includes a public information arm. A mixture of state and federal management is a likely outcome. Governmental handling of needed compensation systems, regulation, interstate compacts, and funding will develop.

FACTS TO CONSIDER

Regardless of the size or duration of state-operated or supported projects, or the technical certainty of weather modification, certain basic truths and needs will always exist. Several relate to state policy. The state may choose to ignore these (a policy option), but they will exist.

First, the weather changes, regardless of the success or size of the project, will not be detectable by the user public. A key for the wise use of new complex technology without social conflict is considered to be 'observability of the technology'. Because precipitation enhancement is not 'observable', there is an *absolute need for scientific and technical monitoring and evaluation of the projects* to assess the effects. Several states like Illinois, North Dakota, South Dakota, Colorado and Utah have performed such evaluations of in-state projects. However, in all instances, more in-depth monitoring of the operational projects is needed. Adequate meteorological monitoring and evaluation require the collection of quality data. For example, one needs: surface precipitation from dense networks and raingages; hail insurance data; detailed records of the modification operations showing (for each event) the reasons for seeding and where it was done; associated radar data; and satellite data.

The questions about this meteorological monitoring and assessment, then, are:

— Who performs these activities? (Credibility of those performing these assessments is important.)

— Who pays for this activity? (The federal government presumably should contribute, but will not assume the responsibility in the foreseeable future.)

— Who sees that the job gets done? (What state entity should be responsible for this activity?)

Environmental monitoring is also a recommended activity. Obviously, the above described atmospheric monitoring is the first stage of environmental monitoring that is done. However, monitoring of 'downwind weather effects' should be considered. The degree of monitoring of the second-order impacts on plants, animals, and water resources greatly depends on the size and duration of the project. Basically, available information suggests that environmental monitoring need not begin unless a project is expected to last more than 5 years. Questions then are: (a) who controls it; (b) who performs it; and (c) who pays for it?

A third area is *economic monitoring and evaluation.* Questions to be addressed include: 'Is the change detectable and at what level in crop yields, runoff, income variations, and activities of impacted business?' Again, the basic state policy questions are: (a) who performs it; (b) who pays; and (c) who manages it?

There are other key needs. Past experience has shown that there should be a *project information system* available to provide the results of project operations, as well as to describe for laymen the monitoring and evaluation results. The questions for such a program of public information are: (a) who pays; (b) what is the state's role; and (c) who does it? So far in most states, this has been done on an ad hoc basis, largely based on information supplied locally by the project sponsors and occasionally general information coming from groups like the Cooperative Extension Service.

Past planned weather modification projects lasting 4 or 5 years (or more) have shown that mechanisms which produce community interaction and participation in the program are needed. It is particularly important to get community involvement (not just sponsors) in on the decisions to proceed or stop (temporarily or permanently) the project. It is important to realize that a precipitation enhancement program operated by a state is a community or public decision, not just a series of individual decisions. Obtaining such community (regional) interactions raised several questions like:

— Who organizes such efforts?

— Who monitors public attitudes?

— Who develops an extension committee? (The extension service?)

— Who pays for the activities?

— Who and how does someone respond to valid claims of loss? (Are the courts enough?)

TRANSLATING NEEDS TO STATE POLICIES

The societal and environmental impacts from planned weather modification and their call for involvement of the local, state, and federal government units, reflect on a need for state policy. It is clear that states need a law that controls and monitors cloud seeding projects. The Council of State Governments has offered a model law (Counc. State Gov., 1979), and the means whereby a control law was enacted in Illinois is a useful model (Ackermann et al., 1974). The federal government requires only the filing of project records.

There must be great care exercised in awarding permits for weather modification projects, particularly in areas of different cropping patterns where different weather-crop interactions exist, when hail and rain changes are uncertain, or when droughts develop.

The state also must consider and ultimately define its role in managing, performing, and paying for the recommended monitoring and evaluation of: weather changes; public attitudes; environmental impacts; economic impacts.

These are activities that must be done on the state level, at least in the foreseeable future, and possibly through existing state regulatory mechanisms. A state can choose to ignore these activities.

The state must also consider and define its role to ensure: public participation in local project decisions; public information system exists and functions; developing a means for compensation of those harmed.

It is clear that there will be valid claims of loss and damage when rain is changed up or down. Possibly a contigency fund should be made available, possibly by contributions by those supporting the program.

The state must also establish a posture (policy) towards research and development. The federal role in research and development for weather modification has been large. However, states have a potential role: (a) in providing some funding and control; (b) for gathering data through their operational projects useful in evaluation research; and (c) for enabling the 'piggyback' type efforts to evolve where scientific data are gathered and scientific study is on top of operational projects (WMAB, 1978).

An ultimate necessity, for securing state involvement in these recommended areas, is to involve the state in joint funding of the operational projects. Approaches are already identified in at least four states (North Dakota, Utah, Illinois and South Dakota) whereby local entities (counties or special districts) chose to be involved in regional weather modification projects and then raised funds adequate to pay all or part of the project cost. In these instances, the state can then help by matching funds. Such state support will help ensure better projects, more total effort, and better quality control of the programs.

Obviously, a policy approach that a state can consider is one of flexibility. The impacts of small modification projects based on uncertain technology are minor and may call for policies that are very different from the policy needs for larger and long duration projects likely based on more certain technology and wide usage.

The state involvement needs to become larger as the technology improves and usage grows. With growth in usage, efforts should begin to develop legislation to allow for partial state support or organizational support for county-type local projects. This one action will likely take several years to accomplish . It will help provide the basis for greater state involvement in the ultimate activities recommended which include: (a) monitoring and evaluation of projects; (b) developing means for compensating the losers; and (c) ensuring public participation in project decisions. A state plan for weather modification leading to the ultimate implementation of these activities, including operational projects, should be made and support sought. Once in hand, implementation can move ahead in an orderly fashion.

REFERENCES

Ackermann, W.C., Changnon, S.A. and Davis, R.J., 1974. The new weather modification law for Illinois. Bull. Am. Meteorol. Soc., 55: 745—750.

Changnon, S.A., 1980. Weather modification: impacts and implications. In: State Government, Winter Issue. Council of State Government, Lexington, KY, pp. 10—18.

Changnon, S.A., Davis, R.J., Farhar, B.C., Haas, J.F., Ivens, J.L., Jones, M., Klein, D., Mann, D., Morgan, G.M., Jr., Sonka, S.T., Swanson, E.R., Taylor, C.R. and Van-Blokland, J., 1977. Hail Suppression Impacts and Issues. Illinois State Water Survey, Urbana, IL, 430 pp.

Counc. State Gov., 1979. Suggested State Legislation, Vol. 37. Council of State Government, Lexington, KY.

WMAB, 1978. The Management of Weather Resources, Vol. 1. Report to Secretary of Commerce, Weather Modification Advisory Board. (H. Cleveland, Chairman). U.S. Government Printing Office, Washington, DC, 229 pp.

Agricultural Water Management, 7 (1983) 23—35
Elsevier Science Publishers B.V., Amsterdam — Printed in The Netherlands

UTILIZATION AND ASSESSMENT OF OPERATIONAL WEATHER MODIFICATION PROGRAMS FOR AUGMENTING PRECIPITATION

LEWIS O. GRANT

Department of Atmospheric Science, Colorado State University, Fort Collins, CO 80523 (U.S.A.)

(Accepted 28 January 1983)

ABSTRACT

Grant, L.O., 1983. Utilization and assessment of operational weather modification programs for augmenting precipitation. *Agric. Water Manage.,* 7: 23—35.

Operational weather modification programs to augment precipitation will continue routinely during the next decade in some areas, and intermittently during drought periods in other areas. The benefit to cost relationship for such programs is very favorable if even only modest changes in precipitation are produced. Evaluations and assessments are needed in any specific region to both confirm and establish the amount of precipitation change and to provide the basis for improvement in program performance.

This paper identifies the various interactive components that should be considered in any weather modification program. It then briefly discusses evaluation and assessment approaches and describes recently developed methodologies that can greatly strengthen such assessments.

INTRODUCTION

Extensive operational cloud seeding programs for augmenting precipitation are underway in many areas of the United States. These programs reflect the efforts by interested citizen groups to improve the weather in ways that provide economic benefits. It is clear that increases in the mean precipitation, at least in the western United States, would be economically beneficial. In nearly all areas of the country increased precipitation can be beneficial during drought periods.

Numerous studies have shown that even small increases in precipitation of as little as a few percent would, in many cases, produce benefits exceeding the costs. Such benefit is readily apparent to most water users in the water-short West, and to groups experiencing drought. This obvious nature of the potential benefits provides the impetus for many of the programs of operational weather modification. The farm operator, for example, is frequently receptive to operational programs of weather modification because the cost for the program is frequently much less than two dollars per hectare. This cost is minor in comparison with other operating costs which may run to

0378-3774/83/$03.00

around \$250/ha for semi-intensive agriculture to well over \$500/ha for intensive agriculture. It is also a small cost for a return of several quintals of additional grain production or kilograms of beef that can be expected with even a 5 to 10% increase in precipitation. Thus, the user of weather modification services frequently concludes that, even if rainfall is not increased, the dollar loss in attempting the programs is small and thus worth the risk. He concludes that this is a minor gamble in relation to most other production variables that the farm operator takes in producing and marketing a farm crop (market prices; variable effectiveness of herbicides, insecticides, fertilizers in a particular year; changes in interest rates, etc.). The farm operator is also conditioned to being no more able to assess the immediate value of many of his other agricultural practices in any specific year than he is to assess the returns from weather modification.

The problem, however, is more complex. It is not just a matter of whether there may be substantial, some, or no precipitation increases from a weather modification program. One cannot rule out the posssibility that there may be adverse effects in the form of precipitation decreases or, in some cases, the creation of severe weather. Additionally, in contrast to other farm practices, modification of the weather can affect a considerably larger area than just an individual farm. It is essential when undertaking a weather modification program that up-to-date knowledge of the weather and weather modification potential of a specific region be considered by experienced weather modification engineers, operators, or researchers. For some types of weather modification, such as seeding for some wintertime orographic cloud systems, the technology has a reasonably sound basis. For other applications, the 'state of the art' may be so marginal, or cloud information for a geographical region of concern so incomplete, that background studies and measurement programs are a necessity before a reasonable weather modification treatment program should be considered. A pre-program design can give visibility to the various interacting processes and requirements. An assessment phase once a program is underway should provide: (a) measurements on estimates of the benefits being obtained; (b) a system for monitoring various societal and environmental impacts; and (c) a basis for continual upgrading and improvement of the program. The following portions of this paper discuss features that should be considered in pre-program design plans and in ongoing program assessments.

DESIGN CONSIDERATIONS FOR OPERATIONAL WEATHER MODIFICATION PROGRAMS

The main considerations for operational weather modification programs in the past have been: (a) who wants it; (b) how much it will cost; (c) who will pay for it and how the funds will be raised; (d) what method of seeding should be used; and (e) which operator will be hired to conduct the program? These are, of course, all important considerations. They are, however, not all of the factors that need to be considered.

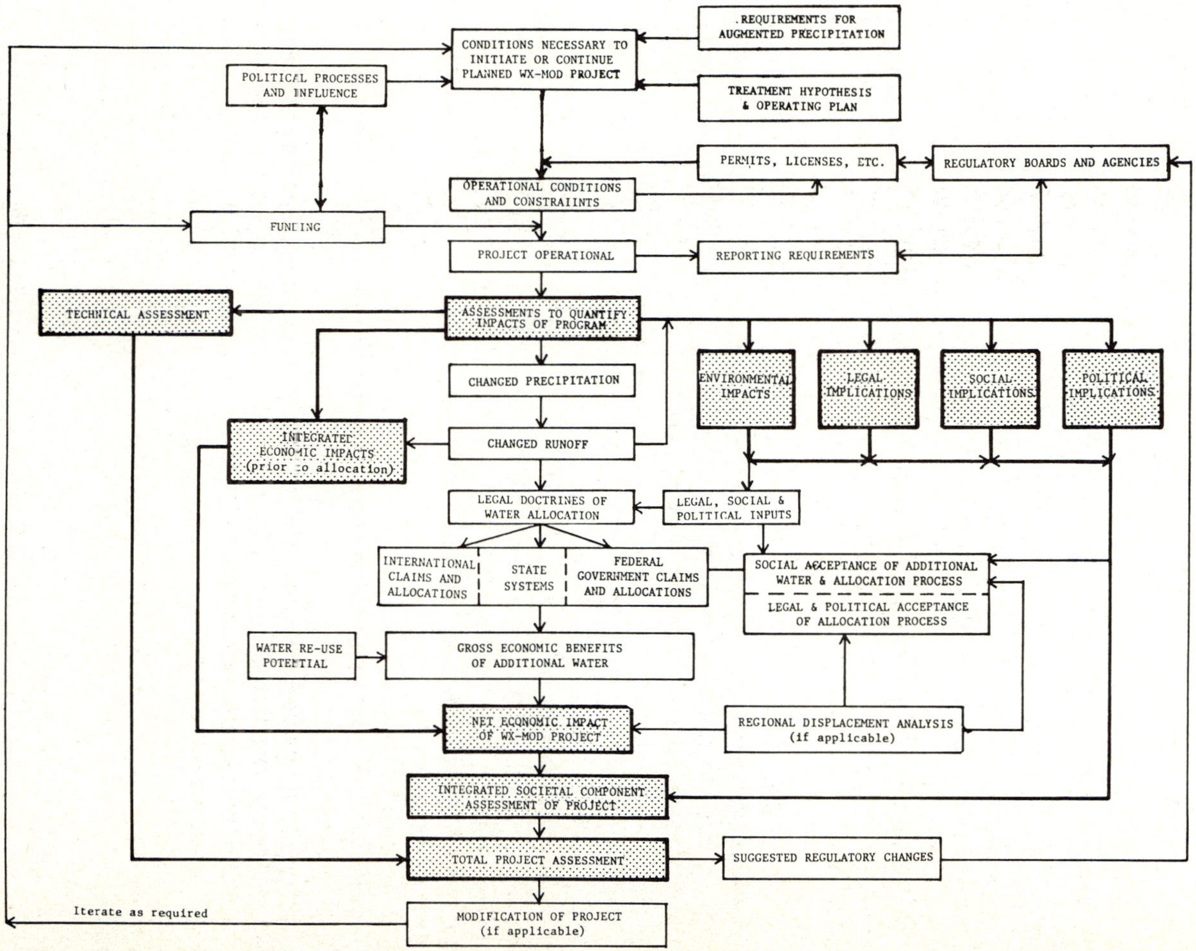

Fig. 1. Weather modification operational programs: flow chart of components.

Fig. 1, developed from a weather modification decisional process suggested by Jones et al. (1979), shows the various factors and processes that may be directly or implicitly involved in an operational weather modification program. The blocks in the diagram identify the factors or processes involved. The arrows show how these factors or processes interact or change with time. It can be noted that as the flow progresses down the central section it includes considerations of: the conditions necessary for a program; operational conditions and constraints; an operational program; assessments to quantify the impacts; specification of the amount of augmented precipitation; specification of the amount of runoff; legal aspect of ownership of 'new' water; gross economic benefits; net economic impact; integrated societal impacts; total project assessment; and design changes for upgrading of the program.

The relationship of these main components with necessary interacting processes (political, funding, social, and program improvement) and requirements (need for water, technical feasibility, permits, licenses, reports, and various types of assessments) are shown. Clearly a successful weather modification program requires more than the realization of a need for water and an airplane to do cloud seeding. The program cannot be successful if the funding, and technical, political, social, economic and assessment aspects are not considered and placed in perspective.

All of the aspects shown in Fig. 1 should be addressed during the design of a program and monitored while it is being conducted. In some programs certain features or processes may not apply. After identification and consideration they can easily be bypassed and dropped. In other programs, there are likely other factors or processes that need to be identified. Some of the program aspects and processes can be addressed in only a tentative way in operational programs. The decision process depicted in Fig. 1 provides a checklist for identifying the components and their interaction.

ASSESSMENT OF OPERATIONAL WEATHER MODIFICATION PROGRAMS

The types of evidence required to evaluate operational weather modification programs are frequently different than required in research programs. Research cloud seeding programs are designed to provide fundamental knowledge. In contrast, operational cloud seeding programs are designed to provide immediate beneficial economic gains. However, operational programs can provide fundamental knowledge, and research programs may cause changes in the weather. In the field of weather modification, the need for the developing fundamental knowledge, as well as water, from actual operating programs is substantial.

The recent final report of the Weather Modification Advisory Board (WMAB, 1978) requested by Congress in 1976 emphasizes the potential importance of weather modification to the nation's water supplies and delineates the main types of research that will be required over the next 20 years to realize the potential. This program of research has not been initiated, but a start has been made at implementing a few of the recommendations.

Thus, despite the positive conclusion and specific program recommendation in the study commissioned by Congress, it does not appear that a significant national program of weather modification research will be undertaken in the near future. The failure to develop such a national effort is the result of many factors. They largely reflect other pressing national interests, the economic recession, and the failure of federal agencies to come forward with a well designed and coordinated program. These obstacles will likely be overcome only when the next drought period arrives or the general water supplies become more critical. In the absence of such a strong national research thurst, it behooves supporters of local and state operational weather modification efforts to incorporate particularly strong assessment components to their operational programs.

In the above discussion both 'evaluation' and 'assessment' have been used. The interpretation of the results of all weather modification programs are frequently referred to as 'evaluations'. This language has led to misunderstandings between the scientific community and weather modification users and operators. Scientific evaluation places emphasis on very solid physical evidence and does not necessarily incorporate considerations of the economic consequences. In contrast, weather modification users and operators use of the term 'evaluation' has been loose and typically incorporates a value component in which less stringent proof is required, particularly if the possible economic returns can be great. For operational weather modification programs, the interpretation of results should probably be referred to as assessments: this infers more of a value judgment appraisal that can include sound but less stringent scientific proof. In this chapter the interpretation of results of operational weather modification programs are referred to as assessments, while rigid scientific interpretation of weather modification programs are referred to as evaluations. The assessment approaches suggested in this paper for operational programs are, however, considerably more stringent than usually used and draw heavily on the great progress toward weather modification evaluation that has recently been made by statisticians.

The shaded blocks in Fig. 1 identify the factors that should be considered for assessment in operational weather modification programs. The interconnecting lines suggest how these various assessment components should phase into the program and how their results should feed back into the overall program assessment. The main components include: technical assessment; economic assessment; assessment of the legal, social, and political impacts; and assessment of the environmental impacts.

These components are briefly discussed in the following sections. The emphasis is on technical assessment.

Technical assessment

Technical assessment should include assessment of: the efficiency with which the program was operated; the verification of the expected chain of events following the treatment; and the verification of the expected changes in precipitation.

A technical assessment should be considered as a vital component of any weather modification program.

Operational efficiency

Prior to the start of a weather modification operational program, consideration of the experience and competency of the technical operating team and the hypothesis and seeding criteria to be used must be based on the best information available. Once the program is started, relevant measurements and analyses are needed to test the treatment hypothesis and the efficiency with which the treatment is made. All too often program users assume that the operations will be carried out efficiently, and all weather systems which meet the seedability criteria established in advance will be seeded efficiently. Weather systems are complex and treatment using the proper seeding materials must be made over a large area with great precision both in time and space. Proper seeding requires highly-efficient organization, sophisticated monitoring systems, and highly-skilled operators. It is likely, even with the best presently available monitoring and seeding delivery systems, that precise coverage of a high percentage of the clouds cannot be accomplished. This fault may constitute as great or greater a constraint on the results of many operational weather modifications than do the scientific limitations.

A first step in the assessment of operational efficiency is the documentation of clouds that do or do not meet the seedability criteria established for the program (including ones that may produce severe weather and should be avoided). This step should include careful documentation of the time, location and seeding requirements for each system so identified. Once the cloud systems that meet the program criteria have been identified, further assessment can be carried out to define how well the seedable cases were identified on a real time basis, and how well they were treated in accordance with the program design. The results of these analyses can also be valuable in refining the climatology of the seeding potential for the region.

Analyses of program operating efficiency should be included as a direct program cost. This type of analysis is similar to that carried out as part of the management input for farming and business enterprises. It essentially involves the assessment of how well the program or business is run. All too often reports of weather modification operations show how many hours generators were run or how many hours aircraft flew without addressing the questions of How well were the seedable clouds identified?, What portion of the seedable clouds were actually treated?, and How good was the timing for the treatment of the seeded clouds?, etc. Such assessment should be considered as an integral part of sound, operational weather modification programs. These analyses should be carried out by objective criteria established in advance, and preferably by individuals who are not involved in the conduction of operations.

Physical evaluation

Confidence in the results of operational seeding is greatly increased if verification is obtained that each of the physical changes expected from the seeding actually occurred in accordance with the program seeding hypothesis. This procedure basically involves verification that: (a) the proper concentration of appropriate seeding material reached the proper levels in the cloud; (b) ice crystals were produced at the time, location and in the concentrations expected; and (c) the ice crystals grew and/or interacted with other cloud particles and fell out in the manner expected. In the case of dynamic seeding, the verification needs to extend to cloud size, rate of growth, vertical motions, etc.

Clearly, this type of assessment is not feasible to any substantial extent in practical operational weather modification programs. In such programs it must be assumed that, if efficient operations are carried out according to the best available technology, the sequential movement of atmospheric aerosol and cloud changes will follow. This type of verification has not been fully carried out to date even in the most sophisticated research programs. Some assessment of the progression of the seeding effect should, however, be incorporated in operational programs when possible.

While the limitations of physical assessments of operational programs are substantial, operational programs provide a significant opportunity for 'piggy back' research that can provide insight into all of the basic scientific and technological questions of weather modifications. The operational program can provide the treatment effect required for such research and at least part of the observational base. These contributions can substantially reduce the cost of the research. In turn, the superimposed research can provide the operational program with a physical evaluation of the efficiency of the seeding process, and a basis for upgrading the seeding hypothesis and the technology being used. This cooperative 'piggy back' approach offers a substantial opportunity for the development of weather modification technology (Grant and Mielke, 1982). This concept is particularly important in the current period when both research and operational program resources are limited.

Verification of precipitation changes

The main goal of technical assessment of operational weather modification programs is the documentation of the expected small change in precipitation in relation to the large, natural variability. This goal is difficult to achieve but must be routinely established for operational programs of weather modification.

Weather modification researchers and a small, but highly competent, group of statisticians have made great progress during the past 10 years in the treatment of this problem, both for research and operational programs (Gabriel and Changnon, 1981; Gabriel and Hsu, 1981; Mielke et al., 1982). Research on evaluation procedures has provided the basis for experimental designs that can appropriately include procedures like randomization, physical stratification of the data sets, and statistical procedures for handling complex problems, such as unknown data distributions, "uncontrolled" natural precipitation variability, disproportionate allocations of non-seed and seed experimental units, extreme events, differential treatment effects, etc. (Mielke, 1980a, b; Gabriel and Petrondas, 1981; Mielke et al., 1982; Mielke and Medina, 1983). Full use of these sophisticated techniques can be incorporated into operational evaluations but place some constraints on program operations. These constraints are frequently, or perhaps generally, not fully acceptable to the program supporters, however. Compromises are possible that permit full scientific integrity while still permitting most of the operational program goals. Even in the operational programs where nothing but full operations are acceptable to the supporters, the statistical tools that have been developed for research programs may still be employed, but results must be interpreted with more caution. The degree of bias that remains in the sample from not having an independent randomly selected non-seeded sample, can be studied with re-randomization techniques. When positive results are obtained with such methods, the remaining uncertainties are not large and are likely not substantially different, and maybe less, than those involved in many other aspects of farming or business activities.

The compromises in operational programs, but which will still permit scientific integrity should be given full consideration by program supporters. In many cases, the losses in seeding opportunities may not be critical, and would be well dispersed through the season because experimental units would typically be on individual days or storms. The gains in confidence in program results and in subsequent improvements in the program, in the long turn, can substantially offset the short-term losses of seeding opportunity. The compromises can include procedures like random seeding, but using a greater 'seed'/'non-seed' ratio than the 50/50 split generally used in research programs. This approach will likely increase the time required for significant results, but time is frequently or generally not a critical factor in operational weather modification programs. As a general rule, the time required for obtaining statistically significant results is only modestly increased until the seed/no-seed ratio is increased to above 70/30. The use of good covariates in the evaluation can substantially reduce the number of episodes required and, in many cases, can largely offset the greater time required for analysis with a large ratio of seeded/non-seeded events.

Another procedure that can strengthen assessment in some operational programs, and still permit fulltime seeding, can involve randomization of the method of seeding rather than randomization between seed and no-seed events. For example, the randomization might be between events to be seeded by aircraft versus ones to be seeded by ground generators. The randomizations might also be between ground seeding events using one type of seeding material or seeding rate versus another type of material. This procedure provides information on relative results but has serious limitations for assessing absolute effects. This approach may have advantages in some cases but may also cause more confusion than clarification. If the results of the different treatments produce similar effects, for example, there is no basis for concluding what their common effect was relative to what it would have been without seeding. Even when there is significant difference between two alternate seeding treatments, the information obtained is still limited. The treatment with the least precipitation may actually have caused an increase, not had any effect at all, or caused a negative effect that makes the favorable appearing treatment look good. The use of randomization between treatments does have value for comparison of treatments but has very limited value for total program assessment.

Even in programs where no randomization can be accepted, meaningful assessments of results are possible and should receive priority attention. There always will be questions, especially within the scientific community, about bias in the results with analyses from non-randomized programs. This doubt should not override all other considerations. The results can provide valuable input for the user in considering what return is being obtained for the investment involved, and for considering ways to improve the seeding program.

The main problem in evaluation of non-randomized programs is that the statistical analyses assume that all observations are independent and have the same distribution. Because all precipitation events may not be independent of each other, the statistical tests employed may not be strictly valid. Yet, as stated by Gabriel and Hsu (1981), "many statistical methods are quite robust to departures from the assumptions under which they were derived, and allow valid inferences even when the underlying variables are not IID (independent and identically distributed) and not normal and equally variable". The question then becomes one of knowing how valid the tests are when applied to any particular set of data. This question has been addressed by various authors including Mielke (1980b), Gabriel and Petrondas (1981) and Hsu et al. (1981). Gabriel and Petrondas (1981) and Hsu et al. (1981) have simulated seeding experiments with long-term precipitation records by arbitrarily designating seed and non-seed intervals. They then tested the statistical significance levels arising from the simulated experiment to see how they compared with those for a valid statistical technique which should have produced *P*-values distributed from 0 to 1. The use of long-term treatment units, like a month as opposed to a day, helps in this type of analysis to mitigate some of the data problems such as dependence between units. Based on studies of this type Hsu et al. (1981) concluded that "at best, we may conclude that the standard [statistical] techniques for comparing operational with historical precipitation are not blatantly invalid". These analyses were based on data from Illinois. They found that in some comparisons the validity of the statistical technique seemed to be confirmed but perhaps indicated a slightly better significance level than justified. On the other hand, other comparisons gave conservative results indicating somewhat worse significance levels than were required.

Gabriel and Petrondas (1981) made similar type analyses using both Illinois and 'world-wide' data. They concluded that the methods tend to be radical and indicate appreciably more significance than they should. They concluded that "perhaps one should discount most published analyses of this type roughly to the extent of doubling the *P*-values stated; i.e., if results are said to be 5% significant, they really should be regarded as only 10% significant."

Even if the significance is discounted to the extent of doubling the *P*-values obtained, as suggested by Gabriel and Petrondas (1981), signficant information is still obtained from the assessment of operational weather modification programs. In some cases the significant levels can still be impressive. In other cases, the lack of any significance can be extremely informative. The scientist's main concern is that he not make a type I error, that is to conclude that there was a seeding effect when there really was none. A main concern of most weather modification operations supporters is that they not make a type II error, which is to conclude that there was no effect when there really was one. The application of good statistical analysis can provide a strong, albeit non conclusive, basis for understanding the results of most operational weather modification programs that are continued in a reasonably similar manner for a few years. These results, especially when combined with parallel physical studies of the chain of events taking place, can, over a period of time, provide impressive information on the likely value of the programs and on the treatment technology being employed.

The comments above on analyses of non-randomized seeding programs basically deal with possible bias resulting from the inadequacies of the historical precipitation data base. There are other types of bias and inadequacies of analysis that can occur. These can and should be avoided in all program analyses. Other biases which should be avoided can include biases involved in the selection of the control areas, selection of the historical period, the collection of data, etc. As for analyses, the sound statistical procedures that have been developed and tested during the past few years must be employed. Good procedures include the use of meteorological covariates, covariates selected by principal component regression, rank tests, 're-randomization' (even with non-randomized data sets), non-parametric methods using residuals, etc.

Other assessments

The section above emphasized technical assessment which is basic to all research or operational programs of weather modification. Such technical analyses are required to determine how much the precipitation has been changed. This need constitutes basic input for socio-economic type evaluations and/or assessments.

Assessments of the societal components constitute the end result. These components basically address questions related to the economic, social, legal, political, and environmental benefits and costs. The magnitude of the economic impact is of immediate and direct interest to operational program supporters. This factor provides adequate incentive for its assessment. Assessments of the other societal impacts, while very critical to the long-term soundness of operating programs, will generally be of secondary interest to operational program supporters. Assessments of non-economic components of societal issues will probably emphasize only those components for which there are political or legal pressures. Details on the societal components of weather modification programs are covered by Changnon (1983, this issue).

SUMMARY AND CONCLUSIONS

There have been and continue to be a significant number of operational weather modification programs for augmenting water supplies. The potential for substantial economic gain from even small precipitation increases provides the incentive for these programs. The incentive is particularly strong in the drier, western parts of the country and in all areas during drought periods. The direct cost of such programs is relatively small in comparison to other costs incurred by most water users. The indirect costs, which could conceivably include decreased precipitation under some weather conditions are not well defined but also must be considered.

Various interacting factors must be considered in evaluating or assessing operational weather modification programs. These include aspects relating to the organizational, legal, assessment and integration components of such programs. At least some significant program assessments should constitute a vital part of all sound operational weather modification programs. Such assessments need to emphasize the technical and economic components to the program, and increasingly the other societal related aspects.

Technical assessment should include assessment related to the operational efficiency of the program, physical studies to verify the occurrence of the chain of events expected from the seeding hypothesis employed, and statistical analyses to establish the changes in precipitation. The costs and resource requirements for analyses of the physical chain of events taking place following seeding are far too great to be included as a routine component of operational weather modification programs. Such analyses, however, are vital to the understanding of program results and to further development of the technology. In the present environment of restricted research funding, physical analyses by independent researchers superimposed on operational programs provide an unique opportunity for learning. Such studies can facilitate scientific research at less cost and provide significant technical input to the operational weather modification program. In such "piggy back' research efforts, the pertinence of the findings to the operational program is direct, and the cost of the research can be significantly less because all treatment costs and some of the observational costs are provided by the operational seeding program. Federal/state/local programs of 'piggy back' weather modification research for this type of research justifies increased consideration.

In summary on the verification of precipitation changes, it is concluded that: (a) randomization is highly desirable; (b) if full randomization is not acceptable to program supports, some 'piggy back' compromises are feasible that can still permit scientific integrity of the analysis; and (c) even when no randomization is possible, good analysis can still provide highly useful and valuable information. It is essential that refined statistical procedures developed during the last few years be employed.

ACKNOWLEDGEMENTS

The author gratefully acknowledges drawing on discussions with and materials prepared by the participants of the working committees for the development of a report entitled *A Program for Federal/State/Local Cooperative Weather Modification Research*.

This report was prepared for the NOAA Weather Modification Program Office under Contract No. NA80RAC0007.

REFERENCES

Changnon, S.A., Jr., 1983. Society's involvement in planned weather modification. Agric. Water Manage., 7: 15—21.

Gabriel, K.R. and Changnon, S.A., Jr., 1981. Piggyback weather experimentation: superimposing randomized treatment comparisons on commercial seeding operations. Tech. Rep. 81/2, Deparment of Statistics, University of Rochester, Rochester, NY, 11 pp.

Gabriel, K.R. and Hsu, C.-t., 1981. Evaluation of the power of randomization tests, with applications to weather modification experiments. Tech. Rep. 81/11, Department of Statistics, University of Rochester, Rochester, NY, 26 pp.

Gabriel, K.R. and Petrondas, D., 1981. On using historical comparisons in evaluating cloud seeding operations. Tech. Rep. 81/12, Department of Statistics, University of Rochester, NY, 6 pp. and appendices.

Grant, L.O. and Mielke, P.W., Jr., 1982. A program of federal/state/local cooperative weather modification research. Final Report to NOAA, Weather Modification Program Office. Department of Atmospheric Science, Colorado State University, Fort Collins, CO, 70 pp.

Hsu, C.-t., Changnon, S.A., Jr., Huff, F.A. and Gabriel, K.R., 1981. The assessment of statistical-physical techniques for the evaluation of weather modification operations. SWS Contract Report 286, Vol. 3 OSET Final Report, Meteorology Section, Illinois State Water Survey, Urbana, IL, 90 pp. and appendices.

Jones, E.B., Grant, L.O., Klein, D.A., Davis, J.R. and Piernot, C., 1979. A societal component approach for a Colorado River Basin weather modification program. Preprints of Seventh Conference on Inadvertent and Planned Weather Modification, Banff., Alta., Canada. American Meteorological Society, Boston, MA.

Mielke, P.W., Jr., 1980a. A suggested piggyback randomized experiment for the Utah wintertime orographic seeding program. A section of the Statistical Evaluation of Cloud Seeding Operations, an appendix of a final report to NOAA entitled A Program for Federal/State/Local Cooperative Weather Modification Research, Department of Atmospheric Science, Colorado State University, Fort Collins, CO.

Mielke, P.W., Jr., 1980b. A suggested piggyback randomized experiment for North Dakota summertime cumulus cloud seeding program. A section of the Statistical Evaluation of Cloud Seeding Operations, an appendix of a final report to NOAA entitled A Program for Federal/State/Local Cooperative Weather Modification Research, Department of Atmospheric Science, Colorado State University, Fort Collins, CO.

Mielke, P.W., Jr. and Medina, J.G., 1983. A new covariate procedure for estimating treatment differences with applications to Climax I and II Experiments. J. Clim. Meteorol. (In press).

Mielke, P.W., Jr., Berry, K.J. and Medina, J.G., 1982. Climax I and II: distortion resistant residual analyses. J. Appl. Meteorol., 21: 788—792.

WMAB, 1978. The Management of Weather Resources. Report to Secretary of Commerce, Weather Modification Advisory Board, (H. Cleveland, Chairman). U.S. Government Printing Office, Washington, DC, 229 pp.

Agricultural Water Management, 7 (1983) 37—52
Elsevier Science Publishers B.V., Amsterdam — Printed in The Netherlands

CONSIDERATIONS IN THE DEVELOPMENT OF A STATE OPERATIONAL WEATHER MODIFICATION PLAN

JAMES W. SCHUELEIN[1], HAROLD L. SPRINGER[1], MICHAEL E. MATHIS[1], RANDAL A. ZIPSER[1] and D. RAY BOOKER[2]

[1] *Oklahoma Water Resources Board (OWRB), P.O. Box 53585, Oklahoma City, OK 73152 (U.S.A.)*
[2] *Aeromet, Inc., P.O. Box 45447, Tulsa, OK 74145 (U.S.A.)*

(Accepted 11 January 1983)

ABSTRACT

Schuelein, J.W., Springer, H.L., Mathis, M.E., Zipser, R.A. and Booker, D.R., 1983. Considerations in the development of a state operational weather modification plan. *Agric. Water Manage.*, 7 : 37—52.

Through a cooperative effort with the Bureau of Reclamation, under the Southwest Drought Research Program, the Oklahoma Water Resources Board is coordinating a plan to develop and manage the atmospheric resources of the state. To avoid the controversy surrounding many cloud seeding programs, the Board is carefully evaluating the potential for such a program and the most scientifically sound method of conducting seeding operations.

In support of the state effort, the Oklahoma Climatological Survey has developed a series of background studies which provide information on the space, time and intensity distribution of daily rainfall and derived variables such as soil moisture, runoff, drought and flood indices. A study is also being conducted to evaluate the number of cloud seeding opportunities, based on model-predicted thermodynamic and kinematic variables.

Aeromet, Inc. has designed an operational weather modification program for the state which specifies methods and tools that can be used to increase the available water resources for the state while protecting the interest of the general public. Designed to develop a capability for managing rainfall over a number of years, the plan proposed a program based on the current state of knowledge about modifying the atmosphere and the safeguards that are required for public safety and economic welfare.

The design has a strong provision for measuring the effects of seeding which, over time, will determine the particular seeding methods that work best and those that do not.

Before a final design can be adopted by the OWRB for the long term, an initial phase is proposed in which cloud physics measurements would be conducted with aircraft. This would also enable Oklahoma project planners and managers an opportunity to test and evaluate techniques and instrumentation proposed in the design study.

INTRODUCTION

For many years the State of Oklahoma has met the constantly increasing demands for water through the development of new sources of supply.

0378-3774/83/$03.00

From the drilling of new wells to a vast system of lakes and reservoirs, new sources of water have been developed to sustain the thriving Oklahoma economy and to maintain the comfort and well-being of a growing population.

Consistent with the authority granted in Oklahoma Statutes (Title 82), the Oklahoma Water Resources Board (OWRB) continues to explore a variety of new and emerging technologies both structural and non-structural to augment the state's water resources including desalination, chloride control, artificial recharge of the state's ground water resources, wastewater reuse, water conservation and interbasin transfer.

Supported by the Bureau of Reclamation (U.S. Department of the Interior) through its Southwest Drought Research Program, the OWRB and its contractors at Aeromet, Inc. and the Oklahoma Climatological Survey (OCS) are currently investigating the feasibility of implementing a state plan for management of the state's atmospheric water resources. This plan is being designed as part of a continuous rainfall enhancement program and, if shown to be feasible, would be incorporated as a part of the state's comprehensive strategy for managing Oklahoma's water resources.

SOUTHWEST DROUGHT RESEARCH PROGRAM

The Southwest Drought Research Program is a cooperative effort between the Bureau of Reclamation (BuREC) and the southwestern states, developed in response to a recommendation by the U.S. Senate Committee on Appropriations (Report No. 95-1069 on the Energy and Water Development Appropriation Bill, 1979, dated August 7, 1978). For the purpose of this research effort, the southwest has been defined by the BuREC as the state of Arizona, Colorado, Kansas, New Mexico, Oklahoma, Texas and Utah. Kansas and New Mexico elected not to participate in the program.

As a part of the cooperative effort, each state acting through the appropriate agency responsible for weather modification activities, performs multidisciplinary research tasks leading toward a regional capability for applying weather modification to lessen the impacts of drought. Each participating agency shares, with the BuREC, both the costs of the research and the responsibilities for developing the work tasks and evaluating the products. As a result of this program, the states have increased their knowledge of weather modification technology and its appropriate application (and limitations). The program was designed by the BuREC to be flexible to accommodate the level of technical expertise and experience of the individual states. Some involved in the program are preparing weather modification contingency plans, while others are acquiring information needed to prepare plans in the future. It should be emphasized that no actual seeding was undertaken as a part of the program (BuREC, 1981).

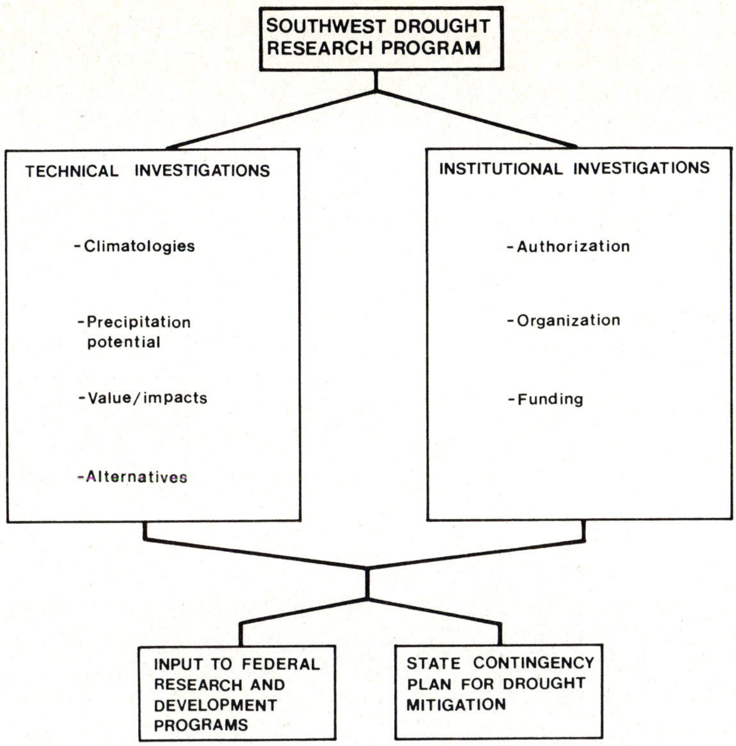

Fig. 1. Categories of investigation in the Southwest Drought Research Program. After BuREC (1981).

Drought research studies are being conducted under two broad categories (Fig. 1). Technical investigations include developing climatologies for drought and non-drought periods, assessing the potential for weather modification during drought and non-drought periods, estimating probable social and agricultural benefits and probable social and environmental impacts of cloud seeding for drought mitigation, and evaluating alternatives to cloud seeding for drought mitigation.

Institutional investigations consider the issues of authorization for conducting cloud seeding for drought mitigation, the issues of organization for decision making and for communicating the need for, and results of cloud seeding, and the issues regarding funding to support cloud seeding for drought mitigation (BuREC, 1981).

OKLAHOMA CLIMATE

Oklahoma is divided into two basic climate regimes, the humid east and semi-arid west. Summers are long and hot, while winters are shorter and less vigorous than those of the plains states lying farther north. Moist

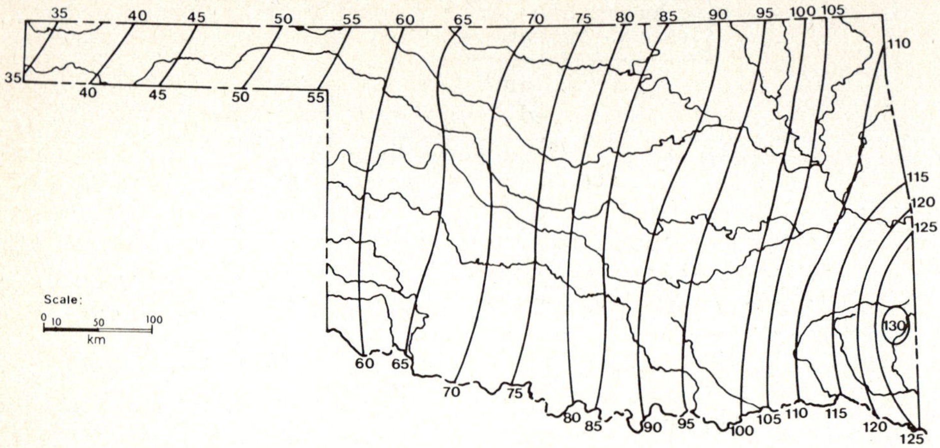

Fig. 2. Mean annual precipitation (cm) for Oklahoma. Base maps with major rivers. Isohyets from Pflaum and Eddy (1982). After Aeromet (1982).

air currents from the Gulf of Mexico temper the weather during most of the year, but cool, moist air masses from the Pacific and cold, dry Canadian air masses influence Oklahoma's winter temperatures (OWRB, 1980).

Maximum precipitation occurs in the spring, when thunderstorms frequently spawn damaging hail and/or tornadoes. May is usually the wettest month and rainfall decreases through the summer until fall, the second wettest season (OWRB, 1980).

Average annual precipitation decreases sharply across the state from east to west as shown in Fig. 2 (Pflaum and Eddy, 1982) ranging from over 130 cm in the southeast to less than 35 cm in the western end of the panhandle. A great deal of effort has been expended by Oklahoma project sponsors in analyzing Oklahoma rainfall and related atmospheric and geophysical measurements as a basis for estimating the opportunities for, and benefits of, cloud seeding.

The OCS has prepared a series of climatologies of different weather variables by month and time of day, including rainfall amount, number of days with rainfall, and radar echo frequency. Studies of this type have played an important role in the development of an operational weather modification plan for the state and will remain important policy considerations among decisionmakers in the event a program moves from the development to implementation phase.

On the pages that follow OCS background studies will be reviewed with a brief discussion of some of their implications; a design study, proposed by Aeromet, Inc. suggesting a plan for managing Oklahoma's atmospheric water resources will follow; and finally a brief analysis of future considerations concerning additional research needed before an operational program can be implemented in the state.

BACKGROUND STUDIES

Analysis of where and when rainfall occurs provides information concerning the number of days seeding might be attempted, assuming the presence of rain indicates favorable seeding conditions. Unfortunately rainfall and other climate records are not always consistent. For example, one reporting station may record rainfall days as those occuring with a minimum value greater than 0.01 in (0.254 mm) while another may record those occurrences with rainfall greater than 0.1 in (2.54 mm). Another problem effecting rainfall day statistics, and rainfall records in general, is that the distribution of reporting stations is not uniform across the state. Even where there is a fairly uniform distribution of stations, precipitation from isolated showers could easily miss observations (Aeromet, 1982).

Despite these problems, rainfall day statistics provide a rough measure of how frequently seeding might be attempted. The appendix provides a summary of days with rainfall greater than 0.01 in by month and climate division across the state. It is interesting that the east-west variation in rainy days does not seem as great as the variation in monthly rainfall amounts. The variation between climatic divisions, within any given month is particularly small. This suggests that for any given month the number of opportunities for cloud seeding to increase rainfall might be relatively independent of location in Oklahoma (Aeromet, 1982).

Another means of estimating rainfall occurrence is available through the use of weather-radar echo data. A number of National Weather Service (a division of the U.S. Department of Commerce, National Oceanic and Atmospheric Administration) stations routinely record the occurrence of echo returns registered on their radar scopes. As a part of the recording procedure, the observer records an intensity code on the face of the scope, which has been divided into squares. Referred to as manually digitized radar (MDR), these data provide a means of estimating the distribution of rainfall in both time and space, since the intensity of the echo is directly related to the precipitation rate.

The OCS has examined MDR data recorded in Oklahoma and the surrounding states covering 53 grid blocks drawn over Oklahoma and the Texas panhandle for the period 1 April to 30 September during the years 1974 to 1980. They then summarized the results of their echo-intensity-category into calendars of MDR echo frequency for each of the 53 blocks.

Figs. 3 through 5 represent a sample of these calendars across the state for all intensity categories combined. Each calendar is identified in the title by a code designating the block number and the radar where the echo data were derived. The map on the lower right shows the geographic location of the block (i.e. Fig. 3 is the block in the middle of the Oklahoma panhandle). Local standard time runs along the top of each calendar and months are indicated along the left side. Frequency occurrence (%) is indicated in the body of the calendar by various symbols, while the approximate time of sunrise and sunset are shown by the dark curved lines.

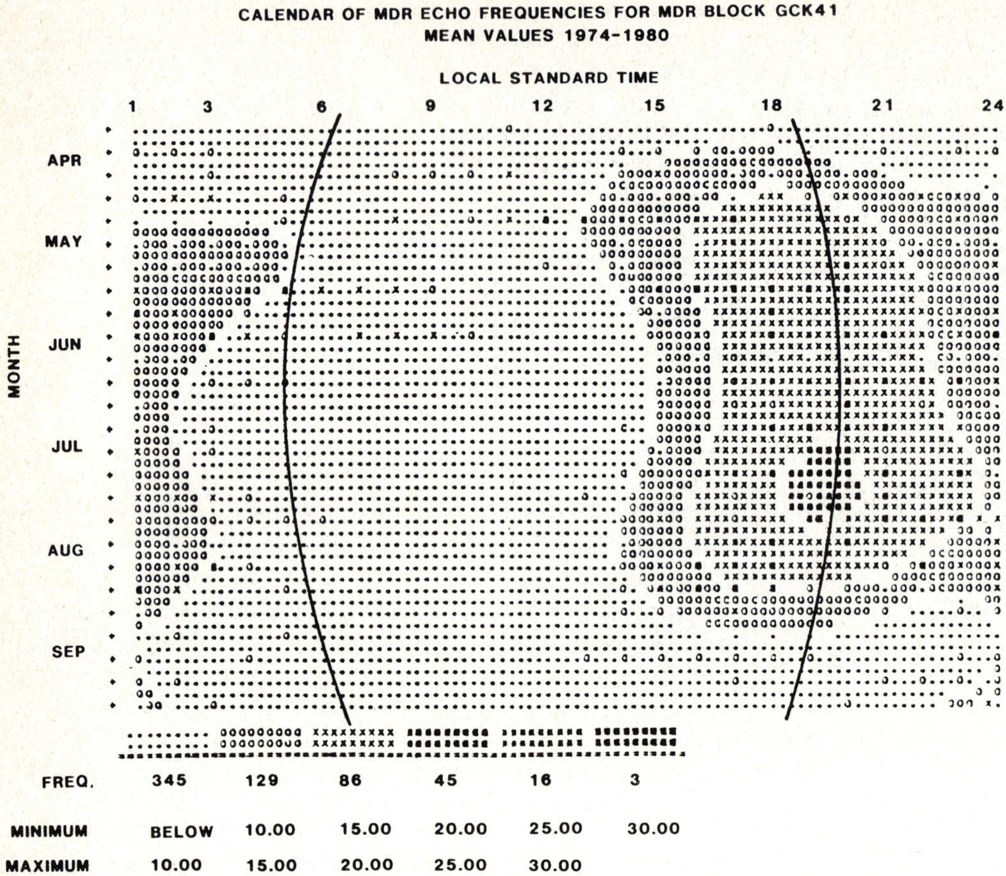

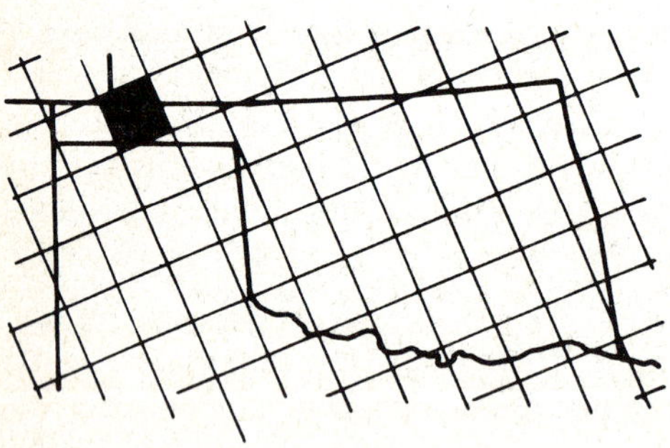

Fig. 3. Calendar of MDR echo frequency data for panhandle climatic division of Oklahoma.

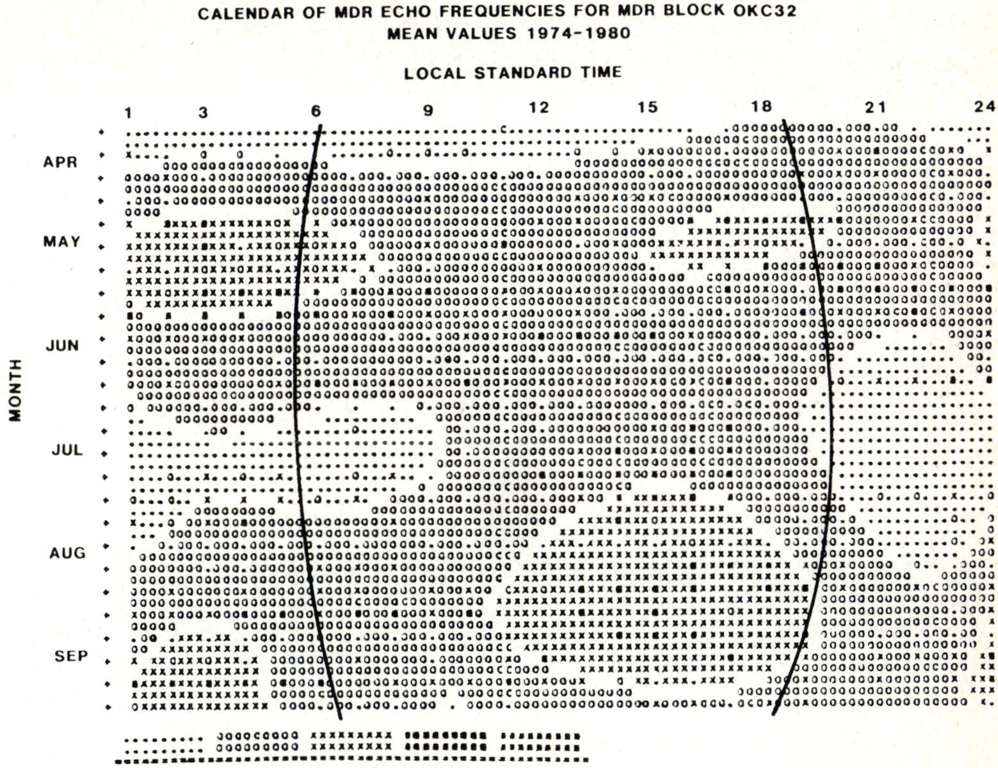

FREQ	215	167	148	71	23	
MINIMUM	BELOW	10.00	15.00	20.00	25.00	30.00
MAXIMUM	10.00	15.00	20.00	25.00	30.00	

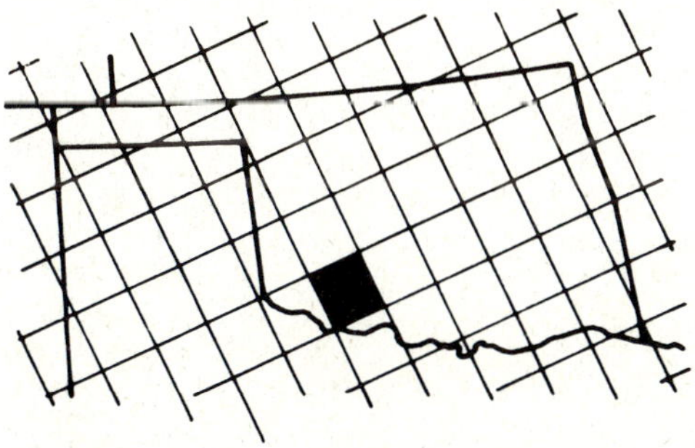

Fig. 4. Calendar of MDR echo frequency data for southwest climatic division of Oklahoma.

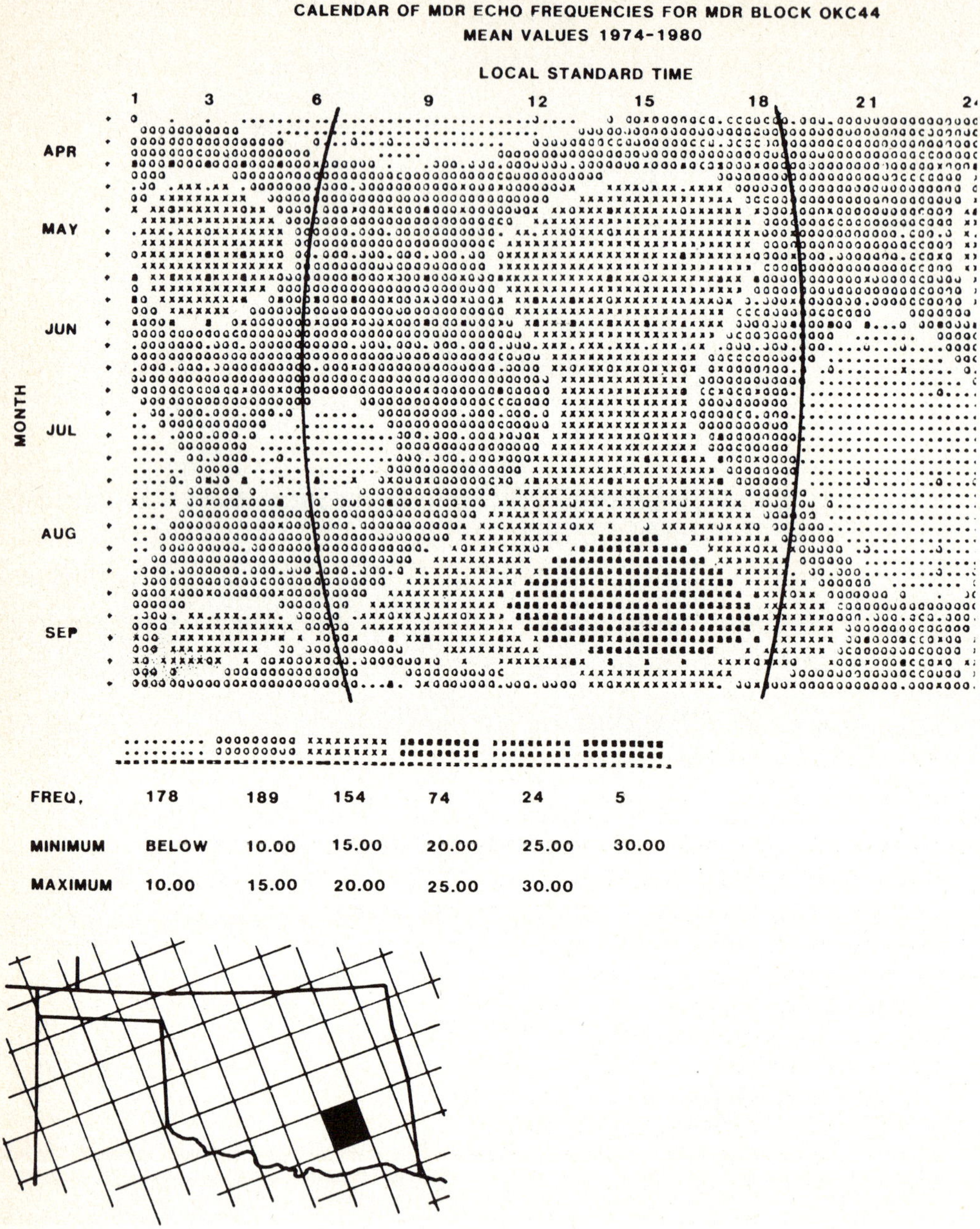

Fig. 5. Calendar of MDR echo frequency data for southeast climatic division of Oklahoma.

Since rainfall and MDR intensities are directly related it seems fair to draw the following conclusions by climatic division for the figures shown:

(1) Panhandle (Fig. 3): Rain occurs most frequently during the afternoon and at night. Rain is least likely during the early dawn and morning hours, while the diurnal variation is more pronounced than in any other section of the state.

(2) Southwest (Fig. 4): The climatology indicates peaks in rainfall limited to the early morning hours (midnight to sunrise) and the afternoon hours during May and September. August has a maximum from just before noon to sunset. While there is a definite lack of rainfall occurrence after sundown through the early dawn hours during July.

(3) Southeast (Fig. 5): There is a marked diurnal variation throughout the calendar. All months show maximum rainfall frequency during the afternoon with peak values in August and September. Rainfall is also prominent for three or four hours after midnight during parts of May and June and most of September.

An analysis of the radar-echo data indicates without night seeding, many opportunities will be missed (Aeromet, 1982).

While the efficacy of using cloud seeding to increase rainfall in times of drought has not been established, the background research being performed by the OCS has already provided a great deal of valuable information toward the design of an operational program as evidenced by the discussion on radar-echo data just presented. Other studies soon to be completed by the OCS include: (a) a rainfall climatology demonstrating the space, time and intensity distributions of daily rainfall; (b) a drought-flood climatology, which will reveal space, time and intensity distributions of derived variables such as soil moisture, runoff and drought and flood indices; and (c) a cloud-model climatology prepared utilizing the Great Plains Cloud Model (GPCM) of Hirsch (1971). This report will present distributions of thermodynamic, kinematic, and model predicted variables including values for cloud base temperature, cloud top temperature and maximum vertical motion, which are of interest to cloud physicists.

The cloud physics associated with the cloud model-implied climatology are currently the subject of a close examination at the OCS (Eddy, 1982). An illustrative example of the GPCM for Oklahoma City (OKC) rawinsonde data analysis indicates during the summer season of 1980 (a particularly hot, dry summer throughout Oklahoma and much of the rest of the south central U.S.), 51 days out of 92 (55%) had model predicted cloud base temperatures greater than -10°C, cloud top temperatures less than -45°C, and maximum vertical motions of over 20 m s^{-1}. This compares to a 5 year average (1976–1980) of only 42% of the days which fall in this category. The criteria expressed here are those which have been identified by some cloud physicists (Todd and Howell, 1981) as logically associated with potential seeding (Eddy, 1982).

OPERATIONAL DESIGN

Supported in part by the background studies of the OCS, Aeromet, Inc., utilized their experience in the operational aspects of weather modification to put together a design plan of operation that would best suit Oklahoma. Highlights of their plan follows.

SEEDING HYPOTHESIS

Aeromet, based on their experience and consultation with others familiar with the latest discoveries and innovations in the field of weather resources management, recommends a combination of both cloud top using dry ice (CO_2); and cloud base using silver iodide (AgI) seeding in Oklahoma.

The following sequence of events is likely to follow seeding:

(1) An average ice crystal concentration of about ten crystals per liter will be produced in the supercooled water cloud at temperatures warmer than −10°C. This concentration will be higher than in similar clouds left unseeded.

(2) Ice crystals will grow rapidly to a size where they begin to fall and collect supercooled cloud droplets. The particles will continue to grow to a size of about 1 mm, in concentrations of about one per 10 l. This concentration will be greater in seeded clouds than in similar unseeded clouds at the same time in the life cycle.

(3) These greater concentrations of 1 mm size particles in seeded clouds will result in earlier and lower first radar echoes.

(4) The graupel particles will fall through the freezing level, melt, and continue falling as rain drops. More rain will fall into the downdraft and downdrafts will be more vigorous in seeded clouds than in unseeded ones.

(5) As a consequence, the downdrafts in seeded clouds will be more likely to trigger secondary showers as a result of the stronger outflow.

(6) Rainfall below the cloud base and at the level of the lowest radar scan will be greater. Therefore, rainfall estimates made by use of radar or aircraft-mounted instruments, will be greater in seeded clouds than in similar unseeded ones; and finally.

(7) Rainfall at the ground will be greater from seeded than unseeded clouds so that rainfall estimated by rain gages, combined with radar data, will be greater downwind of seeded compared to similar unseeded clouds.

As a result of heat released by freezing of ice particles, it is likely some of the seeded clouds will grow taller than if left unseeded. If so, some of the additional rainfall downwind of the seeding event may be due to the increased size of the cloud in addition to microphysical effects. While this effect is not included as a part of the seeding hypothesis, it should be suspected as a possible effect of seeding and radar data should be examined to determine if the effect is present (Aeromet, 1982).

MEASUREMENT OF EFFECTS

Sooner or later every project sponsor wants to know how effective the seeding has been. According to the 1978 report of the Weather Modification Advisory Board, "the overriding problem for operational cloud seeding is that, except in a few cases, the predictability of the results of the seeding is not to the point where operations can be conducted with confidence" (WMAB, 1978a).

To avoid this problem both Aeromet, and the OCS have made recommendations for an evaluation procedure to be included as an integral part the operational plan design.

Aeromet's study points out two good reasons for utilizing a randomization method for seeding treatment. First, a number of people in the target area may not be convinced that cloud seeding works to their benefit. Without adequate data to point out the results of the seeding, this group may demand the project to be stopped. Second, comparative data (between seeded and unseeded clouds) are needed to improve the seeding method. Experience has shown that many clouds should not be seeded, in effect decreasing rainfall.

For theses reasons, they suggest an automatic randomizing sustem using a microprocessor, which can be programmed with a fixed series of random numbers and give a signal that activates the seeding device. When the computer gives the signal indicating no-seed, the system dispenses a placebo, or null treatment. The pilot and persons making observations and measurements are unaware of which clouds are actually seeded and which are treated with a placebo — so it is impossible to bias the data, even unconsciously (Aeromet, 1982).

The microprocessor can also monitor the resulting action to verify that no mechanical malfunction prevented the instrumentation from performing as instructed. Using a dry ice hopper similar to that used by HIPLEX (High Plains Experiment performed by the BuREC), the microprocessor can record the temperature in the dump chute to verify actual seeding.

In agreement with the Statistical Task Force of the Weather Modification Advisory Board (WMAB, 1978b), Aeromet suggests seeding two out of every three clouds, while the remainder remain unseeded (Aeromet, 1982).

It would also be advisable for the seeding aircraft to make additional penetrations after some seeding runs to compare the populations of ice crystals and larger ice particles in seeded and unseeded clouds. Both seeded and unseeded clouds must be penetrated in the same way to prevent bias. A random selection process is also suggested so seeded and unseeded clouds have the samce chance of having a post seed penetration without the crew knowing whether the cloud was seeded or not (Aeromet, 1982).

ROLE OF THE EVALUATOR

It is very important for all interested citizens to know that the information they receive about the effects of seeding in their area is unbiased. The role of the person or group which interprets the comparative data is very important to the long term health of the program because it is the foundation upon which public confidence is built.

The evaluator must be well qualified in statistics and the science of cloud seeding, yet must have no personal interest in the outcome of the cloud seeding project. He/she must also be able to communicate the evaluation results to the interested public effectively and objectively, while maintaining the respect and trust of the scientific community (Aeromet, 1982).

He/she must also be provided data from seeding operations and related measurements, and maintain an ongoing analysis as those operations are in progress. When the data indicate trends that are important to the operator as well as the interested public, they should be reported.

Aeromet warns against the temptation to drop randomization during drought to maximize rainfall. They point out there is insufficient evidence that seeding increases rain during drought to proceed without measuring effects. Rainfall could actually be decreased during drought — and for this reason, a continuation of the randomization procedure is encouraged during drought.

At the same time they recognize the need to suspend seeding during very wet periods, during harvest and at other times when additional rainfall is not desirable. It is however recommended that equipment and personnel be kept in place and observations continued during temporary suspensions to gather information needed to design longer term seeding methods (Aeromet, 1982).

OPERATIONS

As with any effort to modify the atmosphere, an operational program has an obligation to apply certain safeguards to protect the public interest and safety.

The State, through the OWRB, should manage the operations at a policy and funding level. However it would not be desirable for the state to conduct any of the operations or perform the primary evaluation, since a conflict of interest could develop or at least be perceived by some who would object to cloud seeding. The OWRB would be advised by two advisory groups, a Scientific Advisory Committee and a Citizens Advisory Group. The Scientific Advisory Committee would consist of (perhaps five) well known scientists in the theory and practice of weather modification and the estimation of seeding effects. Disciplines represented on the committee might include operations, measurements, evaluation statistics and government.

The cloud seeding contractor has a fundamental conflict of interest in that his business depends on a decision that the seeding is effective, and as a result should not conduct any part of the evaluation (Aeromet, 1982).

Since the Aeromet design study, based on the current state of knowledge, is oriented toward enhancement of precipitation from cumulus clouds, seeding operations would necessarily be confined to April to September, when cumulus clouds are common.

It is strongly recommended to employ weather modification operations as a water resources management tool every year, whether there is a drought or not — in fact Aeromet would encourage the state to make a commitment for at least 10 years. (This is not presently feasible under Oklahoma statutes which prohibit the commitment of funds for more than one fiscal year.)

An important consideration in support of this policy decision is the fact that significant benefits can almost always be gained from increased rainfall. At the same time, qualified personnel and well calibrated equipment could not be maintained with an intermittent operation.

Among other considerations discussed by Aeromet, is the potential threat of severe weather primarily during the spring in Oklahoma. Seeding should not be conducted in or around clouds likely to produce severe weather. Recent work at the Illinois State Water Survey on the severe weather threat to seeding-opportunity days in western Kansas indicates the number of days lost can exceed 80% depending on the restrictions in suspension criteria (Achtemeier, 1979).

According to the Aeromet design for Oklahoma, the project's chief meterologist will be ultimately responsible for recognizing severe weather situations and suspending seeding. The project forecaster will closely monitor weather conditions for developing severe weather indicators, including official forecasts. A National Weather Service severe weather warning should be treated as an absolute condition for suspending seeding in the area affected. At this time it is felt that a severe weather watch would be treated as a particularly sensitive indicator but not necessarily mean that seeding be suspended. At the same time, the Illinois work points out that a considerable number of the tornadoes which were reported in western Kansas during the study period occurred in the absence of appropriate weather watches (Achtemeier, 1979). This emphasizes the need for project personnel to be able to recognize threatening conditions and have firm procedures established for suspending operations.

CONSIDERATIONS FOR LONG-TERM PROJECT

Aeromet concluded there is insufficient information about the microphysical makeup of clouds in Oklahoma to submit a final design for the long term. As a result, they recommend the most economical way to get the information required for the final design would be to conduct a preliminary study as Phase I, seeding with aircraft equipped with sufficient instruments to collect the cloud physics data needed to answer scientific questions about cloud seeding in Oklahoma (Aeromet, 1982).

Project leaders at the OWRB would also welcome this opportunity to resolve differences of opinion between its contractors and other members of the scientific community who have expressed an interest in the Oklahoma program. It might also provide a way to evaluate several small commercial projects currently underway in the state.

Information on the size, number, and distribution of ice crystals in Oklahoma cumulus clouds will shed light on prospects for testing warm phase seeding and other seeding agents in some sections. Considering the wide variation in climate across the state, it is felt by some that more than one seeding hypothesis may be appropriate.

The initial phase of the Oklahoma project would also enable the contractor to test the operation of aircraft in restricted airspace over Oklahoma; to experiment with night time operations seeding at cloud top using low light level cameras; and to evaluate the automatic randomization technique using the microprocessor system — all relatively new mechanisms which may require a trial period for testing prior to implementation on an operational basis. (Aeromet, 1982).

SUMMARY AND CONCLUSIONS

Supported in part by funding from the Bureau of Reclamation, the State of Oklahoma has laid the ground work for the implementation of a statewide operational weather modification program to enhance precipitation during the spring and summer seasons.

Background studies compiled by the Oklahoma Climatological Survey provide links between the user interest and the scientific community and yield estimates between non-seeded climatological expectations and those anticipated from seeded clouds.

The Aeromet operational design recognizes the urgent need Oklahoma often has for water, but bases their proposed action on the current state of knowledge about modifying the atmosphere and the safeguards that are required for public safety and economic welfare.

The design also has a strong provision for measuring the effects of seeding. Indications are the primary responsibilities for evaluating a state-run program should rest with an independent contractor with no ties to the Oklahoma project who can relay his/her findings convincingly to the scientific community and the general public alike.

Before the OWRB can recommend a final project design for the state, additional information is needed characterizing the physical make-up of Oklahoma clouds. The cloud physics studies suggested in Phase I of Aeromet's Oklahoma Precipitation Enhancement Project design are the next logical step leading to the successful implementation of a project design for the long term.

Now is a good time for Oklahoma to move ahead with developing the capability for managing her atmospheric water resources. The tools required for observing atmospheric conditions and measuring the effects of treatment have been developed and tested. Our understanding of cloud processes has advanced to the point we can move ahead as long as we proceed with caution and a keen sense of public responsibility.

As the competition for Oklahoma's available water resources continues to increase, water resources planners and managers will look to a number of different alternatives to augment state water supplies. One alternative that holds particular promise, both from an economic and technological view point is precipitation enhancement through weather modification. The Oklahoma Water Resources Board feels a keen sense of responsibility to aggressively manage the atmosphere if we can increase rainfall in areas where it is needed for the common good — within the limits of known scientific methods and with adequate safeguards to protect the public interest.

ACKNOWLEDGEMENTS

The Oklahoma Climatological Survey is located at the University of Oklahoma in Norman, OK, under the direction of State Climatologist, Dr. Amos Eddy.

Aeromet, Inc. is a private consulting firm engaged in aeronautical and atmospheric research headquartered in Tulsa, OK. Dr. D. Ray Booker is president of the company.

REFERENCES

Achtemeier, G.L., 1979. Planned weather modification and the severe weather threat in the central High Plains. J. Appl. Meterol., 18: 348—354.

Aeromet, Inc., 1982. Oklahoma Precipitation Enhancement Program (OPEP), Design study Part I. Final Report to the Oklahoma Water Resources Board, 81 pp. (Unpublished).

BuREC, 1982. Summary Report, Southwest Drought Research Program. Division of Atmospheric Resources Research, Bureau of Reclamation, U.S. Department of Interior, Denver, CO, 30 pp. (Unpublished).

Eddy, A., 1982. Climatological background studies for operational weather modification. Paper presented to the American Water Resources Association International Symposium on Hydrometereology, June 1982, Denver, CO (Unpublished).

Hirsch, 1971. Computer modeling of cumulus clouds during project cloud catcher. Rep. 71-7, Institute of Atmospheric Science, South Dakota School of Mines, Rapid City, SD, 66 pp.

OWRB, 1980. Oklahoma Comprehensive Water Plan. Publ. 94, Oklahoma Water Ressources Board, Oklahoma City, OK, 284 pp.

Pflaum, J.C. and Eddy, A., 1982. A course of action for Oklahoma. Weather Modification Series, Vol. 3. Oklahoma Climatological Survey, Norman, OK, 18 pp.

Todd, C. and Howell, W., 1981. The relevant physics. In: A. Eddy (Editor), A Seminar Series. Weather Modification Series, Vol. 1. Oklahoma Climatological Survey, Norman, OK, pp. 100—150.
WMAB, 1978a. The Management of Weather Resources, Vol. I. Report to Secretary of Commerce, Weather Modification Advisory Board (H. Cleveland, Chairman). U.S. Government Printing Office, Washington, DC, 229 pp.
WMAB, 1978b. The Management of Weather Resources, Vol. II. Report to Secretary of Commerce, Weather Modification Advisory Board (H. Cleveland, Chairman). U.S. Government Printing Office, Washington, DC, 92 pp.

APPENDIX

Average number of days with rainfall greather than 0.01 in (0.254 mm)

Month	Division		
	PH	NC	NE
April	6	7	8
May	9	10	10
June	8	7	9
July	8	7	7
August	9	6	7
September	5	6	7
October	4	5	6
Total	49	48	53
	WC	C	EC
April	7	9	10
May	10	10	10
June	8	9	8
July	7	7	7
August	7	6	7
September	6	7	7
October	5	6	6
Total	50	54	55
	SW	SC	SE
April	7	9	10
May	10	9	10
June	8	8	8
July	7	7	7
August	6	6	8
September	6	7	7
October	5	5	6
Total	49	51	56

Data in the table were estimated from NOAA climatological summaries for 28 weather reporting stations in and around the state of Oklahoma. (After Aeromet, 1982.)

Introduction to Part 2

SOIL, PLANT AND ATMOSPHERE EFFECTS

The most serious losses to agricultural production in the Great Plains region are caused by insufficient water for crops during drought periods — whether short or protracted in length. Almost every year contains evidence of some degree of drought somewhere in the Great Plains. The history of the Great Plains is marked by droughts and there is reason to believe that severe droughts will occur in the future. Similar conditions prevail in other regions of the world.

Through mankind's efforts, the weather may be modified by planned or inadvertent activities. The increasing concentration of carbon dioxide in the atmosphere may modify the climate. Although some propose that this modification will have negative effects, others suggest that increased production and improved water use efficiency will result, thereby reducing the impact of and perhaps even reducing the frequency and severity of drought. This interesting and somewhat speculative topic is discussed in this section. Man's attempts to modify the weather on a large scale are discussed in Part 1 while some procedures for modifying the microclimate are suggested in this part. In addition, principles and techniques for managing the soil and crops for proper utilization of whatever water is available are described here. Included are various conservation and tillage practices, proper selection of plant species, planting at proper populations, management of the soils to improve the ability of roots to extract water, and the proper scheduling of irrigation.

BLAINE L. BLAD
Department of Agronomy, University of Nebraska
Lincoln, NE, U.S.A.

Agricultural Water Management, 7 (1983) 55—72
Elsevier Science Publishers B.V., Amsterdam — Printed in The Netherlands

INCREASING ATMOSPHERIC CO_2: EFFECTS ON CROP YIELD, WATER USE AND CLIMATE

B.A. KIMBALL and S.B. IDSO

U.S. Water Conservation Laboratory, 4331 E. Broadway, Phoenix, AZ 85040 (U.S.A.)

Contribution from Agricultural Research Service, U.S. Department of Agriculture

(Accepted 15 February 1983)

ABSTRACT

Kimball, B.A. and Idso, S.B., 1983. Increasing atmospheric CO_2: effects on crop yield, water use and climate. *Agric. Water Manage.*, 7: 55—72.

Probable effects of increasing global atmospheric CO_2 concentration on crop yield, crop water use, and world climate are discussed. About 430 observations of the yields of 37 plant species grown with CO_2 enrichment were extracted from the literature and analyzed. CO_2 enrichment increased agricultural weight yields by an 36% . Additional analysis of 81 experiments which had controlled CO_2 concentrations showed that yields will probably increase by 33% with a doubling of atmospheric CO_2 concentration. Another 46 observations of the effects of CO_2 enrichment on transpiration were extracted and averaged. These data showed that a doubling of CO_2 concentration could reduce transpiration by 34%, which combined with the yield increase, indicates that water use efficiency may double.

Several theoretical models have predicted that the doubling of atmospheric CO_2 concentration will increase the earth's temperature by 2—3° C, which could seriously disrupt agricultural production. More recent empirical evidence suggests that the warming may only be about 0.25° C, so the primary effects on agriculture are likely to be the beneficial increases in crop yields and water use efficiency.

INTRODUCTION

Prior to the Industrial Revolution, the CO_2 concentration in the air was probably somewhere in the range of 260 to 290 parts per million (Bolin et al., 1979). Today, however, the CO_2 content of the air is nearly 340 ppm, or 25% greater, due largely to mankind's burning of fossil fuels. Moreover, its concentration is predicted to double well before the middle of the next century (NAS, 1979), possibly even as early as 2025 (Gribbin, 1981).

This phenomenon would be largely of only academic interest, were it not for the fact that CO_2 affects the climate of the Earth as well as the primary productivity and water use of its flora. Greater CO_2 concentrations will probably boost agricultural production with less water consumption, which will be a boon to Earth's ever-expanding population. On the other hand, increased CO_2 has been predicted to cause a catastrophic atmospheric warming due to CO_2's 'greenhouse' properties, the proportions of which

Earth has never before experienced. It behooves us to look objectively at the whole question in critical review.

CO_2 AND AGRICULTURAL YIELD

Ever since De Saussure (1804) demonstrated that peas exposed to high CO_2 concentrations grew better than control plants in ambient air, scientists have been enthralled with the prospect of vastly increasing crop productivity by means of CO_2 enrichment. For many years, the only obvious place of practicality for such aerial fertilization of plants was the glasshouse or greenhouse; but with the progression of industrialization and its concomitant utilization of fossil fuels, the vast amounts of carbon dioxide dumped into the atmosphere as a waste product of our society have turned the globe itself into a gigantic field trial of the technique.

Effects of CO_2 on crop growth and development are myriad; but the most important ones could probably be lumped together under the categories of biological nitrogen fixation, photosynthesis, water use efficiency and yield component production. Lemon (1977), Wittwer (1978, 1980) Allen (1979), Kramer (1981), and Rosenberg (1981) have reviewed these several subjects. However, the most comprehensive review of CO_2 effects on ultimate harvestable yield has been presented by Kimball (1982) who examined more than 70 reports about effects of CO_2 enrichment on the economic yields and growth of 24 crops and 14 other species, and extracted more than 430 observations.

Elevated CO_2 concentrations have had an overwhelmingly positive effect on yield (Kimball, 1982). Of 437 separate enriched samples, only 39 yielded less than their respective controls. Of this group, 20 were flower crops whose yields were measured by number of flowers rather than by weight. Frequency distributions of the logarithms of the relative yield increases (ratios) are plotted in Fig. 1a for all of the crops and in Fig. 1b for all of the crops except flowers (Kimball, 1982). The mode ratio (the ratio with the highest frequency of observation) was about 1.1, considering all observations (Fig. 1a), or about 1.2, considering only the weight yield ratios and smoothing the curve (Fig.1b). The mean yield ratio and the 99.9% confidence intervals for the means were computed and plotted as the bracketed dots in Fig. 1. The large number of data points made the confidence intervals relatively short, so considering all the mature agricultural crops, the average relative yield increase was 1.28, or 1.36 excluding flowers.

Means for several classes of crops and their respective 95 and 99.9% confidence intervals, as computed by Kimball (1982), are presented in Table I. Effects of CO_2 enrichment on flower yields were generally smaller than effects on other crops, but this was not surprising because flower yields were the number of blooms per plant. Mean yield ratios of the other crop categories with more than three observations were 1.23, 1.32, 1.42, 1.54, and 1.52 for fruit, C3 grain, leaf, legume seed, and root crops, respectively

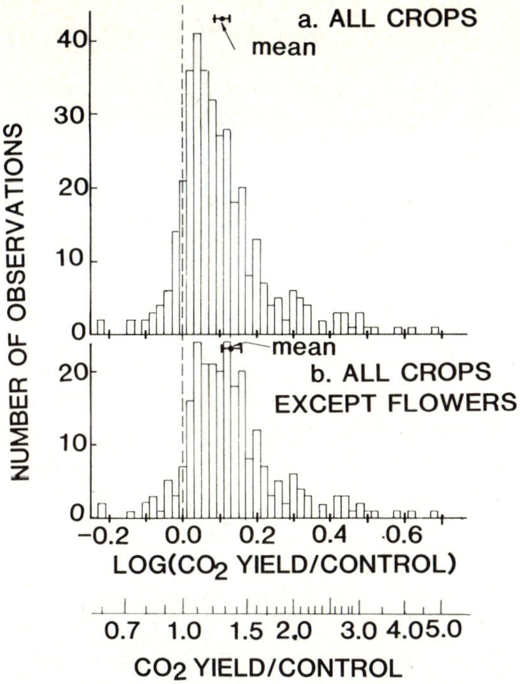

Fig.1. Frequency distribution of the logarithms of the relative yield increases (ratios) of CO_2-enriched plants to their respective unenriched controls for: (a) all mature agricultural crops; and (b) all mature agricultural weight crops excluding flowers. The points and associated brackets are the means and 99.9% confidence intervals for the means computed from the logarithms of the yield ratios. From Kimball (1982).

(Table I). The fruit yield ratio was dominated by tomato (72 of the 97 observations), so the relatively low fruit yield ratio is partly due to the fact that many of the tomato experiments were done in greenhouses that were ventilated and not enriched with CO_2 for a portion of the time, whereas relatively more of the experiments with other crops were done with growth chambers that were continuously enriched.

Also included in Table I are the mean yield ratios with confidence intervals for 79 observations from experiments that were terminated before the plants matured. Without exception these experiments were performed using controlled CO_2 enrichment for their duration, and the resultant mean yield ratio of 1.60 was significantly higher than the mean yield ratios of 1.28 or 1.36 for the mature plants. When these immature plant results were pooled with the mature, the overall resultant mean CO_2/control yield ratio was 1.33 with a 99.9% confidence interval from 1.27 to 1.40.

The results presented in Fig. 1 and Table I indicate it is highly probable that crop yields will increase with increasing concentrations of atmospheric CO_2. However, these results could not easily be used to predict quantitatively how much yields will increase with a doubling or tripling in CO_2 concen-

TABLE I

Mean relative yield increases (ratios) of CO_2-enriched to control crops and associated confidence intervals for various classes of crops

Crop	No. of observations	Lower limits		Mean	Upper limits	
		– 95%	– 99.9%		95%	99.9%
Mature agricultural crops (marketable yield)						
Fiber	2	—	—	2.59	—	—
Flower	110	1.06	1.08	1.12	1.16	1.18
Fruit	97	1.15	1.18	1.23	1.28	1.32
Grain (C_3)	34	1.16	1.23	1.32	1.42	1.50
Grain (C_4)	2	—	—	2.98	—	—
Leaf	72	1.24	1.31	1.42	1.53	1.62
Legume seed	24	1.20	1.34	1.54	1.77	1.99
Root and tuber	17	1.02	1.23	1.52	1.88	2.26
All mature	358	1.22	1.24	1.28	1.32	1.35
All mature except flowers	247	1.28	1.31	1.36	1.42	1.45
Immature crops (total plant weight or height)						
Agricultural crops	56	1.39	1.49	1.64	1.81	1.94
Non-agricultural, herbaceous	12	1.06	1.22	1.38	1.58	1.80
Woody plants	11	1.23	1.42	1.62	1.84	2.11
All immature	79	1.41	1.48	1.60	1.72	1.81
All mature plus immature	437	1.27	1.30	1.33	1.37	1.40

Adapted from Kimball (1982)

tration. First, many workers did not report what concentrations were used. Second, even more used widely varying concentrations, particularly those whose greenhouses had to be ventilated for temperature control on hot sunny days. Still other workers enriched their plants only for specific stages of growth. When only those experiments that used controlled and monitored CO_2 concentrations for nearly 100% of the daytime hours for the entire lifetimes of the plants were considered, most of the experiments were eliminated. Further restricting to those experiments with weight yields (excluding flowers) and concentrations of less than 1 200 ppm (an arbitrary cut-off point to focus on current rates of increase rather than on where a yield plateau occurs), only 38 mature agricultural crop experiments and 43 immature plant experiments were left (Kimball, 1982). The average rate of yield ratio increase with increasing CO_2 concentration (slope) was determined for each of these 81 experiments.

Results of the statistical analysis of the slopes of the yield ratio lines are presented in Table II from Kimball (1982). Mean slope was 0.000 96/ppm, with a 95% confidence interval from 0.000 62 to 0.001 36/ppm, for the mature agricultural crops. The mean slopes for the immature crops were not significantly different from that for the mature crops. Pooling the data, the overall mean slope was 0.001 00 with a 95% confidence interval from 0.000 85 to 0.001 17.

TABLE II

Means and confidence intervals for the change in relative yield increase per 1 000 ppm change in CO_2 concentration (slopes) for those experiments having controlled CO_2 concentrations. Also the subsequent predicted relative mean yield increase and confidence intervals for a 660 ppm CO_2 concentration (atmospheric doubling) and for a 1 000 ppm CO_2 concentration (atmospheric tripling) based on the overall mean slope (from Kimball, 1982).

Item	No. of observations	Lower limit		Mean	Upper limit	
		−99.9%	−95%		+95%	+99.9%
Slopes [change in relative yield increase/change in CO_2 concentration (ppm)] × 10^3						
Mature agricultural crops	38	0.41	0.62	0.96	1.36	1.73
Immature agricultural crops	31	0.61	0.87	1.26	1.72	2.16
Immature non-agricultural crops	6	− 0.39	0.04	0.42	0.97	2.35
Immature woody	6	− 0.14	0.34	0.75	1.29	2.58
All mature plus immature	81	0.74	0.85	1.00	1.17	1.30
Yield increase at 660 ppm CO_2		1.24	1.28	1.33	1.39	1.43
Yield increase at 1 000 ppm CO_2		1.50	1.57	1.67	1.78	1.87

Using the overall mean slope in Table II and a present-day CO_2 concentration of 330 ppm, the yield ratios at future concentrations of 660 ppm (doubling) and 1 000 ppm (tripling) were predicted. At a CO_2 concentration of 660 ppm, mean yields are predicted to be 1.33 times greater than now with a 95% confidence interval from 1.28 to 1.59. At 1 000 ppm, the increases are twice those at 660 ppm for a predicted mean yield ratio of 1.66 with a 95% confidence interval from 1.57 to 1.78. Thus, a doubling of the earth's CO_2 concentration will probably increase agricultural yield by about 33%, with a 95% confidence range from 28 to 39%.

CO_2 AND CROP WATER USE

In addition to increasing yields, the increasing atmospheric CO_2 concentration is likely to have at least one other major beneficial effect on agri-

TABLE III

Percentage reduction in transpiration of various species with a doubling of CO_2 concentration

No. of Observations	Species	Reduction in transpiration (%)	References
1	Barley (*Hordeum vulgare*	9	Akita and Moss (1972)
2	Carnation (*Dianthus caryophyllus*)	15	Enoch and Hurd (1979)
3	Corn (*Zea mays*)	16	Moss et al. (1961)
4	Corn	33	Moss et al. (1961)
5	Corn	68	Pallas (1965)
6	Corn	68	Pallas (1965)
7	Corn	66	Pallas (1965)
8	Corn	68	Pallas (1965)
9	Corn	27	Akita and Moss (1972)
10	Corn	24	Akita and Moss (1973)
11	Corn	40	Carlson and Bazzaz (1980)
12	Cotton (*Gossypium hirsutum*)	34	Pallas (1965)
13	Cotton	29	Pallas (1965)
14	Cotton	34	Pallas (1965)
15	Cotton	25	Pallas (1965)
16	Cottonwood (*Populus deltoides*)	28	Regehr et al. (1975)
17	Cottonwood	38	Carlson and Bazzaz (1980)
18	Dandelion (*Taraxacum officinale*)	9	Akita and Moss (1972)
19	Foxtail (*Setaria viridis*)	41	Akita and Moss (1972)
20	Simson weed (*Datura stramonium*)	23	Carlson and Bazzaz (1980)
21	Pigweed (*Amaranthus retroflexus*)	38	Akita and Moss (1972)
22	Pigweed	35	Carlson and Bazzaz (1980)
23	Ragweed (*Ambrosia artemisiifolia*)	40	Carlson and Bazzaz (1980)
24	Silvermaple (*Acer saccharinum*)	33	Carlson and Bazzaz (1980)
25	Sorghum (*Sorghum bicolor*)	68	Pallas (1965)
26	Sorghum	58	Pallas (1965)
27	Sorghum	61	Pallas (1965)
28	Soybean (*Glycine max*)	34	Pallas (1965)
29	Soybean	53	Pallas (1965)
30	Soybean	41	Pallas (1965)
31	Soybean	22	Egli et al. (1970)
32	Soybean	19	Egli et al. (1970)
33	Soybean	17	Egli et al. (1970)
34	Soybean	25	Van Bavel (1972)
35	Soybean	24	Van Bavel (1972)
36	Soybean	43	Van Bavel (1972)
37	Soybean	24	Carlson and Bazzaz (1980)
38	Soybean	30	Rogers et al. (1981)
39	Sunflower (*Helianthus annuus*)	12	Rogers et al. (1981)
40	Sycamore (*Platanus occidentalis*)	25	Rogers et al. (1981)
41	Tomato (*Lycopersicon esculentum*)	21	Pallas (1965)
42	Tomato	33	Pallas (1965)
43	Tomato	36	Pallas (1965)
44	Velvet leaf (*Abutilon theophasti*)	38	Carlson and Bazzaz (1980)
45	Wheat *Triticum aestivum*)	11	Akita and Moss (1972)
46	Wheat	8	Akita and Moss (1973)
	Average	34 ± 3	

culture — an increase in water use efficiency. Increased CO_2 concentration has been found to decrease stomatal apertures and reduce transpiration. Table III lists 46 observations of the reduction in transpiration of 18 different species obtained with a doubling of CO_2 concentration in short-term growth chamber experiments. The average reduction was 34%, which coupled with yield increases of about 33% (Table II), indicates water use efficiency may double with a doubling of CO_2 concentration. In longterm growth chamber experiments, Gifford (1979a, b) and Sionit et al. (1980) found that the relative yield increase with CO_2 enrichment of waterstressed wheat was as large as well-watered wheat. Gifford found little change in the actual amount of water used with CO_2 enrichment, and Sionit et al. did not report water use, but the 32 to 100% increase in yield found by both groups means there was a large increase in water use efficiency.

The actual long-term consumptive water use of CO_2-enriched crops under field conditions needs to be investigated. Larger plants due to CO_2 enrichment will have a larger leaf area tending to increase transpiration. They will also probably have a larger, more vigorous root system and be able to extract more water, resulting in less water stress and higher yields, but possibly a higher total consumptive water use. Any decreases in transpiration will also make more energy available for soil evaporation. Nevertheless, it is our belief that once the crop canopy becomes closed, the higher stomatal resistances with increased CO_2 concentrations will cause higher total canopy resistances, resulting in less total water consumption. Coupled with the increase in yield, the increases in water use efficiency are likely to be dramatic and probably will have a significant beneficial impact on irrigated and dry-land agriculture.

CO_2 AND CLIMATE

Carbon dioxide has made the term 'greenhouse effect' into a household word. Unfortunately, the phraseology is inappropriate, because CO_2 does *not* warm the surface of the Earth in a manner analogous to the way in which a greenhouse keeps its temperature elevated, but the terminology has been established by popular convention. The reason CO_2 is referred to as a 'greenhouse gas' is because it is fairly transparent to solar radiation but largely opaque to thermal radiation. Thus, the rays of the sun penetrate the atmosphere and heat the Earth's surface. When this absorbed solar radiation is reradiated back towards space as thermal radiation, CO_2 in the atmosphere intercepts a certain fraction that would otherwise escape unimpeded to space. The intercepted radiation is subsequently reradiated, with a portion of it finding its way back to the surface and increasing Earth's temperature.

Qualitative aspects of the carbon dioxide-climate connection have been well understood for over a century, but there has always been some concern about the numbers involved. The first extensive calculations of the magnitude of the carbon dioxide effects on climate were made by Arrhenius (1896). He estimated that a doubling of the CO_2 content of the atmosphere could elevate average surface air temperature by as much as 8 K. As a result of his work, and subsequent contributions by Chamberlin (1899), the carbon dioxide theory of climatic change came to be widely accepted. By the early 1950's, however, its adherents had dwindled considerably; but then, in the mid-50's, Plass (1956a, b; 1957) revived the concept and started it on its way to the position of preeminence that it occupies today.

Many investigators of the past two decades have calculated the change in surface air temperature to be expected from an increase in CO_2 concentration. The standard number that is used for comparative purposes in this context is the change to be expected from a doubling of the CO_2 concentration from a nominal value of 300 ppm to 600 ppm. In initiating the modern period of investigation, Plass (1961) obtained a 3.6 K increase for such a concentration doubling. Following him, there were numerous other calculations, ranging from a low of 0.7 K (Weare and Snell, 1974) to a high of 9.6 K (Moller, 1963). Manabe and Wetherald (1967) started the modern 'radiative-convective approach' used by most modelers today, although many different assumptions and significant differences still exist among the various models in use today. However, there has come to be a consensus among modelers that a doubling of the atmospheric CO_2 concentration would probably lead to a global warming on the order of 2 to 4 K (Reck, 1982).

An important new aspect of CO_2-climate investigations of recent years has been the application of empirical approaches to the question. Whereas the numerical modeling approach has been to introduce a compositional or radiative perturbation into a mathematical model of the atmosphere and then compute how the model atmosphere responds, the empirical approach identifies environmental circumstances which provide natural experiments illustrative of how the real atmosphere responds to real perturbations (Idso, 1980a).

Due to difficulties inherent in dealing with the real world, empirical estimates of CO_2 effects on climate were essentially nonexistent until the last few years. But then, two independent studies appeared which cast considerable doubt upon the previous consensus estimate of the numerical modeling approach.

The first of these studies was an investigation by Newell and Dopplick (1979) into large scale constraints on tropical ocean surface temperatures. Analyzing empirical observations of Newell et al. (1978), they concluded that the sea at low latitudes reaches a limiting temperature close to 303 K, based on the balance between radiative energy input and evaporative heat loss. This fact, coupled with certain simple energy balance considerations, led them to conclude that the change in air temperature over tropical waters corresponding to a given change in absorbed radiant energy would be about 0.03 K $(W\ m^{-2})^{-1}$, and that for the globe as a whole it would be about 0.1

K (W m^{-2})$^{-1}$. Applying this sensitivity factor to the radiative perturbation caused by a doubling of the atmospheric CO_2 concentration, a temperature increase of only 0.25 K was obtained.

Idso (1980b) soon followed with his independent analysis of the problem, in which empirical observations were employed to evaluate what he called the 'surface air temperature response function' by three separate and completely unrelated natural experiments (Idso, 1982a). The first of these 'events' was the vertical redistribution of airborne dust which occurs between summer and winter at Phoenix, AZ (Idso and Kangieser, 1970). Idso and Brazel (1978) demonstrated that the 'thermal blanketing' effect (Idso, 1974) of this dust was much stronger in winter than in summer, and they presented empirical data to indicate that it raised the winter surface air temperature of the city by 2.4 K over what it would have been for the type of vertical distribution of dust that exists in summer. Subsequently, Idso (1981a) measured the actual change in thermal radiation from the atmosphere caused by this vertical redistribution of dust between summer and winter and found it to be 13.9 W m^{-2}. Dividing this radiant energy flux differential into the 2.4 K temperature differential, the surface air temperature response function of the atmosphere over Phoenix was found to be 0.173 K (W m^{-2})$^{-1}$.

The second event used to evaluate this atmospheric parameter was the advent of the southwestern United States' summer monsoon, which annually immerses Phoenix in humid air from the Gulf of Mexico at high levels and the Gulf of California at low levels sometime near the end of June or the beginning of July (Hales, 1974). Idso (1982a, c) demonstrated that for extremes of surface vapor pressure ranging from 4 to 20 mbar over this transition period, surface minimum air temperature at Phoenix increases by 11.4 K. Using a newly developed equation which gives a very accurate representation of the thermal radiation from the atmosphere as a function of surface air temperature and vapor pressure (Idso, 1981b), this change in vapor pressure from 4 to 20 mbar was found to create a radiant energy flux differential of 58.1 W m^{-2}, which when divided into the 11.4 K temperature differential yielded a surface air temperature response function of 0.196 K (W m^{-2})$^{-1}$.

The third naturally occurring phenomenon used to obtain this function was the annual variation in solar radiation received at the surface of the Earth. Bennett (1975) had evaluated the summer-winter differential of this flux for 105 stations in the United States. Assuming an average surface albedo of 0.30, Idso (1982a, c) divided the resultant absorbed solar energy differential to obtain a result exactly identical to the mean of the two previous determinations of the surface air temperature response function for the *interior* United States. For 15 stations on the west coast, however, the result was only half as great. Assuming these stations to be greatly influenced by the Pacific Ocean, Idso (1982a, c) took the lower result for those stations to be an upper limit for the water surfaces of the planet.

Apportioning these two numbers between the appropirate percentages of land and water surfaces yielded an upper limit for the mean global surface air temperature response function of 0.113 K $(W\ m^{-2})^{-1}$. Multiplying this number by the calculated radiative perturbation of a doubling of the atmospheric CO_2 content thus led Idso (1980b) to conclude that the surface air temperature would warm by only 0.25 K, or less.

The similar empirical results of Newell and Dopplick (1979) and Idso (1980b) posed a clear challenge to the credibility of the numerical model results, which were a full order of magnitude greater. Consequently, criticisms were leveled at Newell and Dopplick by Watts (1980) and at Idso by Schneider et al. (1980) and Leovy (1980). The latter two criticisms were basically philosophical in nature and, hence, were initially answered on an equivalent basis (Idso, 1980c). Watts' remarks were somewhat more specific, however, requiring Newell and Dopplick to reply in more detail.

Newell and Dopplick (1981) contended that both the theoretical and empirical approaches to the problem obtained essentially the same result for the radiative perturbation associated with a doubling of the atmospheric CO_2 concentration and, hence, that the major difference between the two approaches derived from their differing evaluations of the responses characteristics of the atmosphere, specifically, the *surface* air temperature response function. This latter point of emphasis is particularly important, because the empirical approaches have been criticized for dealing only with the surface and failing to account for the fact that energy removed from the surface as latent heat had to reappear somewhere else in the atmosphere when the water vapor within which it was trapped later condensed. Newell and Dopplick admitted that this was so, but noted that this result was just the point — the heat did not reappear at the surface, but had its main influence higher in the atmosphere, as had been shown years before by Newell et al. (1974). They also noted that it was by no means obvious that this release of latent heat would cause an increase in temperature even at cloud level, since part of it would probably be lost via radiation to space and other portions of it could participate in a number of water-cloud-circulation feedback processes, the ultimate consequences of which are not known. We additionally note that another possibility is a resultant change in lapse rate. There is no a priori reason why a temperature increase at high levels must lead to a temperature increase at low levels. Indeed, the eruption of Mt. Agung provided a good example of just the opposite effect, where aerosols that penetrated the tropopause apparently produced a warming of several kelvin in the tropical stratosphere (Newell, 1970, 1971) and a cooling of about 0.5 K in the tropical troposphere (Newell and Weare, 1976).

The second significant criticism leveled against Newell and Dopplick (1979) by Watts (1980) was that they ignored the effects of the great thermal inertia of the oceans. In countering this argument, Newell and Dopplick first noted that Newell and Chiu (1980) had just completed a comparison of the first harmonics of sea surface temperature and surface energy balance components on a seasonal time scale for the subtropical Pacific, finding the analysis to yield a sensitivity factor of 0.04 K (W $m^{-2})^{-1}$, which was almost identical to the response function of Newell and Dopplick (1979). They then proceeded to test the idea of a longer time scale effect by comparing the present climate with the climate of 18 000 years ago.

The basis for this comparison arose from the findings of Delmas et al. (1980) and Oeschger (1980) that the atmospheric CO_2 content at 18 000 BP (before present) was only about half that of the present, as estimated by analyses of air trapped in ice cores. Coupled with the estimate of Newell et al. (1975) that rainfall about that time was perhaps 10% less than at present, and the calculations of Verneker (1972) that incoming solar radiation at 20°N was essentially invariant between then and now, a good case could be made for a reduced input of radiant energy to the subtropical gyre (circulation zone) of the Atlantic at 18 000 BP as compared to the present. However, data obtained from three cores at that site (McIntyre et al., 1976) indicated that the mean annual temperature difference between then and now was only about 0.1 K, and maybe not even that much. Newell and Dopplick (1981) thus rightly noted that "these data do not support Watts' contention that at times longer than the oceanic response time of 20 years, an influence of altered CO_2 or any other process that alters radiation at the surface will be observed in the oceanic temperature."

An even stronger argument against the ocean thermal inertia argument of the numerical modelers was provided by Idso (1982b, c). In an attempt to empirically derive a value for the surface air temperature response function that would include all direct, indirect, and feedback processes of the total Earth-ocean-atmosphere system — both known and unknown — and that would also apply to long time scales and be valid over a wide range of forcing function values, he compared the present climate with that of the Earth without an atmosphere. For the no-atmosphere case, Idso calculated that the equilibrum surface temperature of the planet would be 254.4 K. Since the current mean global surface air temperature is about 288 K, the total 'greenhouse' effect of the atmosphere as now constituted is seen to be 33.6 K.

What is the value of the forcing function that has brought about this surface air temperature response? It is merely the current mean rate of thermal radiant energy emission from the atmosphere to the Earth's surface, since in the initial synthetic equilibrium state, this energy flux is non-existent. Calculating the clear sky component of this flux by the equation of Idso (1981b) and modifying the result to accommodate the effects of clouds as described by Idso (1980b), Idso (1982b, c) obtained a forcing flux of 348 W m^{-2}. Dividing this number into the 33.6 K temperature increment for which it is responsible, he obtained a value for the mean global surface air temperature response function of 0.10 K (W $m^{-2})^{-1}$, thus vindicating the original analyses of Newell and Dopplick (1979) and Idso (1980b).

The excellent agreement among the several different empirical approaches to the problem (Table IV) indicates that the surface air temperature response

TABLE IV

Empirical evaluations of Earth's surface air temperature response function

Investigators	Nature of "natural experiment"	Value of response function
Newell and Dopplick (1979)	Variation of tropical sea surface temperature with available energy	0.03 K $(W\ m^{-2})^{-1}$ over water 0.1 K $(W\ m^{-2})^{-1}$ for globe
Newell and Chiu (1981)	Comparison of first harmonics of temperature and energy balance components of subtropical Pacific	0.04 K $(W\ m^{-2})^{-1}$ over water
Newell and Dopplick (1981)	Variation of temperature of Atlantic subtropical gyre with CO_2 and water vapor variations since 18 000 BP	So small as to elude accurate evaluation
Idso (1982a, b, c)	Vertical redistribution of dust from summer to winter over Phoenix, AZ	0.173 K $(W\ m^{-2})^{-1}$ over land
Idso (1982a, b, c)	Temporal variation of vapor pressure with advent of monsoon season at Phoenix, AZ	0.196 K $(W\ m^{-2})^{-1}$ over land
Idso (1982a, b, c)	Annual variation in mean air temperature caused by yearly cycle of solar radiation at 100 stations in the United States	0.185 K $(W\ m^{-2})^{-1}$ over land $<$ 0.092 K $(W\ m^{-2})^{-1}$ over water $<$ 0.113 K $(W\ m^{-2})^{-1}$ for globe
Idso (1982a, b, c)	Transition from equilibrium state of Earth without an atmosphere to present condition	0.10 K $(W\ m^{-2})^{-1}$ for globe

function is a very stable characteristic of the Earth-ocean-atmosphere system; and this fact implies that it can be used to look at much bigger atmospheric disturbances than merely a simple doubling of the atmospheric CO_2 content. Consequently, Idso (1982b, c) went on to estimate what would happen to the mean surface air temperature of the planet if its effective emittance as measured at the surface were increased from its present value of 0.893 to the maximum value of unity. His analysis indicated that under such extreme circumstances, a temperature increase of just over 4 K was the most that could occur in contrast with the 40 K increase predicted by the numerical models.

One standard of discrimination which may help to decide whether a 4 or 40 K increase is more realistic is the past temperature history of the Earth. The consensus of the Carbon Dioxide and Climate Research Program Conference held in Washington, DC, on 24—25 April 1980 seemed to be that the Earth has probably never been more than 2.5 K warmer than now for the last few million years (Schmitt, 1980), and Matthews and Poore (1980) have suggested that $\delta^{18}O$ data in ocean cores imply a constant ocean temperature reaching back over 50 million years. In light of this immense period of time over which there is no evidence for global surface air temperature change of anything of the magnitude the theoretical numerical models are predicting for the CO_2 'problem', it would seem prudent to consider those predictions as unsubstantiated.

AGRICULTURAL PRODUCTIVITY AND CLIMATE CHANGE

Although we consider the predictions of the numerical models to be unsubstantiated, the question of what will happen to agriculture if they are right needs to be asked. Wittwer (1980) and Rosenberg (1982) have both addressed this question and both concluded that there is sufficient resiliency in agriculture to adapt to the predicted changes. Wittwer notes that during the past century, American farmers have coped with year-to-year variations of 30 mm in rainfall in eastern Kansas and with two week changes in the growing season in Minnesota. From 1915 to 1945 Indiana farmers experienced a +0.1°C per year trend and a total change of +2°C during the past century.

Rosenberg (1982) considered the results of one particular model that predicted increases in average annual temperature of 3—4°C and decreases in soil moisture of 20—50 mm at a latitude of 35—45° N for a hypothetical continent. This may or may not really apply to the Great Plains of North America, but Rosenberg went on to show the change in extent of hard red winter wheat production on the Plains between 1920 and 1980. In 1920 production ranged from northern Texas to central Nebraska and from eastern Colorado to central Illinois. Since then production has adapted to a 88 mm decrease in mean annual precipitation, a 4.2°C decrease in mean annual temperature, and a 10 day decrease in growing season and has spread northwestward across Montana to the Canadian border. It has also adapted to a 100 mm decrease in precipitation, a 0.7°C decrease in temperature, and a 10 day increase in growing season while spreading southward to central Texas. We also note that equivalent yields are obtained in western Nebraska and central Texas, the latter having 300 mm more precipitation but an average annual temperature that is 8.5°C warmer. Thus, agriculture has already demonstrated a capacity to cope with changes of the size predicted.

We discussed earlier how doubling CO_2 concentration increases stomatal resistance and has decreased transpiration an average 34% in short term experiments (Table III). Thus plants already have a mechanism to compensate for the predicted reductions in available soil moisture. Rosenberg's (1982) data show a range in annual precipitation from 357 to 845 mm over the current hard red winter wheat production area. At least half of this is lost as transpiration, so doubled CO_2 has the potential to save more than the predicted soil moisture decreases of 20—50 mm. Indeed, it seems more likely that wheat production can spread into more arid areas, and areas that presently use a summer fallow schedule may be able to crop every year, especially if the model predictions prove to be false.

FUTURE RESEARCH NEEDS

The doubled atmospheric CO_2 concentration in the next century is likely to affect plant production and management under both well-watered and drought conditions. Considerable information is available about crop production with CO_2 enrichment in greenhouses, but relatively little about open-field agriculture. At the recent conference on Rising Atmospheric Carbon Dioxide and Plant Productivity sponsored by the American Association for the Advancement of Science and held in Athens, GA, 23—28 May 1982, many areas of research needing further investigation were identified. Several of these concerned obtaining a better understanding of basic plant physiological processes, so that both present and future plant production can be better modeled, predicted, and managed. It seems likely that higher fertility levels will be required to sustain higher rates of crop growth at the higher CO_2 levels, so additional information is needed about the effects of mineral nutrients, particularly nitrogen, on plant growth and development and on stomatal responses to CO_2 concentration and water stress. Most CO_2 enrichment studies have been performed using high-value greenhouse crops (Table I: tomato, chrysanthemum, and lettuce), so the genetic information base needs to be greatly expanded. Studies of the effects of CO_2 enrichment on many more agronomic and forest species are needed. These studies should include members from the different photosynthetic classes of plants: C_3, C_4, and CAM. Laboratory measurements have shown that photosynthesis of C_4 plants changes relatively less with CO_2 concentrations than that of C_3 plants, so corn (C_4) may become less profitable than C_3 grains. C_3 weeds may become relatively more of a problem in C_4 crops, so more studies of crop-weed competition are needed. Also identified at the Athens conference (and probably of most interest to attendees at this conference) is the need to determine what effects the future higher CO_2 concentration will have on productivity, water use, and water use efficiency of field-grown crops under both well-watered and drought conditions.

With respect to the carbon dioxide-climate question, perhaps the most objective statement that can be made at this time is that it remains just that — a question. Those investigators whose work is based on general circulation models of the atmosphere admit that there are large uncertainties in their results, yet they generally feel that CO_2 effects on climate will be dramatic. If they are correct, we need to take steps now to curtail the release of CO_2 into the atmosphere because Keeling (1982) has indicated that much more than a mere doubling of CO_2 concentration will occur with unrestricted burning of fossil fuels. Indeed, if the ultimate 4- to 6-fold increase he eventually sees as occurring comes to pass, and if the modelers are correct, the Earth will experience a temperature increase on the order of 4—12 K, which may well produce the doomsday scenario they envison.

On the other hand, those investigators who take a more empirical approach to the problem find only a nominal effect of increasing CO_2 on climate and claim little room for error. If they are correct, we may *want* to increase the CO_2 content of the atmosphere; for it appears almost certain that such action would have a tremendous direct effect on plant productivity, boosting it several tens of percent world wide. With our expanding world population, this enhanced agricultural capacity could well mean the difference between war and peace in the years ahead.

Obviously, both of these assessments cannot be right, for they are mutally exclusive; and in light of their ponderous consequences, it is incumbent upon us to determine which is the more correct. We admit to being strongly biased in favor of the nominal effect assessment; but others are equally ardent supporters of the opposite point of view. Thus, more research must be done to break this deadlock. In particular, climate models must be improved to the point where they incorporate all of the important feedback processes that operate in the real world, many of which are presently omitted (Idso, 1982c, d); and they should demonstrate an ability to simulate real-world "natural experiments" (Idso, 1980a).

REFERENCES

Akita, S. and Moss, D.N., 1972. Differential stomatal response between C_3 and C_4 species to atmospheric CO_2 concentration and light. Crop Sci., 12: 789–793.

Akita, S. and Moss, D.N., 1973. Photosynthetic responses to CO_2 and light by maize and wheat leaves adjusted for constant stomatal apertures. Crop Sci., 13(2): 234—237.

Allen, L.H., Jr., 1979. Potentials for carbon dioxide enrichment. In: B.J. Barfield and J.F. Gerber (Editors), Modification of the Aerial Environment of Crops. ASAE Monogr. 2, American Society of Agricultural Engineers, St. Joseph, MI, pp. 500—519.

Arrhenius, S., 1896. The influence of the carbonic acid in the air upon the temperature of the ground. Philos. Mag. (Ser. 5), 41: 237—276.

Bennett, I., 1975. Variation of daily solar radiation in North America during the extreme months. Arch. Meteorol. Geophys. Bioklim. Ser. B, 23: 31—57.

Bolin, B. et al. (Editors), 1979. The global carbon cycle (Scope 13). John Wiley, New York, NY, p. 5.

Carlson, R.W. and Bazzaz, F.A. 1980. The effects of elevated CO_2 concentrations on growth, photosynthesis, transpiration, and water use efficiency of plants. In: J. Singhjag and A. Deepek (Editors), Environmental and Climatic Impact of Coal Utilization. Academic Press, New York, NY, pp. 609—623.

Chamberlin, T.C., 1899. An attempt to frame a working hypothesis of the cause of glacial periods on an atmospheric basis. J. Geol., 7: 545—584, 667—685, 751—787.

Delmas, R.J., Ascencio, J-M. and Legrand, M., 1980. Polar ice evidence that atmospheric CO_2 20,000 BP was 50% of present. Nature, 284: 155—157.

De Sassure, T., 1804. Recherches chemiques sur la vegetation, Paris. Transl. by A. Wieler from Chemische Untersuchungen über die Vegetation, Engelmann, Lepzig, 1890, 22 pp., as cited by D.C. Dalrymple, 1973, A global review of greenhouse food production, U.S. Dep. Agric. Econ. Res. Ser., 44 pp.

Egli, D.B., Pendleton, J.W. and Peters, D.B., 1970. Photosynthetic rate of three soybean communities as related to carbon dioxide levels and solar radiation. Agron. J., 62(3): 411—414.

Enoch, H.Z. and Hurd, R.G., 1979. The effect of elevated CO_2 concentrations in the atmosphere on plant transpiration and water use efficiency. A study with potted carnation plants. Int. J. Biometeorol., 23: 343–351.

Gifford, R.M., 1979a. Carbon dioxide and plant growth under water and light stress: implications for balancing the global carbon budget. Search, 10: 316–318.

Gifford, R.M., 1979b. Growth and yield of CO_2-enriched wheat under water-limited conditions. Aust. J. Plant Physiol., 6: 367—378.

Gribbin, J., 1981. The politics of carbon dioxide. New Sci., 90: 82—84.

Hales, J.E., Jr., 1974. Southwestern United States summer monsoon source — Gulf of Mexico or Pacific Ocean? J. Appl. Meteorol., 13: 331—342.

Idso, S.B., 1974. Thermal blanketing: a case for aerosol-induced climatic alteration. Science, 186: 50—51.

Idso, S.B., 1980a. Human actions and nature's reactions. Am. Q. Assoc. Abstr., 6: 105—106.

Idso, S.B., 1980b. The climatological significance of a doubling of Earth's atmospheric carbon dioxide concentration. Science, 207: 1462—1463.

Idso, S.B., 1980c. Carbon dioxide and climate. Science, 210: 7—8.

Idso, S.B., 1981a. An experimental determination of the radiative properties and climatic consequences of atmospheric dust under nonduststorm conditions. Atmos. Environ., 15: 1251—1259.

Idso, S.B., 1981b. A set of equations for full spectrum and 8—14 μm and 10.5—12.5 μm thermal radiation from cloudless skies. Water Resour. Res., 17: 295—304.

Idso, S.B., 1982a. A surface air temperature response function for Earth's atmosphere. Boundary-Layer Meteorol., 22: 227—232.

Idso, S.B., 1982b. An empirical evaluation of Earth's surface air temperature response to an increase in atmospheric carbon dioxide concentration. In: R.A. Reck and J.R. Hummel (Editors), AIP Conf. Proc. 82, Interpretation of Climate and Photochemical Models, Ozone and Temperature Measurements. American Institute of Physics, New York, NY, pp. 119—134.

Idso, S.B., 1982c. Carbon Dioxide: Friend or Foe? An inquiry into the climatic and agricultural consequences of the rapidly rising CO_2 content of Earth's atmosphere. IBR Press, Tempe, AZ, 92 pp.

Idso, S.B., 1982d. Shortcomings of CO_2-climate models raise questions about the wisdom of their energy policy implications. Appl. Energy (Submitted).

Idso, S.B. and Brazal, A.J., 1978. Climatological effects of atmospheric particulate pollution. Nature, 274: 781—782.

Idso, S.B. and Kangieser, P.C., 1970. Seasonal changes in the vertical distribution of dust in the lower troposphere. J. Geophys. Res., 75: 2179—2184.

Keeling, C.D., 1982. The oceans and terrestrial biosphere as future sites for fossil fuel CO_2. In: R.A. Reck and J.R. Hummel (Editors), AIP Conf. Proc. 82, Interpretation of Climate and Photochemical Models, Ozone and Temperature Measurements. American Institute of Physics, New York, NY, pp. 47—82.

Kimball, B.A., 1982. Carbon dioxide and agricultural yield: an assemblage on analysis of 430 prior observations. WCL Rep. 11, U.S. Water Conservation Laboratory, Phoenix, AZ, 48 pp.

Kramer, P.J., 1981. Carbon dioxide concentration, photosynthesis, and dry matter production. Bioscience, 31: 29—33.

Lemon, E.R., 1977. The land's response to more carbon dioxide. In: N.R. Anderson and A. Malahoff (Editors), The Fate of Fossil Fuel CO_2 in the Ocean, U.S. Office of Naval Research, Symposium Series in Oceanography, Plenum Publishing Corp., New York, NY, 97—130.

Leovy, C.B., 1980. Carbon dioxide and climate. Science, 210: 7.

Manabe, S. and Wetherald, R.T., 1967. Thermal equilibrium of the atmosphere with a given distribution of relative humidity. J. Atmos. Sci., 24: 241—259.

Matthews, R.K. and Poore, R.Z., 1980. The tertiary S180 record: An alternative view concerning glacio-eustatic sea level fluctuations. Geology, 8: 501—504.

McIntyre, A., Kipp, N.A., Be, A.W.H., Crowley, T., Kellogg, T., Gardner, J.V., Prell, W. and Ruddiman, W.F., 1976. Glacial North Atlantic 18,000 years ago: a CLIMAP reconstruction. In: R.M. Cline and J.D. Hays (Editors), Investigation of Late Quaternary Paleoceanography and Paleoclimatology. Geol. Soc. Am. Mem., 145: 43—76.

Moller, F., 1963. On the influence of changes in CO_2 concentration in air on the radiative balance of the Earth's surface and on the climate. J. Geophys. Res., 68: 3877—3886.

Moss, D.N., Musgrave, R.B. and Lemon, E., 1961. Photosynthesis under field conditions. III. Some effects of light, carbon dioxide, temperature, and soil moisture on photosynthesis, respiration, and transpiration of corn. Crop Sci., 1: 83—87.

NAS, 1979. Carbon dioxide and climate: A scientific assessment. National Academy Press, Washington, DC, 22 pp.

Newell, R.E., 1970. Modification of stratospheric properties by trace constituent changes. Nature, 227: 697—699.

Newell, R.E., 1971. The global circulation of atmospheric pollutants. Sci. Am., 224: 32—42.

Newell, R.E., and Chiu, L.S., 1980. Climatic changes and variations: a geophysical problem. Proc. NATO Advanced Study Institute on Climatic Variations and Variability: Facts and Theories, Erice, Sicily. Reidel, Dordrecht, The Netherlands.

Newell, R.E. and Dopplick, T.G., 1979. Questions concerning the possible influence of anthropogenic CO_2 on atmospheric temperature. J. Appl. Meteorol., 18: 822—825.

Newell, R.E. and Dopplick, T.G., 1981. Reply to Robert C. Watts' "Discussion of 'Questions concerning the possible influence of anthropogenic CO_2 on atmospheric temperature." J. Appl. Meteorol., 20: 114—117.

Newell, R.E. and Weare, B.C., 1976. Factors governing tropospheric mean temperature. Science, 194: 1413—1414.

Newell, R.E., Vincent, D.G. and Boer, G.J., 1974. The General Circulation of the Tropical Atmosphere and Interactions with Extratropical Latitudes, Vol. 2. MIT Press, Cambridge, MA, 371 pp.

Newell, R.E., Gould-Stewart, S. and Tanaka, M., 1975. Decreased global rainfall during the past Ice Age. Nature, 253: 33—34.

Newell, R.E., Navato, A.R. and Hsiung, J., 1978. Long-term global sea surface temperature fluctuations and their possible influence on atmospheric CO_2 concentrations. Pure Appl. Geophys., 116: 351—371.

Oeschger, H., 1980. The history of atmospheric CO_2 as revealed from ice core studies. WMO Project on Research and Monitoring of Atmospheric CO_2. Rep. No. 1, Appendix G. World Meteorological Society, Geneva.

Pallas, J.E., 1965. Transpiration and stomatal opening with changes in carbon dioxide content of the air. Science, 147: 171—173.

Plass, G.N., 1956a. Effect of carbon dioxide variations on climate. Am. J. Phys., 24: 376—387.

Plass, G.N., 1956b. The influence of the 15-micron carbon dioxide band on the atmospheric infrared cooling rate. Q. J. R. Meteorol. Soc., 82: 310—324.

Plass, G.N., 1957. The carbon dioxide theory of climatic change. In: Recent Research in Climatology. Committee on Research in Water Resources, University of California, La Jolla, CA, pp. 81—92.

Plass, G.N., 1961. The influence of infrared absorptive molecules on the climate. Ann. N.Y. Acad. Sci., 95: 61—71.

Reck, R.A., 1982. Confidence in climate models including those with particles. In: R.A. Reck and J.R. Hummel (Editors), AIP Conf. Proc. 82, Interpretation of Climate and Photochemical Models, Ozone and Temperature Measurements. American Institute of Physics, New York, NY, pp. 23—33.

Regehr, D.L., Bazzaz, F.A. and Boggess, W.R., 1975. Photosynthesis, transpiration, and leaf conductance of *Populus deltoides* in relation to flooding and drought. Photosynthetica, 9: 52—61.

Rogers, H.H., Beck, R.D., Bingham, G.F., Curie, J.D., Davis, J.M., Heck, W.W., Rawlings, J.O., Riordan, A.J., Sionit, N., Smith, J.M. and Thomas, J.F., 1981. Response of vegetation to carbon dioxide: Field studies of plant responses to elevated carbon dioxide levels. Re. 005, Carbon Dioxide Research Division, U.S. Department of Energy/Agricultural Research Service, U.S. Department of Agriculture, Washington, DC, 108 pp.

Rosenberg, N.J., 1981. The increasing CO_2 concentration in the atmosphere and its implication on agricultural productivity. I. Effects on photosynthesis, transpiration and water use efficiency. Clim. Change, 3: 264—279.

Rosenberg, N.J., 1982. The increasing CO_2 concentration in the atmosphere and its implication on agricultural productivity II. Effects through CO_2-induced climatic change. Clim. Change, 4: 239—254.
Schmitt, L.E. (Editor), 1980. Carbon Dioxide Effects, Research and Assessment Program No. 011. Proc. Carbon Dioxide and Climate Research Program Conference, 24—25 April 1980, Washington, DC. U.S. Department of Energy, Washington, DC.
Schneider, S.H., Kellogg, W.W., and Ramanathan, V., 1980. Carbon dioxide and climate. Science, 210: 6—7.
Sionit, N., Hellmers, H. and Strain, B.R., 1980. Growth and yield of wheat under CO_2 enrichment and water stress. Crop Sci., 20: 687—690.
Van Bavel, C.H.M., 1972. Water use efficiency in plant growth and ambient carbon dioxide level. Tech. Rep. 42, Texas Water Resources Institute, Texas A & M University, College Station, TX, 115 pp.
Verneker, A.D., 1972. Long-period global variations of incoming solar radiation. Meteorol. Monogr. 34, American Meteorological Society, Boston, MA, 128 pp.
Watts, R.G., 1980. Discussion of 'Questions concerning the possible influence of anthropogenic CO_2 on atmospheric temperature'. J. Appl. Meteorol., 19: 494—495.
Weare, B.C. and Snell, F.M., 1974. A diffuse thin cloud structure as a feedback mechanism in global climatic modeling. J. Atmos. Sci., 31: 1725—1734.
Wittwer, S.H., 1978. Carbon dioxide fertilization of crop plants. In: U.S. Gupta (Editor), Crop Physiology. Oxford/IBH, New Delhi, India.
Wittwer, S.H., 1980. Carbon dioxide and climatic change: an agricultural perspective. J. Soil Water Conserv., 35: 116—120.

Agricultural Water Management, 7(1983) 73—88
Elsevier Science Publishers B.V., Amsterdam — Printed in The Netherlands

POTENTIAL FOR PLANT ENVIRONMENT MODIFICATION

BILLY J. BARFIELD[1] and JOHN M. NORMAN[2]

[1] *Department of Agricultural Engineering, University of Kentucky, Lexington, KY (U.S.A.)*
[2] *Department of Agronomy, East Campus, University of Nebraska, Lincoln, NE 68583-0817 (U.S.A.)*

Published as Paper No. 7187 Journal Series Nebraska Agricultural Experiment Station

(Accepted 20 January 1983)

ABSTRACT

Barfield, B.J. and Norman, J.M., 1983. Potential for plant environment modification. *Agric. Water Manage.*, 7: 73—88.

Plants have a remarkable ability to cope with diverse environments. However, the whims of man can lead to plants being so misplaced that environmental modification may be a useful enterprise for improving productivity. Unfortunately, the potential for modifying canopy environments on large scales under drought conditions is limited. The canopy radiation balance can be modified with misting, crop covers or albedo changes, and canopy temperatures modified by area fogging to cool the air or intermittent sprinkling to cool the leaves. Perhaps the most common environmental modification derives from windbreaks, which reduce air movement. The last promising method for large areas is CO_2 enrichment.

Comprehensive plant-environment models can be valuable tools for assessing the impact of various environmental modification techniques because of the complexity of plant-environmental relations. Two examples of model applications are: (a) CO_2 enrichment; and (b) retarding spider mite development with irrigation. The results suggest that field CO_2 enrichment is not practical. Retarding spider mite development by 3 to 6 days may be possible.

INTRODUCTION

Plants have a remarkable ability to cope with diverse environments. This adaptability is essential because plants lack the mobility necessary to avoid extreme environments. Because the productivity of cultivated plants can be severely affected by brief exposure to extreme environmental conditions, modifying the plants' environment to minimize damage seems worthwhile. Actually, such environmental modification is commonplace on small scales: sprinkling the lawn, covering the tomatoes on a cold night, the misting of house plants or the careful placement of certain house plants in sunny or shaded locations. Large scale plant environment modification for the ameli-

0378-3774/83/$03.00

oration of extreme high temperature or drought effects has met with limited success. The following includes summarization of the present state of knowledge, illustration of several possible methodologies, and discussion of some potential benefits.

Plant factors affected by drought

Environmental modifications that minimize the effect of drought on productivity reductions can focus on two factors: (1) plant-part temperatures; (2) and plant water status. Obviously plant water status and plant temperature are interrelated. The temperature of any particular plant part is an integrated result of both environmental and plant factors (Fig. 1). Energy budgets for leaves, flowers and fruits have been measured, and equations describing these energy exchanges are available (Raschke, 1960; Norman, 1979).

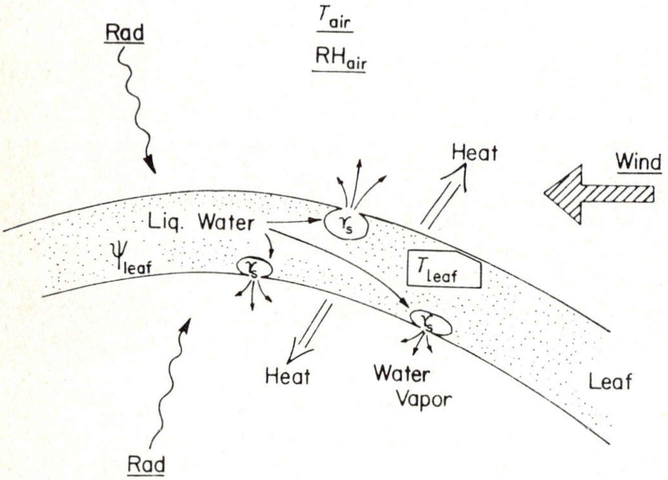

Fig. 1. Diagram of some of the factors involved in a leaf energy budget, where ψ_{leaf} is the leaf water potential and γ_s is the stomatal resistance (Norman, 1982).

The temperatures of various plant parts affect many important plant processes: for example, phenological development, photosynthesis, growth and maintenance respiration, translocation, water uptake, stomatal conductance and disease resistance. As a plant grows, all plant parts are not at the same temperature nor are they necessarily near air temperature. Leaves, which have numerous stomata, routinely experience temperatures from 5°C or more below air temperature (high air temperature and high stomatal conductance) to 5°C or more above air temperature (low stomatal conductance). Flowers without stomata may experience temperatures 10°C or more above ambient air temperature. In many grasses, like corn or sorghum, meristematic tissue remains below ground for much of the vegetative development period so soil temperatures probably are more important than air temperatures in early plant development. Sterility in reproductive organs is common at very high temperatures. Nontranspiring plant parts exposed to high solar radiation levels are particularly vulnerable to high temperature damage when wind speeds are low. Presumably sprinkling these organs could reduce vulnerability to high temperature damage and might be desirable even when the crop has ample available water for transpiration.

If plant water stress occurs in hot environments, elevated plant temperatures can result from stomatal closure. Water stress can also directly affect numerous plant processes independently of temperature and stomatal effects; notable examples are cellular expansion, nitrate reductase activity, respiration, and possibly photosynthesis (Hsiao, 1973).

The temperature of plant parts also can exhibit a major influence on the development rate of pests that may be damaging to a crop. In some cases, modification of the plant environment might retard the development of a pest sufficiently to reduce damage to tolerable levels.

Environmental factors to be modified

The main environmental factors that influence plant growth and development include solar radiation (or net radiation), air temperature, air water vapor content, wind speed, soil temperature and soil water availability. The solar radiation absorbed by a canopy has a major influence on plant temperature and water use. In fact, transient water stress can be experienced by plants with adequate available soil water when excessive transpiration rates would be required to dissipate the absorbed solar radiation. The combination of high incident solar radiation with high air temperature, moderate wind speeds and low relative humidities leads to transient stress conditions. Since the absorbed solar radiation is dependent on leaf spectral properties, there may be some potential for reducing transient water stress by modifying leaf reflectance (Baradas et al., 1976).

Well watered crops tend to have foliage temperatures above ambient air temperatures at lower air temperatures (perhaps less than 30 to 35°C), while at higher air temperatures (above 30 to 35°C) the canopy temperature is usually below air temperature. This relation between crop temperature and air temperature occurs primarily because increases in temperature cause even greater increases in vapor-pressure-deficit (VPD) between the crop and the air, and thus greatly increase transpiration if adequate water is available. Relative humidity of the air also affects the VPD and thus the relation between canopy and air temperature. Misting or sprinkling can increase the air relative humidity and decrease VPD while also decreasing the air temperature.

Wind speed also affects crop temperature. At very low wind speeds (near free convection), crop temperature is always above air temperature even in well watered canopies. As wind speeds increase to moderate levels, the crop temperature decreases by an amount that depends on the VPD and may be depressed well below air temperature. At high wind speeds the crop temperature approaches the air temperature. Thus evaluating the effect of wind breaks on crop temperature and water use is a complex process that involves all of the environmental variables. If soil water is not available in sufficient quantities, stomatal closure results, VPD becomes less important, and absorbed solar radiation is dissipated mainly by sensible heat convention through elevated canopy temperatures. Obviously all of the environmental factors interact with each other so that it is difficult to assess the impact of altering one factor without considering all factors. Some predictions of the effects of water availability, VPD, solar radiation and wind speed on canopy temperature departure from air temperature from the plant-environmental model, Cupid (Norman, 1982) are shown in Fig. 2.

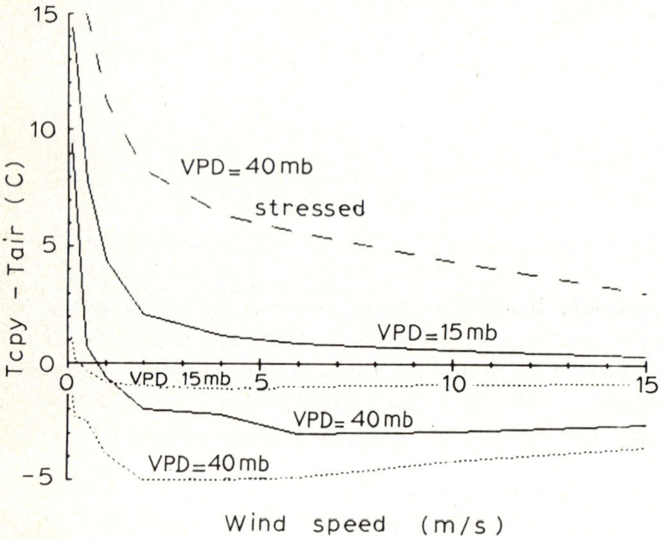

Fig. 2. Predictions, from the plant-environment model Cupid, of the departure of canopy temperature from air temperature as a function of windspeed, vapor pressure deficit of the air and solar radiation (——, full sun; ...., 1/3 full sun) for stressed (– – –) and unstressed (———, ...) corn, 2 m tall.

METHODOLOGIES FOR MODIFYING THE ENVIRONMENT

Proposed methodologies for modifying the environment and improving crop production have ranged in complexity from simple individual crop covers to computer controlled sprinkler systems for evaporatively controlling air temperature. Techniques for modifying the environment to enhance production can be categorized into those techniques that alter the energy balance of a crop and its environment, and those techniques that modify the mass balance of the crop.

A good framework to conceptualize the potential for modifying the environment of crops is derived from the energy and mass budget of a crop canopy (Fig. 3). One can alter the energy balance or mass balance of the box by altering any one of the input or output terms. As a result of the alteration, there will be a corresponding, typically nonlinear, response in the plant. Examples of techniques for making these modifications are given in Table I and discussed briefly below. Detailed case studies are given in a subsequent section.

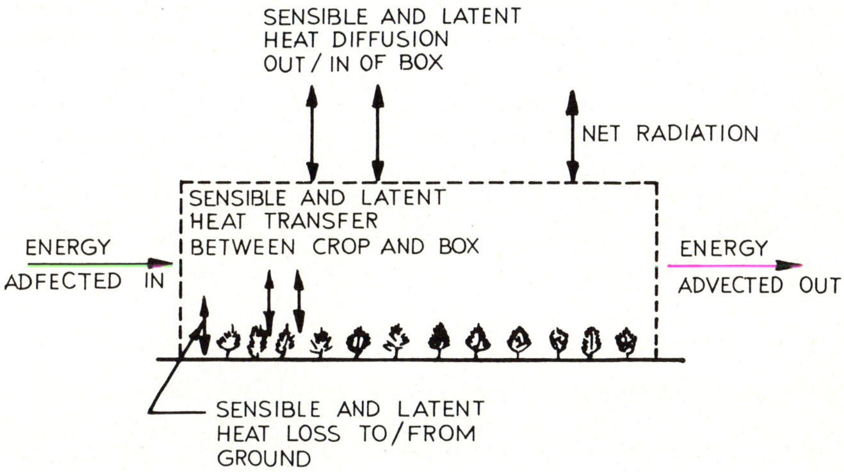

Fig. 3. Box-mass balance for a crop.

TABLE I

Techniques previously used to modify the energy and mass balance of the crop environment shown in Fig. 3

Point of Modification	Technique used
1. Modifying fluxes at the top of the box	a. Misting — modifies radiation flux divergence b. Porous crop covers — modifies net radiation flux, reduce diffusion of heat and mass through the cover
2. Modifying advective fluxes	a. Line source mist nozzles — increases vapor content while cooling the air moving into the box b. Windbreaks — reduces air movement into the box
3. Modifying fluxes at the crop surface	a. Intermittent sprinklers — wets the crop and soil surface. Reduces the internal diffusion resistance to lower crop temperature b. Albedo modification — varies the amount of radiation reflected and absorbed at the surface
4. Internal generation and sources	a. Area fogging — reduces the air temperature by evaporatively cooling the air b. CO_2 Enrichment — increasing the CO_2 concentration of the canopy air

Misting for evaporative cooling

Misting for evaporative cooling can be accomplished by sprinkling on either an area basis or from a line source. Temperature reduction from cooling results from the use of sensible heat in the air to evaporate the water drops. Although the total energy of the resulting system remains unchanged, the loss in sensible heat results in a decrease in temperature. The efficiency of the process increases as the drop size decreases.

Nurnberger et al. (1976) and Merva and Vanderbrink (1979) used an elevated line source to generate mist which subsequently flowed over a plowed field. They verified the use of turbulent heat and mass diffusion equations to predict downwind changes in air temperature and vapor pressure. Using this model, they proposed that the potential for temperature reduction downstream of an elevated line source ranges from 10°C at windspeeds of 0.5 m/s to 2.2°C at windspeeds of 4 m/s. Although the magnitude of the reduction depends on the relative humidity, no relative humidities were given.

Sprinkling to increase evapotranspiration

Evaporation from a leaf with a dry surface occurs due to evaporation within the substomatal cavity. Because the stomatal opening imposes a diffusion resistance to movement of water between the leaf and atmosphere, the amount of latent heat leaving the leaf surface is less than that of leaf with a wet surface. This resistance will cause the temperature of a leaf with a dry surface to be higher than a wetted leaf. Chessness et al. (1979) summarize the results of several studies showing a leaf temperature reduction of up to 10°C with sprinklers.

Another possible benefit of keeping the leaf surface wet is a reduction in transpiration demand. On very hot, dry, windy days when the transpiration demand is high enough to cause water stress in well watered plants, the wetting of the leaf surface will reduce the transpiration demand. An example simulation of the potential benefit is shown in Fig. 4 using the Cupid model of Norman (1982).

Intermittent sprinkling can also have a potentially beneficial effect on the plant by cooling the soil surface. This cooling can result in a lowered radiation load on the plants and in some cases can reduce the incidence of insects like spider mites which initially infest the lower parts of the canopy before moving upward (Toole, 1982). Details of a case study of the spider mite problem are presented in a subsequent section.

Shelterbelts

Shelterbelts (or windbreaks as they are commonly known) are barriers, that shelter crops growing on the lee side of the shelterbelt. These shelter-

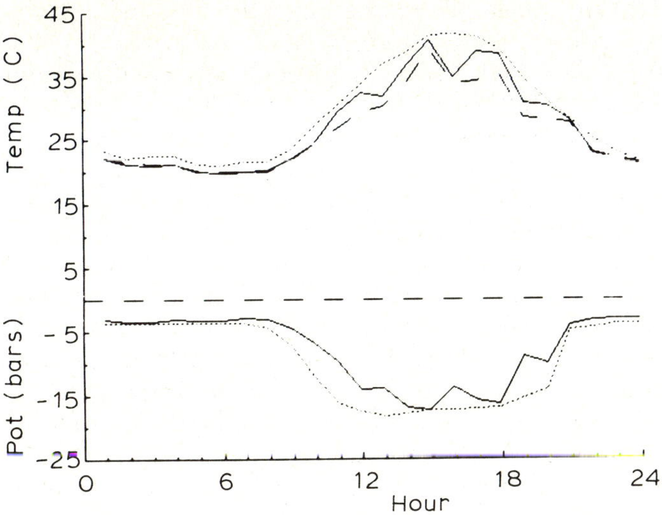

Fig. 4. Predictions of the reduction in corn tassel temperatures, canopy temperatures, and canopy leaf water potential with irrigation of 1 mm every 3 h (from the model Cupid). The dotted lines (....) represented tassel temperature and canopy water potential without irrigation. The solid lines (——) represent tassel temperature and canopy water potential with 1 mm/3 h of irrigation, and the dashed line (— — —) is the average canopy temperature with irrigation.

belts are typically composed of a combination of tall and low growing vegetation with visible porosities of 40 to 60%. Shelters are typically spaced 10 to 20 times the height of the tallest vegetation. Windbreaks have also been established by growing short crops between narrow strips of tall crops.

Rosenberg (1979) summarizes much of the literature on shelterbelts, which indicated that shelterbelts generally increase yields of horticulture and field crops as well as orchards. The exact mechanisms for reducing plant stress and increasing yields is not well established. Rosenberg proposes the following summary of environmental effects of shelterbelts:

Windspeed and turbulence — Wind speed and turbulence are both reduced in shelterbelts resulting in reduced advection and turbulent diffusion.

Radiation balance — Changed very little except near the barrier itself.

Temperatures — Generally higher during the day in the shelter and cooler at night.

Absolute humidity — Higher throughout the day and night.

Sensible heat and vapor transport — Reduced as a result of reduced turbulence.

Transpiration and evaporation — Initially lower in the shelterbelt after a rainfall event due to suppressed turbulence. Ultimately higher than non sheltered crop because of increased water availability.

Photosynthesis — Generally greater in the shelterbelt.

Snow storage — Another advantage of windbreaks is that they operate as a snow fence resulting in additional water storage in the soil profile at the start of the growing season. However, U.S.S.R. studies cited by Rosenberg indicate that this advantage is quickly dissipated.

Recent research indicates that sheltered soybeans experience higher leaf photosynthetic rates, higher leaf stomatal conductances and higher (nearer to zero) leaf water potentials (Ogbuehi and Brandle, 1981). Furthermore, sheltered soybeans have a much different canopy architecture that permits deep light penetration suggesting further possible increases in canopy photosynthesis (Ogbuehni and Brandle, 1982).

In summary, shelterbelts are apparently cost effective methods of reducing environmental stress and increasing plant productivity in semiarid lands.

Albedo modification

The temperature difference between a crop surface and the ambient air is a function of environmental parameters (net radiation, relative humidity and wind speed) as well as a crop resistance, or

$$T_c - T_a = R_n - \frac{B\,(1 - \mathrm{RH})}{r_a + r_c} \tag{1}$$

where T_c and T_a (°C) are crop and ambient temperatures, r_a (s/m) the air resistance which determines the resistance to diffusion of heat and mass between the crop and free air, r_c (s/m) is the crop resistance which is a measure of the diffusion resistance from within the crop leaves to the leaf surface, RH is the relative humidity, and B is a constant. The parameter r_a is a strong function of windspeed while the parameter r_c is a strong function of the stomatal aperture. From Eq. (1), it can be seen that the temperature of the crop surface increases with an increase in net radiation and relative humidity if r_a, r_c, and T_a are held constant. This fact has led researchers to search for ways to decrease R_n in an attempt to decrease crop temperature. One method for decreasing R_n is to increase the reflected radiation which is accomplished by increasing the reflectivity or albedo of the surface.

One method of increasing the surface albedo would be to spray the surface with a reflectant. Experimental and theoretical studies at Nebraska (Doraiswamy and Rosenberg, 1974; Lemeur and Rosenberg, 1975, 1976) using a kaolinite spray on soybeans indicate that albedo's in the visible wavelengths can be increased by as much as 16 to 20 percent (Blad and Lemeur, 1979). The actual albedo depends not only on the reflectivity of the reflectant, but also on the leaf angle distribution. In erectophile canopies portions of the incoming radiation are reflected deeper into the canopy as well as back to the atmosphere. The more horizontal leaves in the planophile canopy will primarily reflect radiation back to atmosphere. Therefore, increasing the reflectivity of leaves in a planophile canopy will result in a greater increase in the crop albedo than a corresponding erectophile canopy. Since the upper leaves are more likely to be light saturated than the lower leaves, it is possible that the radiation reflected by the erectophile leaves could increase the water use efficiency of the canopy. Experimental verification of the concept is needed.

Experimental studies on the use of reflectants have focused on the albedo of the crop. Evaluations need to be made of their effect on photosynthesis. The reflectants may also cover the stomata of the leaf and potentially limit the diffusion of CO_2 into the leaves. Thus, it is possible that the use of reflectants could reduce crop stress and at the same time reduce photosynthesis. This result would be an acceptable response in the case where lethal stresses might otherwise occur.

Reflectants have been used in the past to prevent sun scalding on fruit and vegetable crops. Both theoretical and field studies of the effects of reflectants on photosynthesis and water use efficiency are needed before their use can be recommended for agronomic crops. To date their use has not been economically justified.

Blad and Lemeur (1979) propose that the radiation balance of the canopy can be altered by varying the plant morphological characteristics like plant color, pubescence, and awns. They summerize research showing that lighter colored isolines of plants reflect more radiation than darker colored isolines and are slightly cooler. Because the increased albedo could lead to reduced photosynthesis, the selection of isolines for increased albedo should be guided by a theoretical analysis of the complex interactions which occur between leaf angle, color, reflectivity, radiation absorption, and photosynthesis. A model like the Cupid model (Norman, 1982) could be very helpful for such a study.

Shading

Artificial shades are commonly used to reduce the radiant energy load on high-valued ornamentals and vegetable crops. Blad and Lemeur (1979) summarize research showing that the temperature under shades is reduced 1 to 3°C compared to nonshaded areas and that the relative humidity is increased. The net effect is to reduce evapotranspiration in the shade which should result in a reduction in water stress. The actual reduction in radiant energy will depend in the porosity of the shade. Because of the high capital cost of shades, their use is probably justified only for high value crops.

CO_2 *Enrichment*

Since carbon dioxide (CO_2) is fixed in photosynthesis to form the materials used for crop growth and development, researchers have proposed that increasing the CO_2 content of the atmosphere will lead to enhanced crop growth and production. Laboratory studies on individual leaves have verified that photosynthesis is increased by CO_2 enrichment, but field studies indicate that the potential for economic field application for agronomic crops is seriously limited. Allen (1979) presents an excellent detailed discussion of the research on the subject. Kimball and Idso (1983, this issue) discuss the effect of CO_2 enrichment on leaf photosynthesis and water use.

Studies on enrichment of CO_2 in the greenhouse environment have lead to the conclusion that CO_2 enrichment can enhance crop production. Allen (1979) summarized the results of greenhouse studies showing that the increase in yields range from near zero to over 500%, depending on the type of crop, researcher and enrichment level. Where ventilation is required for temperature control, it is economically infeasible to enrich the greenhouse atmosphere.

Studies of CO_2 enrichment under open field conditions have not been encouraging. The studies have shown that only a small fraction of the CO_2 released is actually fixed in photosynthesis. Most of the released CO_2 diffuses into the atmosphere. Allen et al. (1971) used a crop canopy diffusion model to predict the efficiency with which CO_2 added to the canopy is fixed for release rates ranging from 225 to 450 kg ha^{-1} h^{-1}. The predicted efficiencies ranged from a low of 2% for a tall open crop like corn with a release rate of 450 kg ha^{-1} h^{-1} to a high of 10% for a short dense crop with erect leaves with a release rate of 225 kg ha^{-1} h^{-1}. In all cases, the CO_2 was assumed to be released at the ground. Similar simulations by Waggoner (1969), Duncan and Barfield (1970) and field studies by Allen (1973, 1974, 1975a) have shown that the release rates must be high to have a significant impact and that the trapping efficiency is very low.

The rapid loss of CO_2 by turbulent diffusion in the field studies led to an evaluation of field barriers. Allen (1973, 1975b) used a 0.81 ha shade cloth tent as a diffusion barrier over soybeans. The shade cloth reduced the eddy diffusivity to about 25% of that without the tent, but CO_2 was still rapidly transported away by turbulent diffusion. Buffer crop vertical barriers were used by Kretchman (1960, 1970) in conjunction with CO_2 release in a large-scale field study with a range of horticultural crops. Yield increases ranged from 3 to 18%.

Much attention has recently been focused on the increased concentration of CO_2 in the atmosphere and its effects on crop production. This topic is discussed by Kimball and Idso (1983).

APPLICATION OF COMPUTER MODELS

The interaction of plants with their environment is so complex that it is frequently difficult to predict the response to a modified environment by using simple reasoning. The use of a computer simulation model can greatly facilitate the prediction process through the capacity for evaluating complex interactions which occur within the plant and between the plant and its environment. These models can be used to point to possibilities which might not otherwise have been considered. Two examples are given here both of which have been tested with field studies.

Example application I — CO_2 Enrichment

Prediction of the effects of CO_2 enrichment is a good example of the use of a rather simple computer model to predict the effects of environmental modification on crop production.

The modeling process begins with a model of single leaf photosyntheses rates per unit area, QP_{leaf}, as a function of absorbed photosynthetically active radiation, PAR, leaf temperature T_{leaf} and CO_2 concentration in the air adjacent to the leaf, χ_a, or:

$$QP_{leaf} = f(PAR, T_{leaf}, \chi_a) \tag{2}$$

This functional relationship may take the form of an empirical relationship as that used by Duncan and Barfield (1970) or a theoretical equation based on diffusion through the stomata and mesophyll as that used by Waggoner (1969), Allen et al. (1971) and Norman (1982). The more theoretical approach is desirable, but the empirical approach also yielded satisfactory results for this application.

The next step involves the integration of the single leaf model into a canopy micrometeorological model. If steady state conditions are assumed, the turublent diffusion equation can be written as:

$$(QP_{leaf})\left(\frac{\partial F}{\partial z}\right) - Q_s = \frac{\partial}{\partial z} K \frac{\partial \chi}{\partial z} \tag{3}$$

where F is the distribution which describes leaf area distribution over the canopy height, Q_s is the source of CO_2 per unit area of horizontal surface, K is the eddy diffusivity, and χ is the CO_2 concentration. Based on experimental data, K can be predicted empirically. Equations (2) and (3) can thus be solved simultaneously to obtain QP_{leaf} versus z for a given source strength. The solution is subject to the following boundary conditions:

$$\chi = \chi_r \quad \text{at} \quad Z = H \tag{4}$$

where χ_r is the measured ambient CO_2 concentration at a reference height H above the canopy and:

$$K \frac{\partial \chi}{\partial z} = F_{CO_2} \quad \text{at} \quad Z = 0 \tag{5}$$

where F_{CO_2} is the rate at which CO_2 is evolved from the soil (usually negligible). Typically, numerical techniques are used for the solution. Once a distribution of QP_{leaf} versus height is obtained, the total canopy photosynthesis can be found by summing over the entire canopy. These computations were performed by Waggoner (1969), Duncan and Barfield (1970) and Allen et al. (1971). All of the predictions showed that the rate of photosynthesis in the field could be enhanced slightly by CO_2 enrichment, but that the percent of added CO_2 which is fixed in photosynthesis is less than 10 percent and typically in the range of 2%. Field studies by Allen (1973, 1974, 1975a) confirm these results.

Example application II — Effect of irrigation on spider mite development

The development rate of most pests is very sensitive to the local temperature environment. The Banks Grass Mite (BGM), *Oligonychus pratensis* (Banks), is a good example with developmental rates increasing rapidly to optima near 36° C and significant rates at temperatures above 40° C (Feese and Wilde, 1977). In principle, reducing the temperatures experienced by the BGM with irrigation could slow mite development rates and possibly delay the occurrence of damaging population sizes until later in the season when their presence poses a less serious threat to the maintenance of yield. Field studies have indicated that irrigation tends to reduce BGM population in corn (Chandler et al., 1979). A detailed simulation of BGM development and damage in corn has been attempted at the University of Nebraska through a joint research effort between Entomology and Agronomy. A detailed plant-environmen model (Norman, 1982), which predicts local environmental conditions near developing mites, is combined with a dynamic mite population age distribution submodel (Toole, 1982).

The comprehensive plant-environmental model Cupid is used to describe the processes and interactions occurring in the soil-plant-atmosphere system with a minimum of input data (Norman, 1979; 1982). Input requirements fall into five classes: (a) ambient environment; (b) plant characteristics; (c) soil characteristics; (d) site factors; and (e) initial conditions. Ambient environmental requirements can be met with remote automated weather station measurements of hourly solar radiation, air temperature, air relative humidity, wind speed and precipitation at several meters above the crop with daily measurements of soil temperature and soil water at 1- or 2-m depth. The most important plant characteristics needed are leaf-area-index, canopy height, rooting depth, and the dependence of stomatal conductance and photosynthesis on light, temperature and leaf water potential. Soil texture, bulk density, soil water release curves, and thermal and hydraulic properties as a function of water content are important inputs. Site factors like latitude, longitude and slope are simple to obtain. When the model is first started, an initial profile of soil temperature and soil water content is necessary.

The structure of Cupid is based on defining the relation between individual foliage elements and their immediate environment, then integrating over the canopy to determine the collective effect of all the elements considered. To accomplish this result, the canopy is divided into layers, and leaves within each layer are divided into leaf-angle classes. Soil and the atmosphere immediately above the canopy also are divided into layers. This structuring enables the boundary conditions for the soil and atmosphere to be defined so that profiles of air temperature, and vapor pressure, leaf temperature, soil water content, intercepted rainfall or irrigation, dew formation and thus leaf wetness duration can be generated.

The spider mite submodel is an *m*th order distributed delay model (Manetsch and Park, 1974) that describes the population growth and maturity distribution of individuals within a population (Toole, 1982). Using data from laboratory studies to estimate survivorship, developmental rates and fecundity, initial field counts of mites, and environmental predictions from Cupid, this submodel estimates the change in mite numbers and their damage impact. Spider mites can affect water vapor diffusion resistances so a feedback exists between mite damage and diffusive resistance.

Predictions from the mite-plant-environment model were compared to field measurements on corn at a sandy, irrigated site in central Nebraska (Tools, 1982) in 1981. The results show model predictions always within one standard deviation of measurements and are particularly encouraging since no empirical 'calibration' factors have been used to improve agreement. Moderately stressed plots had BGM numbers three times higher than well watered plots.

The combined mite-plant-environment model also has been used to predict the impact of several irrigation scenerios:

(A) 36 mm irrigation over 3 weeks in 1980 with severe crop stress

(B) 36 mm irrigation over 3 weeks in 1980 with no crop stress

(C) 120 mm irrigation over 3 weeks in 1980 with no crop stress.

From the predictions in Fig. 5, severe stress with light irrigation would have caused almost a 30% increase in the mite population over a moderate irrigation rate with no crop stress after 3 weeks.

The difference between Simulation A and Simulation B is only the presence of crop stress; the difference between Simulation B and Simulation C is only a wetter soil surface and canopy that arises from more frequent irrigation. Clearly a wet soil surface and an occasionally wet canopy can reduce the exposure temperature of developing mites.

Irrigation may be a useful tool for retarding the development of BGM in corn. Although the greatest benefit of irrigation in retarding mite development arises from avoiding crop stress, which is the basic intent of any irrigation system anyway, additional benefit may be derived from more frequent irrigations with lesser quantities of water per irrigation when the risk of mite damage is high. For example, center pivot irrigation systems are available for sprinkling a crop with about 1 mm of water every 3 h. If such irrigation could keep the crop wet one-third of the day, mite development may be delayed 4 days or more compared with normal irrigation, and perhaps delayed a week from a crop in a stress condition. Whether this practice would be economical remains to be determined.

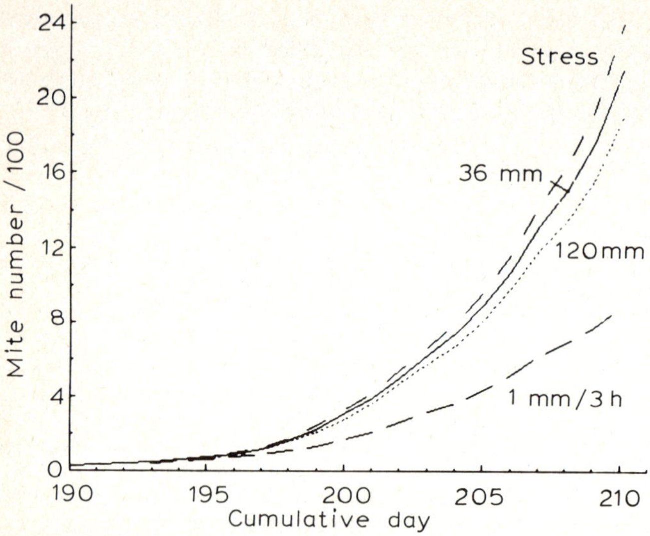

Fig. 5. Predictions of number of Banks Grass Mites per plant with severely stressed corn receiving 36 mm irrigation in 21 days (--------), and unstressed plants receiving 36 mm of irrigation in 21 days (———), 120 mm in 21 days (........) and 1 mm every 3 h for 21 days (- - - - -). Data are from central Nebraska in 1980.

These simulations of spider mite development did not include effects of predator mites that feed primarily on BGM eggs. Including predator mite development would result in greater predicted BGM number reductions with irrigation, because predator mites thrive at somewhat lower temperatures than the BGM. In the field test predator mites were exterminated.

This mite-plant-environment model combines numerous factors that interact in very complex ways. Predictions from this model indicate how mite development depends on plant and environmental factors so that we can explore the most promising practices that will minimize the impact of mite damage.

REFERENCES

Allen, L.H., Jr., 1973. Crop micrometeorology: a wide-row light penetration. B. Carbon dioxide enrichment and diffusion. Ph.D. Thesis, Cornell University, Ithaca, NY, 366 pp. Univ. Microfilm Diss. Abstr. No. 73-14, 716.

Allen, L.H., Jr., 1974. Line source carbon dioxide release. II. Two-dimensional numerical diffusion modeling. Agron. J., 66: 616—620.

Allen, L.H., Jr., 1975a. Line source carbon dioxide release. III. Predictions by a two-dimensional numerical diffusion model. Boundary-Layer Meteorol., 8: 39—79.

Allen, L.H., Jr., 1975b. Shade-cloth microclimate of soybeans. Agron. J., 67: 175—181.

Allen, L.H., Jr., 1979. Potentials for carbon dioxide enrichment. In: B.J. Barfield and J.F. Gerber (Editors), Modification of the Aerial Environment of Crops. ASAE Monogr. 2, American Society of Agricultural Engineers, St. Joseph, MI, pp. 500—519.

Allen, L.H., Jr., Jensen, S.E. and Lemon, E., 1971. Plant response to carbon dioxide enrichment under field conditions: a simulation, Science, 173: 256—258.

Baradas, M.W., Blad, B.L. and Rosenberg, N.J., 1976. Reflectant induced modification of soybean (*Glycine max* L.) canopy radiation balance. IV. Leaf and canopy temperature. Agron. J., 68: 848—852.

Blad, B.L. and Lemeur, R., 1979. Miscellaneous techniques for alleviating heat and moisture stress. In: B.J. Barfield and J.F. Gerber (Editors), Modifications of the Aerial Environment of Crops. ASAE Monogr. 2, American Society of Agricultural Engineers, St. Joseph, MI, pp. 409—425.

Chandler, L.D., Archer, T.L., Ward, C.R. and Lyle, W.M., 1979. Influence of irrigation practices on spider mite densities on field corn. Environ. Entomol., 8: 196—201.

Chessness, J.L., Harper, L.A. and Howell, T.A., 1979. Sprinkling for heat stress reduction. In: B.J. Barfield and J.F. Gerber (Editors), Modification of the Aerial Environment of Crops. ASAE Monogr. 2, American Society of Agricultural Engineers, St. Joseph, MI, pp. 388—393.

Doraiswamy, P.C. and Rosenberg, N.J., 1974. Reflectant induced modification of soybean (*Glycine max* L.) canopy radiation balance. I. Preliminary tests with a kaolinite reflectant. Agron. J., 66: 224—228.

Duncan, W.G. and Barfield, B.J., 1970. Predicting effects of CO_2 enrichment with simulation models and digital computer. Trans. ASAE, 13(2): 246—248.

Feese, H. and Wilde, G., 1977. Factors affecting survival and reproduction of the Banks grass mite, *Oligonychus pratensis*. Environ. Entomol., 6: 53—56.

Hsiao, T.C., 1973. Plant responses to water stress. Annu. Rev. Plant Physiol., 24: 519—570.

Kimball, B.A. and Idso, S.B., 1983. Increasing atmospheric CO_2: effects on crop yield, water use and climate. Agric. Water Manage., 7: 55—72.

Kretchman, D.W., 1969. CO_2 enrichment of outdoor crops. An interim report. Ohio Agricultural Research and Development Center, Department of Horticulture, Wooster, OH, 67 pp.

Kretchman, D.W., 1970. CO_2 enrichment of outdoor crops. Interim report III. Ohio Agricultural Research and Development Center, Department of Horticulture, Wooster, OH, 22 pp.

Lemeur, R. and Rosenberg, N.J., 1975. Reflectant induced modification of soybean (*Glycine max* L.) canopy radiation balance. II. A quantitative analysis of radiation reflected from a green soybean canopy. Agron. J., 67: 301—306.

Lemeur, R. and Rosenberg, N.J., 1976. Reflectant induced modification of soybean (*Glycine max* L.) canopy radiation balance. III. A comparison of the effectiveness of Celite and kaolinite reflectants. Agron. J., 68: 30—35.

Manetsch, T.J. and Park, G.L., 1974. Systems analysis and simulation with applications to economic and social systems. Department of Electrical and Systems Science, Michigan State University, East Lansing, MI.

Merva, G.E. and Vanderbrink, C., 1979. Physical principles involved in alleviating heat stress. In: B.J. Barfield and J.F. Gerber (Editors), Modification of the Aerial Environment of Crops. ASAE Monogr. 2, American Society of Agricultural Engineers, St. Joseph, MI, pp. 373—387.

Norman, J.M., 1979. Modeling the complete crop canopy. In: B.J. Barfield and J.F. Gerber (Editors), Modification of the Aerial Environment of Plants. ASAE Monogr. 2, American Society of Agricultural Engineers, St. Joseph, MI, pp. 249—277.

Norman, J.M., 1982. Simulation of microclimates. In: J.L. Hatfield and I.J. Thomason (Editors), Biometeorology: Integrated Pest Management. Academic Press, New York, NY, pp. 65—99.

Nurnberger, F.V., Merva, G.E. and Harrington, J.B., 1976. Microenvironmental modification by small water droplet evaporation. J. Appl. Meteorol., 15: 858—867.

Ogbuehi, S.N. and Brandle, J.R., 1981. Influence of windbreak-shelter on light interception, stomatal conductance and CO_2 exchange rate of soybeans. Trans. Nebr. Acad. Sci., 9: 49—53.
Ogbuehi, S.N. and Brandle, J.R., 1982. Influence of windbreak-shelter on soybean growth, canopy structure and light relations. Crop Sci., 22: 269—273.
Raschke, K., 1960. Heat transfer between the plant and the environment, Annu. Rev. Plant Physiol., 11: 111—126.
Rosenberg, N.J., 1979. Windbreaks for Reducing Moisture Stress. In: B.J. Barfield and J.F. Gerber (Editors), Modification of the Aerial Environment of Crops. ASAE Monogr. 2, American Society of Agricultural Engineers, St. Joseph, MI, pp. 394—408.
Toole, J.L., 1982. Simulating the population dynamics of Banks Grass Mite. M.S. Thesis, University of Nebraska, Lincoln, NE, 85 pp.
Waggoner, P.E., 1969. Environmental manipulation for higher yields. In: J.D. Eastin, F.A. Haskins, C.Y. Sullivan and C.H.M. van Bavel (Editors), Physiological Aspects of Crop Yield. American Society of Agronomy. Madison, WI, pp. 343—373.

Agricultural Water Management, 7 (1983) 89—114
Elsevier Science Publishers B.V., Amsterdam — Printed in The Netherlands

SOIL MANAGEMENT FOR SEMIARID REGIONS

W.E. LARSON[1], J.B. SWAN[1] and M.J. SHAFFER[2]

[1] *Department of Soil Science, University of Minnesota, St. Paul, MN 55108 (U.S.A.)*

[2] *Agricultural Research Service, U.S. Department of Agriculture, St. Paul, MN 55108 (U.S.A.)*

Contribution of the Minnesota Agricultural Experiment Station, and the Soil and Water Management Research Unit, North Central Region, USDA-ARS, St. Paul, MN 55108

(Accepted 23 January 1983)

ABSTRACT

Larson, W.E., Swan, J.B. and Shaffer, M.J., 1983. Soil management for semiarid regions. *Agric. Water Manage.*, 7: 89—114.

Water related climatic factors and soil physical and chemical properties exert a dominant effect on crop production in semiarid regions. Soil management practices which beneficially affect crop production by increasing water storage or through more efficient water use have been intensively studied at a number of locations with their associated soil conditions and under the climatic conditions encountered. The practical benefits of such research have been extensive. The application of field research to specific sites is restricted by the extreme variations in climatic conditions, by the large differences in soil physical and chemical properties, and by the variation in management practices which occur in the region. Because of these soil and climatic differences, the probability levels of expected responses to management practices generally have not been determined. Part of the difficulty has been due to the inability to precisely define and numerically describe the soil and climatic characteristics involved for sites and years other than those in which the research was conducted.

Computer-based simulation techniques permit the development of long- and short-term site specific soil management recommendations with expected results expressed on a probability basis. The research results from specific geographic locations measured under a specific set of climatic conditions can thus be generalized to a much wider range of soils and expressed on a probability basis using long-term climatic records.

Examples are given where crop yield frequencies are computed as a function of long-term climatic conditions involving various management practices. These results illustrate the impacts of a range of residue management, deep tillage, and fallowing practices on crop production in the semiarid environment. The modeling techniques presented can be used with a range of models and potential management practices.

INTRODUCTION

Soil management in semiarid regions centers around using the limited water supply efficiently. Water conservation involves maximizing infiltra-

0378-3774/83/$03.00

tion and storage of precipitation while minimizing evapotranspiration not directly related to crop yield. Techniques used include crop residue management, weed control, snow management, and runoff and evaporation reduction practices (Papendick et al., 1980a). Other considerations include seedbed preparation and planting, residue management, soil fertility, and weed control. Since tillage plays a major role in most water conservation techniques (Compton, 1942), as well as in other soil management considerations, a major portion of this presentation will deal with tillage and tillage-related practices.

From the standpoint of the practitioner, tillage has remained as much art as science. Hillel (1972) comments that "for the most part prevalent methods of seedbed preparation and of seeding are still based on the accumulated experience of generations and on trial and error methods of experimentation". This is due to the complexity of the problem and to the limited number of experimental sites and limited range in weather. In contrast, tillage recommendations must apply to a large variety of soil types and highly variable environmental conditions. To be most useful, management recommendations must be crop and site specific. Ideally, they should be evaluated over a 25 to 100 year range of climatic conditions, and expressed on a probability basis.

Fruitful research on tillage-soil water-fertility effects and interactions requires integrating what is known into prediction equations or models. Functional relationships should be derived from experimentally determined information at specific sites. These can be built into models capable of extending the relationships to farmers' fields by the use of local soils and weather information.

COMPUTER SIMULATIONS

The development of computer simulation techniques for soil-crop systems has made possible rapid evaluation of management alternatives in the semi-arid environment. Appropriate models have been developed by several research groups (Dutt et al., 1972; Shaffer et al., 1977; Hanks and Hill, 1980; Knisel, 1981; Shaffer and Gupta, 1981; McKinion and Baker, In press; Williams et al., In press. The nitrogen-tillage residue management (NTRM) model developed by USDA-ARS at the University of Minnesota (Shaffer and Larson, 1982; Shaffer et al., In press) is a comprehensive simulator of the soil-crop-water system in one dimension for the unsaturated zone and two-dimensions below the water table. Emphasis is placed on plant and root growth, tillage, water flow, nitrogen transformations, and crop residue interactions.

A mechanistic oriented modeling approach which adequately simulates the interactions of the crop with the physical, chemical, and biological parameters of the system allows extrapolation beyond current data sets. A stochastic approach to model input data and coefficient variability can then be used to generate yield frequency curves and ultimately, confidence bounds for model output.

A suitable model for use in testing management alternatives should be based more on model reliability and validity than on computer costs per run. A good model can be used to evaluate management impacts in the long- and short-term without making a large number of computer runs or making individual runs involving many successive years. This evaluation is done by carefully selecting the input data sets, running representative crop sequences, and using efficient methods of sampling statistical distributions.

The NTRM model was utilized in this study to examine various soil management practices proposed for the semiarid Great Plains, and to demonstrate some applicable modeling techniques, while generating insight into potential methods for alleviating drought impacts.

A transect was taken in a north-south direction from the Dakotas through the Texas-Oklahoma panhandle (Fig. 1). Several methods of soil management were analyzed at three hypothetical sites located near Bismarck, ND (ND), North Platte, NE (NE), and Amarillo, TX (TX). Projected yields were made for grain sorghum and wheat without irrigation. The assumptions were made that the soils were silt loams, adequate crop nutrients were available, and weeds, insects, and diseases were controlled. Simulated site specific, daily values for precipitation, maximum and minimum air temperatures, and solar radiation data were generated for a 100-year period using the Richardson-Nicks climate generator model (Nicks, 1974; Richardson, 1981; Williams et al., In press). The latter three parameters were used to calculate

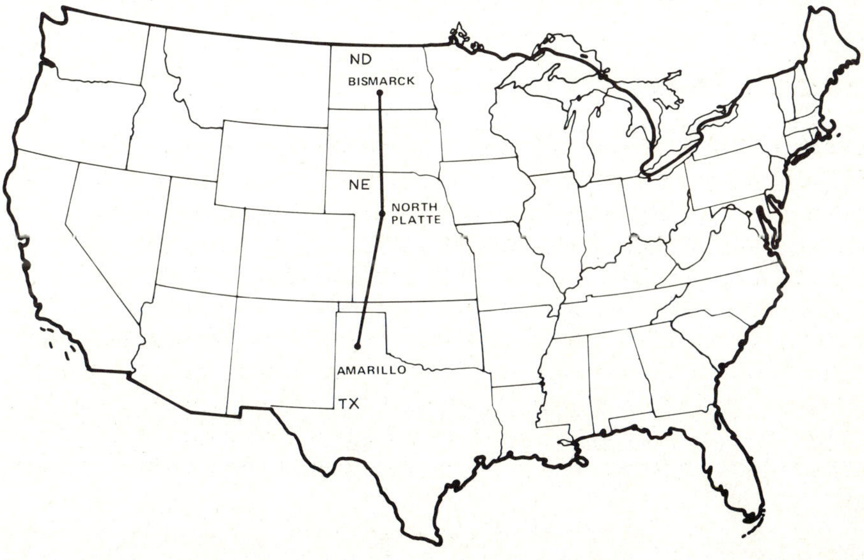

Fig. 1. Location of study sites.

daily pan evaporation (Campbell, 1977). Long-term annual precipitation distribution was simulated for each site (Fig. 2).

Using the 100-year data bases, three sets of representative years were selected, which approximated the values for the mean annual precipitation and two standard deviations on each side of the mean. Final selection of representative years at each site was based on growing season precipitation. The wide data range (Table I) provided definition of the yield response curves suitable for interpolation using a regression approach.

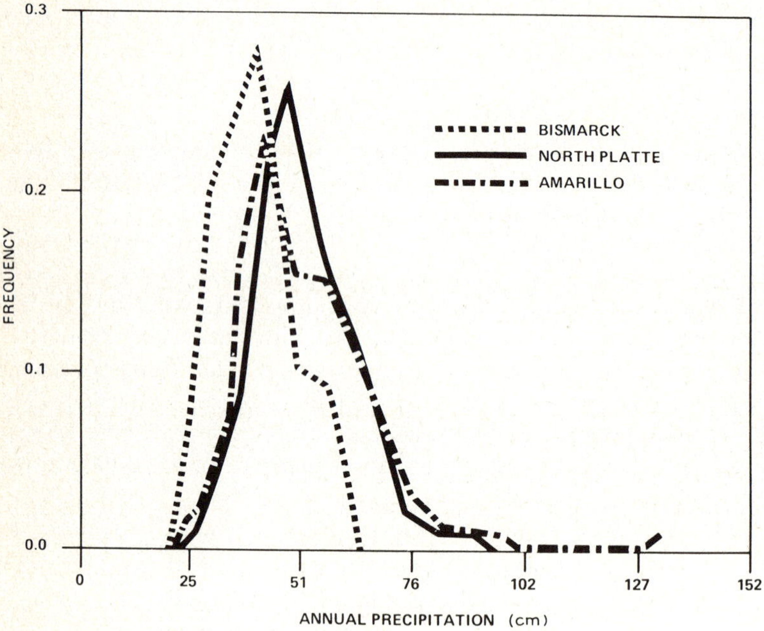

Fig. 2. Long-term distribution of precipitation at the study sites.

TABLE I

Stochastic (simulated) precipitation at the three study sites used as model input

Location	Precipitation (cm)					
	Wet		Average		Dry	
	Annual	Growing season[a]	Annual	Growing season	Annual	Growing season
Bismarck, ND	59.9	38.9	40.6	28.7	24.4	13.5
North Platte, NE	73.9	47.8	51.5	33.3	29.2	21.8
Amarillo, TX	82.8	55.6	51.8	31.8	20.8	12.7

[a]Defined as 1 April to crop maturity.

The computer simulation approach was used to demonstrate effects of continuous wheat vs wheat-fallow, effects of compaction vs non-compaction, and amounts of surface residues on crop yields at the three locations. Additional areas where appropriate models would be of benefit are reviewed and include the effect of: (a) tillage on evaporation; (b) seed placement and seed-soil contact on germination and emergence; (c) snow retention on water storage and yield; and (d) nutrient-water interactions on crop yields. Space limitations prevented demonstrations of the model capabilities in the latter cases.

RESIDUE AND CROP MANAGEMENT

The kinds and amounts of plant residues available are determined by the cropping and management system. Residues of wheat, barley (*Hordeum vulgare* L.) and oats (*Arena satina* L.) predominate in semiarid regions with sorghum (*Sorghum bicolor* L. Moench) or corn (*Zea mays* L.) residues also present.

The amounts of crop residues produced per unit area vary widely between the 29 Major Land Resources Areas (MLRAs) in the Great Plains (Skidmore et al., 1979). Dryland wheat often yields from 1 to 2 t ha^{-1} $year^{-1}$ of straw in the southern Great Plains, and from 3.4 to 4.7 t ha^{-1} $year^{-1}$ in the Central Great Plains (Greb et al., 1979). The number and type of tillage operations determine the percentage of surface residue remaining after tillage.

While a considerable variety of equipment is available, surface residue systems, sweeps or blades are generally operated 12 to 15 cm deep during the initial operation following harvest and shallower during subsequent operations. Uniform spreading of crop residue by the combine is necessary to minimize tillage problems (Unger and McCalla, 1980). When large amounts of residue are present, a disk-type implement may be used in the initial operation to incorporate more of the residue, yet still retain sufficient surface residue for erosion control. Other implements that may be used in heavy residue include skew-treaders or spike-tooth harrows in conjunction with disk plows (Papendick and Miller, 1977).

The use and limitations of summer fallow to store water was reviewed by Black et al. (1974), Greb et al. (1974, 1979), Johnson et al. (1974), Johnson and Unger (1976), Smika (1976) and Willis (1976). While fallow usually increases both soil water content at planting and crop yield, grain production in the Southern Great Plains on a total area basis has been lower with fallow systems (using one-way disk or stubble mulch tillage) than with continuous cropping (Unger, 1981). A major factor contributing to limited soil water storage during fallow in southern areas is the small amount of residue produced. When increased wheat residues due to irrigation were retained on the surface during fallow, significant increases in water storage and grain yields of sorghum were measured (Musick et al., 1977; Unger and Wiese, 1979). Thus, there appears to be opportunity for increased total production through judicious selection of cropping and/or irrigation-dryland sequences.

To investigate the impacts on wheat yields of fallow versus continuous cropping, simulation runs were made using the NTRM model on each of the three study locations. The regression for straw yields versus precipitation at NE are shown in Fig. 3, and illustrate the technique of interpolation between points generated with the NTRM model. To test those results against experiment, a plot was prepared (Fig.4) which compares these data with similar data reported by Allmaras (1983).

Figs. 5, 6, and 7 show the results for the long-term yield frequencies for ND, NE, and TX. The skewed distributions for the case of continuous wheat are replaced with a more normal distribution for wheat-fallow. Thus, one can achieve higher wheat yields and a greater certainty of average yield in any given year with fallow.

The cumulative probability of wheat straw yields in ND equal to or greater than those specified are plotted in Fig.8. Both continuous and wheat-fallow management result in similar straw yields under wet annual climatic conditions. The benefits obtained from the wheat-fallow technique increase rapidly under drier conditions with the maximum obtained at about the 30 to 40% probability levels.

Under conditions of limited available water, increased early growth has resulted in more rapid water use, which later increased drought stress and

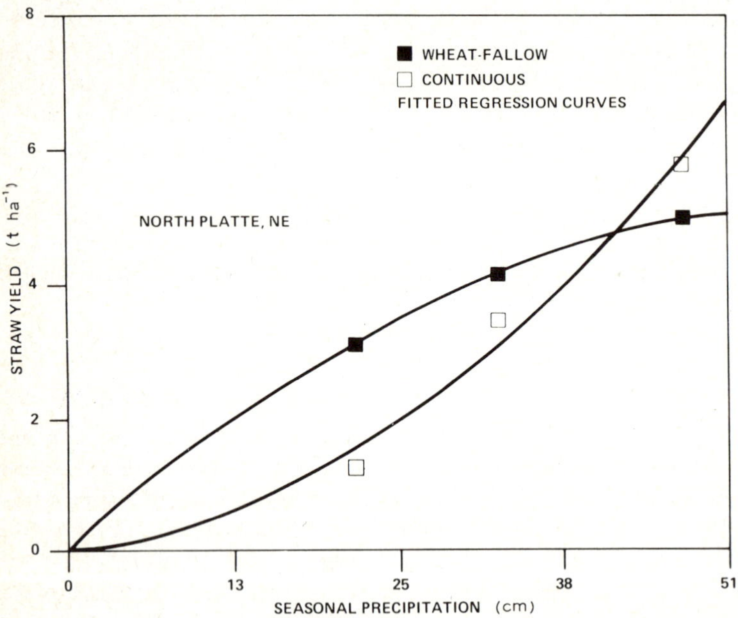

Fig. 3. Simulated data points (NTRM model) and fitted regression curves for wheat straw yield (North Platte, NE).

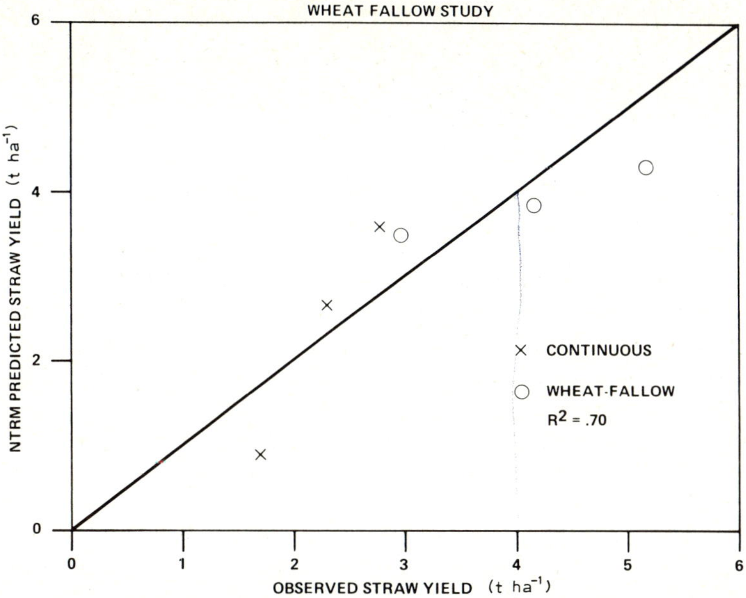

Fig. 4. Observed vs. predicted wheat straw yields.

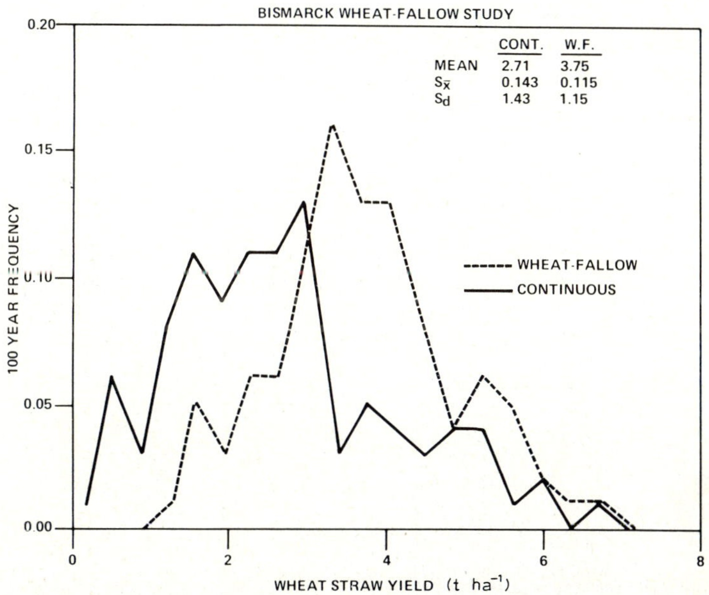

Fig. 5. Long-term distribution of straw yields under continuous and wheat-fallow conditions (Bismarck, ND).

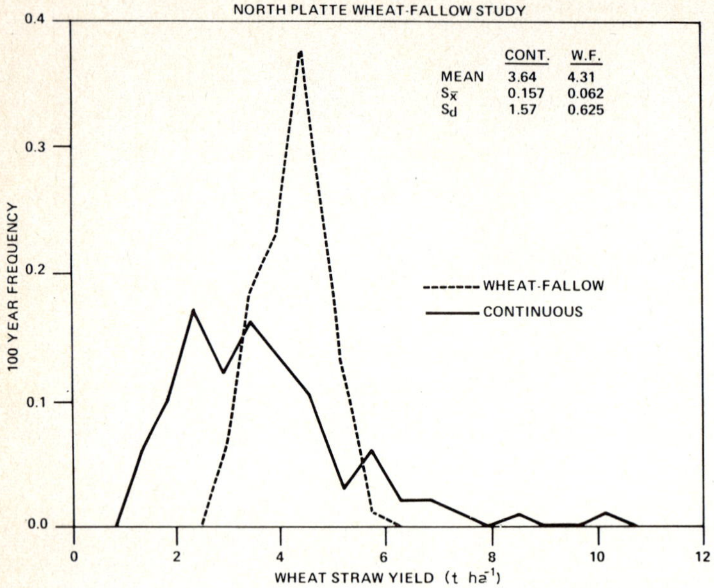

Fig. 6. Long-term distribution of straw yields under continuous and wheat-fallow conditions (North Platte, NE).

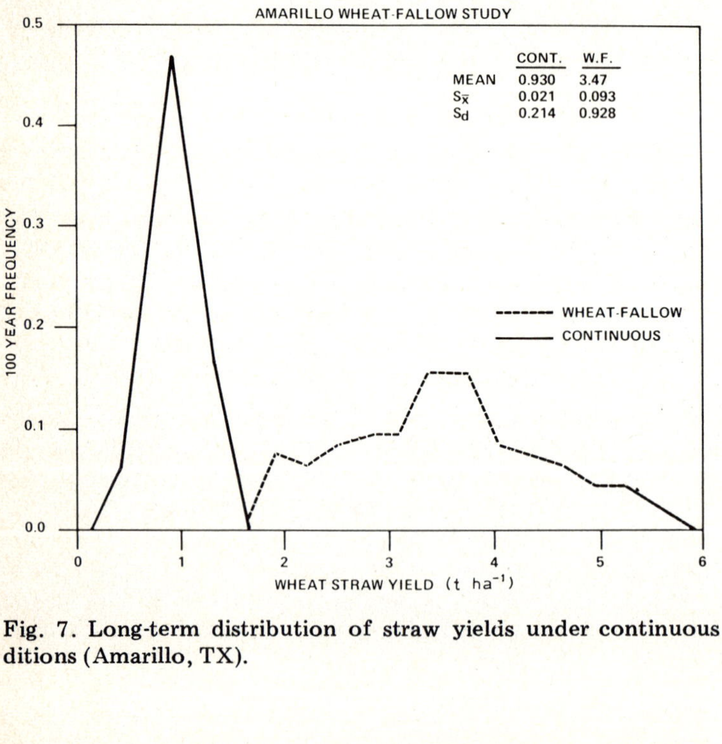

Fig. 7. Long-term distribution of straw yields under continuous and wheat-fallow conditions (Amarillo, TX).

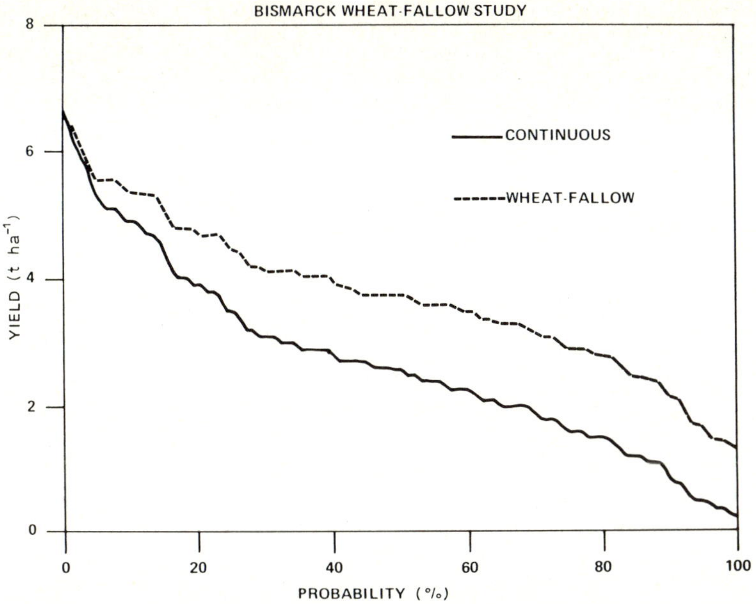

Fig. 8. Probability of wheat straw yields under continuous and wheat-fallow conditions (Bismarck, ND).

decreased grain yields (Viets, 1966; Amemiya). Such observations are in agreement with the greatly increased effect of drought stress on grain yield during the reproductive stages compared to the vegetative stages for grain crops.

Stewart and Musick (1982) cite reviews showing strong linear correlations between cumulative seasonal transpiration and drymatter production when yields are transpiration limited. When transpiration and evapotranspiration are closely associated, drymatter production and cumulative evapotranspiration also show a strong linear relationship. For grain yield, the relationship with seasonal evapotranspiration is often linear but with a different slope than with drymatter. However, based on data of Musick and Dusek (1971) for grain sorghum, Stewart and Musick (1982) argue that the water use yield function is not an explicit relationship and that it can vary considerably depending on various factors which affect both yield and water use. They conclude that limited irrigation, particularly during more critical growth periods, shows considerable promise for increasing water use efficiency. The observed curvilinearity at higher evapotranspiration levels offers some opportunities for other mechanisms of increasing water use efficiencies, presumably including the management of drymatter production rates.

Use of legume green manure crops for N fixation is restricted by the climatic conditions and the large evapotranspiration amounts (Hobbs, 1957; Rogers and Giddens, 1957; Horner, 1960; Horner et al., 1960; Shrader and Voss, 1980). Corn yield decreases following alfalfa are well documented under conditions of insufficient water, a common condition in the western cornbelt (Voorhees and Holt, 1969; Sloan, 1976; Voss and Shrader, 1979).

WATER MANAGEMENT

Tillage

From the literature, Burnett and Hauser (1967) concluded that crop yields were increased most by deep tillage when water was a limiting factor and fertility or salinity was not limiting. The primary benefit of deep tillage was through one or a combination of mechanisms: (a) increased root proliferation which enables the plant to extract water from a larger volume of soil; (b) increased stored soil water due to improved water entry; or (c) increased stored soil water due to a change in particle arrangement. Deep tillage was most effective in increasing crop growth under dry conditions in Canada (Krogman and MacKay, 1980). Deep plowing may assist in reclaiming some salt affected soils (Burnett and Hauser, 1967; Krogman and MacKay, 1980; Karkanis and Cairns, 1981). In portions of the U.S.S.R., "deeper than normal" tillage is extensively used (Papendick et al., 1980a).

To test the hypothesis that deep tillage mitigates the effect of drought on crop yields, simulation runs were made on the silt loam soil under compacted conditions and after removal of the compacted zone by a chisel plow. The compacted zone with a bulk density of 1800 kg m^{-3} was assumed to begin at a 15 cm depth and allowed little root penetration. The initial soil water content was assumed to be 0.32 cm^3 cm^{-3} (field capacity). Fig. 9 illustrates the straw yield responses for the two cases in ND and TX, respec-

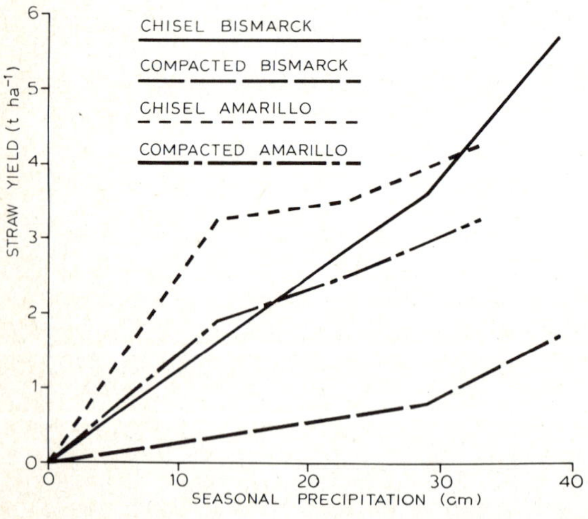

Fig. 9. Simulated long-term wheat straw yields at Bismarck, ND and Amarillo, TX.

tively. The impact of deep plowing on yields is generally similar for TX and NE (not shown) with a rather pronounced increased benefit shown in ND.

Figs. 10 and 11 illustrate the results obtained after applying the stochastic precipitation data to the yield curves in Fig. 9. The mean yields were increased as deeper tillage increased the available water storage in the soil. Deep plowing in ND significantly increased the standard deviation of the mean; although the error is uniformly distributed. The straw yield values in TX are hypothetical since it would be extremely difficult to consistently begin the growing season with a soil water content near field capacity. The yield differences shown between study sites illustrate the importance of weather patterns, planting dates, and their interactions given the same initial set of soil conditions.

Surface roughness (Burwell and Larson, 1969) and surface residue cover of tilled soil (Mannering and Meyer, 1961; 1963) can markedly enhance water infiltration. Van Doren and Allmaras (1978) related the change in soil surface saturated conductivity with the sum of the rainfall kinetic energy, a soil structural stability constant, and the fraction of the soil surface covered by crop residue. The lower the initial soil surface saturated conductivity, the less effective crop residues are in recharging the soil water supply.

Under a range of tillage conditions, Allmaras (1967) found that, on two soils, water storage was positively correlated with cumulative intake; but on a third soil, the results were reversed. The amounts and frequencies of precipitation interacted with the initial profile water content, roughness, porosity, and length of evaporation period between rainfall events to produce differences in distribution and water storage in the soil profiles. In such circumstances, simulation techniques using long-term meteorological data and appropriate soil-water relationships (Linden, 1982) are necessary to assess the probability of given levels of storage expected under different tillage methods.

Residue management

When sufficient amounts were available, surface crop residues have increased water storage and yield (Greb et al., 1979; Smika et al., 1969; Unger, 1978). In the southern Great Plains, Unger (1978) found that 1 or 2 t ha^{-1} of crop residue produced by winter wheat was insufficient to substantially increase yields over conventional stubble mulch tillage. He suggested that low residue production of dryland crops in this region is a major reason for low storage efficiency of precipitation. Crop residue mulches effectively decrease evaporation during the first stage evaporation (Unger and McCalla, 1980). For maximum water conservation over long periods, enough water must be added to penetrate deeply into the soil profile or large amounts of residue must be present.

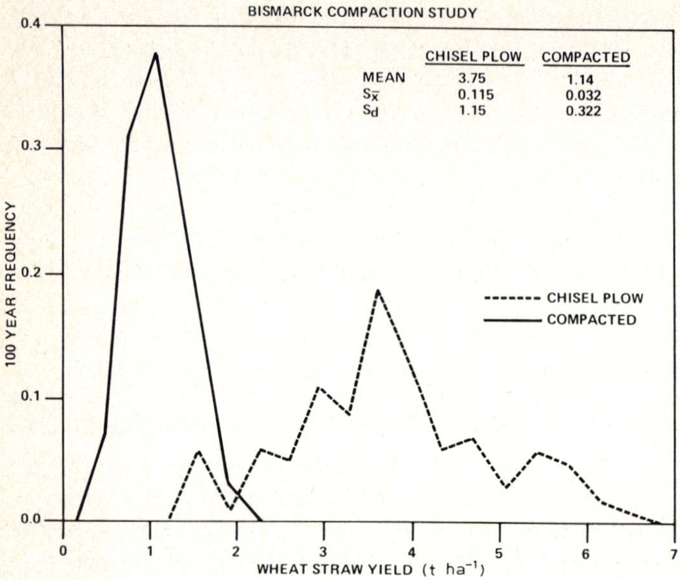

Fig. 10. Long-term wheat straw yield distribution under compacted and chisel plow conditons (Bismarck, ND).

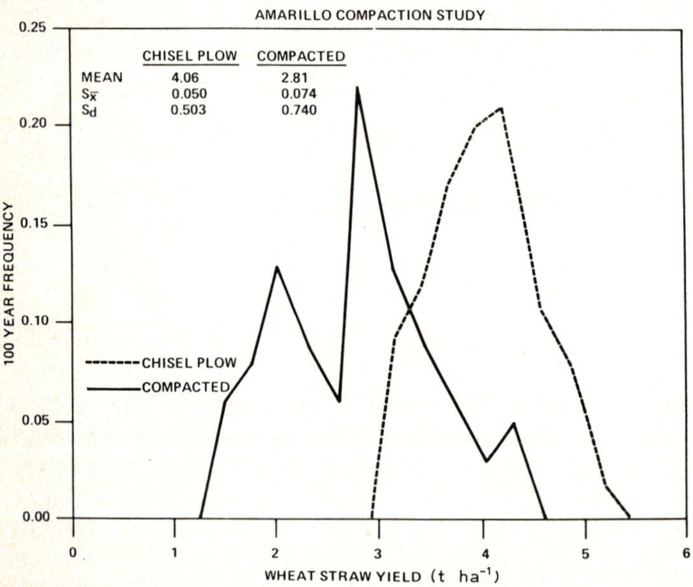

Fig. 11. Long-term wheat straw yield distribution under compacted and chisel plow conditions (Amarillo, TX).

The effect of between growing season residue amounts on grain sorghum yields at three sites and under site specific climate input was predicted using the NTRM model. It was assumed that straw residues were applied at rates of 0, 4, and 8 t ha^{-1} which resulted in spring (1 April) soil water contents of 0.1, 0.21, and 0.32 cm^3 cm^{-3}, respectively. Residue during the growing season was maintained at a rate of 2 t ha^{-1}.

The regression curves in Fig. 12 (TX) were fitted to yield response data calculated by use of the NTRM model. The results provided families of curves which characterize the yield response to precipitation for the residue treatments. The model output was compared with experimental results (data points shown) obtained by Unger (1978) at Bushland, TX which are displayed on the plots for Amarillo, TX. Unger's results have been approximately reproduced by the model illustrating the general agreement of model output with experiment. The curves shown in Fig. 12 (TX) and similar curves for NE (not shown) were used in conjunction with the stochastic precipitation data to produce the yield distribution curves shown in Fig. 13 (TX) and Fig.14 (NE). The distributions represent the responses at each seasonal precipitation value. These figures illustrate the changes in yield distribution, mean yields, and standard deviations for the three assumed residue treatments analyzed over the entire historical record.

The increases in the mean annual yields and decreases in the standard deviations of the mean yields for the residue versus the non-residue treatments indicate a definite long-term increase in plant available water. Thus,

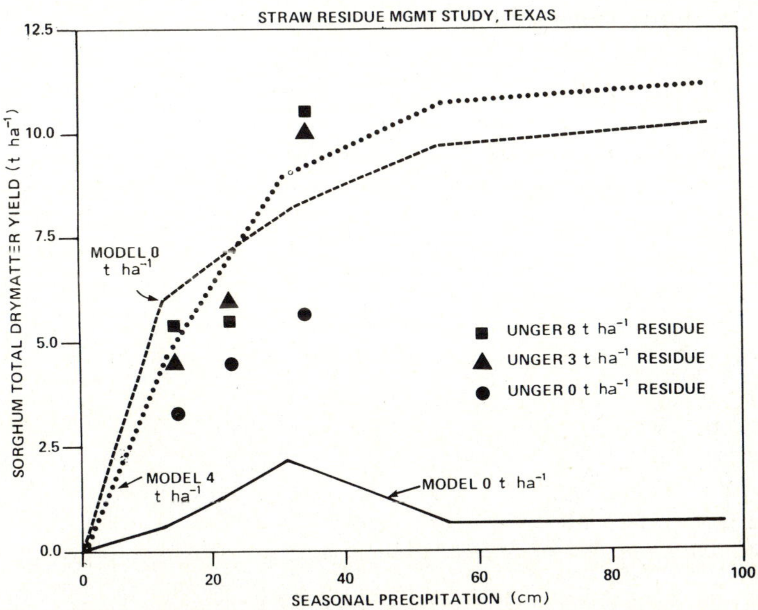

Fig. 12. Predicted and observed sorghum yields at Amarillo, TX.

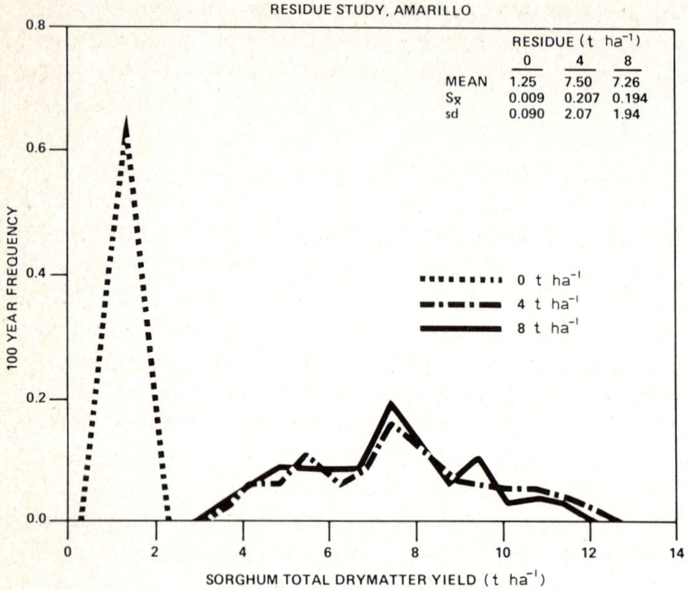

Fig. 13. Long-term yield distribution for sorghum (Amarillo, TX).

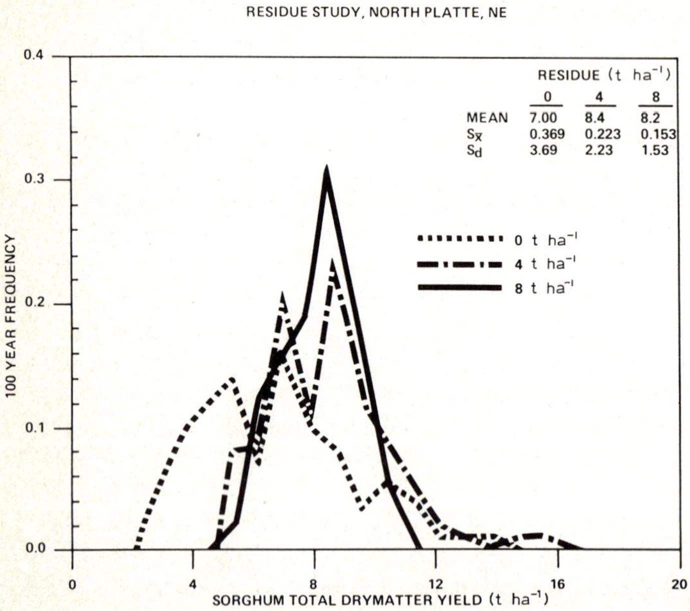

Fig. 14. Long-term yield distribution for sorghum (North Platte, NE).

the chances of obtaining an average yield are increased by maintaining residues on the soil surface between growing seasons. Yield levels are slightly reduced under heavy mulch treatments at NE.

Evaporation reduction

Hillel and Rawitz (1972) point out that the choices for reducing bare soil evaporation depends on the stage of the process of concern. In first stage evaporation, meteorological and surface conditions dominate; while during the second stage, the rate of water supply to the surface is determined by transmission characteristics of the profile. As a consequence, deep tillage to modify the diffusivity of the profile may have little effect on stage one evaporation, but could affect second stage drying. In contrast, surface mulches retard evaporation during the initial stage of drying, providing plants a greater opportunity to utilize water in surface soil layers which is important during germination and establishment phases. Mulches can also enhance internal drainage to greater depths where it is less likely to be lost by evaporation (Hillel and Rawitz, 1972).

Tillage often enhances drying of the tilled layer, but can also reduce water movement from layers below. The net effect of tillage on storage depends on a number of factors including: the duration of the process; the depth, degree, and frequency of tillage; subsequent rainfall (amounts and timing); and the influence of rainfall on reconsolidation (Hillel and Rawits, 1972). A major restriction on water conservation effects from tillage is that by the time the soil surface is dry enough to cultivate, the first stage of evaporation is generally over. Thus, Allmaras's (1967) experimental results illustrate Hillel and Rawitz's (1972) statement that "tillage might thus produce either a larger or smaller loss than would have occurred in the undisturbed soil...".

Aggregate size can affect evaporation rate from seedbeds. Holmes et al. (1960) found that in the laboratory the optimum size of aggregate was 2.5 mm for minimizing the rate of evaporation. Beds of larger or smaller aggregates had higher evaporation rates. Hillel and Hadas (1972) recommended aggregates of 0.25 to 2.0 mm diameter on the soil surface for retarding evaporation from bare soil. Additional research on achieving 'two layer' systems with a coarse layer of large pores overlying a finer structured profile is warranted as this appears to offer the best conditions for both evaporation and infiltration control (Hillel and Rawitz, 1972).

Seed placement

The basic interactions of physiological and environmental factors affecting seed germination and seedling establishment have received little attention until recently. Soil water, temperature, mechanical strength, aeration, and salinity conditions which prevail at the surface are known to be of importance, but their combined and separate effects are incompletely described by the data available (Hillel, 1972; Prihar, 1982). Prihar indicates that since aeration in seedbeds is generally non-limiting, the major concern in seedling emergence reduces to the management of water, temperature, and soil physical impedance. A number of important seedbed factors affecting winter wheat emergence and early growth have been incorporated into an emergence model (Lindstrom et al., 1976).

To achieve rapid germination and early growth, seed placement in moist soil is essential, as is prevention of rapid drying of the layers around the seed (Sepaskhah and Ardekani, 1978; Hillel, 1972). Soil water plays a dominant role in root growth because it not only affects the water availability to the root, but also influences the temperature, mechanical strength, and aeration status surrounding the root (Prihar, 1982). Increasing the normal depth of planting 2.5 to 10 cm can have an adverse effect on cereal germination and seedling emergence (Hart, 1964; Gul and Allan, 1976). De Jong and Best (1979) found that increasing the planting depth not only increased the heating requirement for emergence, but also increased the variability of emergence.

The type of planting equipment is also important for proper seed placement. Shovel and hoe opener drills are generally more satisfactory for planting small grains through a dry surface soil and placing seeds in moist soil than are drills with disk openers (Unger and McCalla, 1980). Double- or triple-disk press drills often provide insufficient penetration for proper seed placement in untilled surfaces when soil bulk density in the upper 5 cm exceeds 1200 kg m^{-3}, or when residue exceeds 3.7 t ha^{-1} on the surface (Lindwall and Anderson, 1977). Greb (1979a) credits improved deep furrow planters (12 to 14 cm) with 8 percent of the recent commercial wheat yield increases.

Good seed-soil contact is also necessary for rapid germination. Hadas and Russo (1974) recommended that seedbeds be prepared so as to ensure high soil water conductivities, and as large a seed-soil water contact area as possible without impairing radicle elongation or aeration conditions. Hadas and Russo (1974) found that seed-soil water contact impedance increased as the wetted soil area and/or soil hydraulic conductivity decreased. They concluded that the mean aggregate size should be one-fifth to one-tenth of the seed's diameter for optimal seed-soil water contact. Such a requirement may conflict with Hillel and Hadas (1972) who recommend 0.25 to 2.0 mm diameter aggregates on the soil surface to retard evaporation from bare soil. Consequently, the row and interrow area may require somewhat different tillage. However, root-soil contact may also be an important factor affecting water uptake, which may restrict the latitude over which aggregate size may be varied. Herkelrath et al. (1977) proposed that as the soil dried, the surface area of roots in contact with the soil decreased, which caused an increase in root membrane resistance. They found that by assuming membrane permeability proportional to the relative saturation of the soil, a much better fit to the water uptake data was obtained. Hadas and Russo (1974) conclude that in the sowing row, a good stand can be assured by reducing the aggregate size around the seed or by compacting the soil area. Any treatment which reduces seed-soil water contact impedance should be adopted since a prolonged imbibitional period due to high contact impedance increases the extent of drying of the topsoil.

Snow retention

Considerable potential exists for increasing water storage through improved snow catchment in portions of the central and northern Great Plains. In one winter, the soil water accumulation from improved snow retention could be as large as the soil water accumulations from summer fallow (Snyder et al., 1980). Research from the U.S.A., Canada and the U.S.S.R. has demonstrated gains in soil water from snow catch by stubble management (Papendick et al., 1980a). Management techniques used were: (a) tall stubble; (b) alternating stubble heights; and (c) use of 15-m strips of standing stubble; all three techniques were effective under the conditions encountered.

Research cited by Papendick et al., (1980a) demonstrated the effectiveness of tall wheat grass barriers at Sidney, MT and Akron, CO. In Montana tall wheat grass barriers with 14.6 m cropping intervals increased water storage to 9.9 cm in the protected area, as compared with 5.3 cm in adjacent unprotected areas with continuous cropping (Black and Siddoway, 1976). With a crop-fallow system, the barriers increased water storage efficiency only during the first winter (9-month) period. With continuous cropping an average of 71 to 80% of between-harvest precipitation was used compared to only 30% for spring wheat-fallow. Sunflower, sunflower and mustard, and tree shelterbelts are used to increase snow retention and water storage in various part of the U.S.S.R. In the U.S.S.R., some type of snow retention measures are applied to 60 to 60 million ha $year^{-1}$, mostly in Kazakhstan and Siberia (CIA, 1974). Snow ridges created by special plows are used in the USSR on a significant acreage (CIA, 1974). Long-term yield increases of 270 kg ha^{-1} are reported (Papendick et al., 1980a). The practice is most efficient when combined with vegetative windbreaks, alternating stubble heights, or tall stubble because a minimum snow depth of about 10 cm is required to form ridges.

A rough soil surface may increase soil water storage because of increased snow catchment and infiltration. Tillage techniques which leave a rough surface often retain a portion of the crop residue on the surface and thus a portion of the benefit is due to the crop residue. In the intermountain dryland region, contour fall chiseling below the frost line reduced runoff and stored 87% of the September—April precipitation compared to only 60% stored by standing stubble (Ramig et al., 1983).

Water and wind erosion estimates for the ten Great Plains states are given in the RCA reports (USDA, 1981) (Fig. 15). Wind erosion is greatest in Colorado, Texas and New Mexico. Average water erosion amounts are greater than wind erosion amounts in Nebraska, Kansas, Oklahoma and North Dakota.

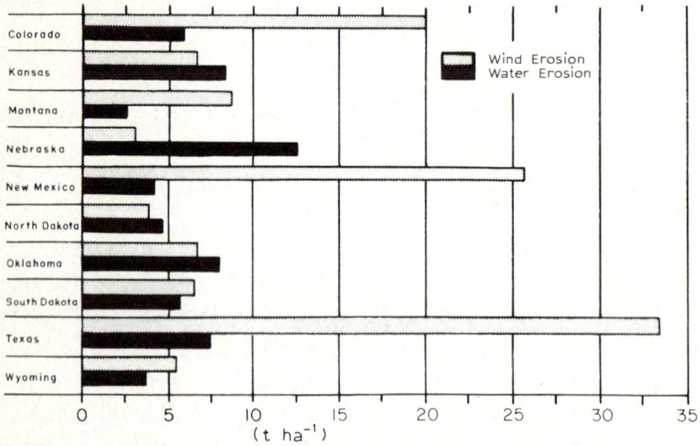

Fig. 15. Average wind and water erosion rates on cropland in the Great Plains states.

In general, tillage practices for wind and water erosion control are similar. Wind erosion control is best achieved through use of any or all four basic management principles: (a) cloddy surface to resist wind force; (b) furrows or ridges perpendicular to the wind direction to slow the wind speed; (c) stripcropping to reduce wind fetch; and (d) adequate vegetative cover to reduce wind speed at the soil surface and to anchor the soil. While wind erosion decreases as the amount of residues maintained on the soil surface increases, the control is more marked once the mulch exceeds 1.1 t ha^{-1} (Greb, 1979b).

Because of weather variability, it is not always possible to produce enough crop residue to protect the soil against wind and water erosion. Allmaras (1983) has given the probabilities of achieving straw yields below a specified maximum value in a wheat-fallow cropping system and in a continuous cropping system of the same wheat type. Of the four regions compared, climatic variability affected straw production most in the central Great Plains winter wheat region. In the southwestern subregion of the central Great Plains, less than 1.21 t ha^{-1} would be produced in 2 of 10 years; 1.98 t ha^{-1} in 5 of 10 years; and 5.02 in 8 of 10 years in a wheat-fallow cropping system. In a continuous wheat system, less than 0.41 t ha^{-1} would be produced 2 of 10 years; 1.65 in 5 of 10 years; and 3.39 in 8 of 10 years. Straw production is usually higher in a wheat-fallow system as compared with a continuous wheat system.

In a wheat-summer fallow management system, at least 50% of the straw at crop harvest is either buried or decomposed by the end of the fallow year (Greb et al., 1970). Allmaras (1983) has calculated that if 0.825 t ha^{-1} of straw remains after wheat seeding, soil loss would be 34% of that from a fallowed bare soil surface exposed to the same rainfall and runoff energy. The same amount of straw after wheat seeding would reduce wind erosion to about 50% of that from a bare and smooth soil surface. Thus, if it is assumed that 50% of the straw is lost by a conservation tillage system, at least 1.65 t ha^{-1} would need to be produced. Management practices that have flexibility to meet the extreme variation in weather and resultant crop production are needed. For example, in years when crop residue production is small, herbicides for weed control, and seeding equipment which conserves the residue on the soil surface might be used. In years of abundant residue availability, a combination of tillage and herbicides might be used.

Emergency tillage to reduce erosion is sometimes necessary when amounts of residue are insufficient or when other management practices have been inadequate. Emergency tillage at right angles to the slope direction is most effective for water erosion control. Emergency tillage for wind erosion control should provide soil roughness by bringing clods to the surface or by construction of ridges and furrows, preferably perpendicular to the prevailing wind direction.

Vegetative barriers grown in rows break the force of the wind. Barriers may be grown with the double objective to trap snow and to protect against wind (Aase et al., 1976). Desirable barrier characteristics include: (a) narrow double rows to insure against wind gaps; (b) plants 0.4 to 1.2 m tall with flexible stalks to minimize breakage; and (c) stalk populations to provide 65 to 75% air porosity (Greb, 1979b). Barriers have been grown in strips 9 m to 18 m apart (Black and Siddoway, 1971). Tall wheatgrass (a perennial), sorghum, and sudan grass have desirable characteristics for barriers. Disadvantages of using vegetative barriers include establishment and maintenance costs, inconvenience in farming operations, and water use by the barrier.

Windbreaks of shrubs or trees are often used for reducing wind velocities at the soil surface and to lower erosion and increase snow trapping. The usual recommendation of spacing for field windbreaks is about 10 times the height of the trees.

NUTRIENT MANAGEMENT

Return of crop residues as opposed to their removal is an important factor affecting the recycling of plant nutrients (Papendick et al., 1980b). Nitrogen and P are the major nutrients of concern in Great Plains crop production, yet most of the N and P taken up by the plant is removed in the grain. For wheat, the N and P in crop residues represent only 19 and

15% , respectively, of the N and P in the grain portion (Holt, 1979). Returning crop residues returns most of the K.

Animal manure adds N and P to the soil and can increase the availability and uptake of soil and fertilizer P by plants in alkaline soils (Olsen and Flowerday, 1971). On alkaline soils, organic matter has increased the concentration of P in the soil solution (Olsen and Barber, 1977).

Because the degree of soil disturbance generally affects mineralization rates of N and P from soil organic matter (Russell, 1961) introduction of reduced or 'no tillage' techniques may slow the release rates of N and P. For example, at Bushland, TX, one-way disk tillage resulted in a more rapid loss of organic matter and higher soil nitrate content than stubble mulch tillage. A wheat-fallow system resulted in a more rapid loss of organic matter and higher soil nitrate content than did continuous wheat (Johnson et al., 1974).

Finally, tillage methods that retain residue on the surface result in slow decomposition of the residues and release of nutrients (Unger and McCalla, 1980).

Fertility-water interactions for dryland crops have been well documented (Viets, 1962; Bauer et al., 1965; C.A. Black, 1966). The effects is clearly demonstrated in data of Ramig and Rhoades (1963) and cited by Black (1966) for winter wheat in western Nebraska (Fig. 16). Yield increased as N fertilizer increased for the treatment with 20.6 cm of available water at planting. However, yields were not greatly different at all N fertilizer levels for the 0 cm available water treatment. Black (1966) observed that for grain crops grown "with a small supply of available water, it may be desirable to limit the rate of exhaustion of water by limiting the growth of vegetative parts of plants, thereby saving a greater proportion of the available water for the stage of grain development".

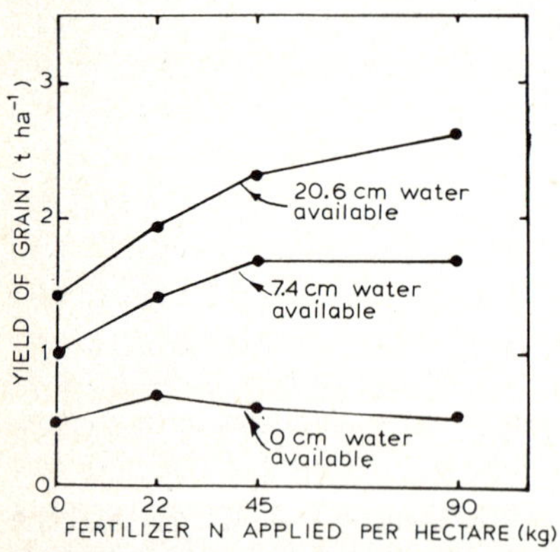

Fig. 16. Yield of grain and total dry matter of winter wheat in western Nebraska with different applications of fertilizer nitrogen and different quantities of available water. The figures in the graphs correspond to the centimeters of available water in the soil at planting time.

Regulating the time of N uptake by date of application appears to be a useful tool for maximizing the efficiency of use of N fertilizer and water.

The interaction of N fertilization and soil water with yield under dryland conditions is well documented (Bauer et al., 1965; Young et al., 1967; Black and Siddoway, 1976). Young et al. (1967) developed equations relating yield of spring wheat and barley to stored available soil water and NO_3—N at seeding, and precipitation and temperature. Smika and Grabouski (1976) concluded that both soil NO_3—N and soil water had to be at similar depths for the NO_3—N to effectively increase grain protein. In an aridic Haplustoll, application over 1 year before seeding gave maximum accumulation of nitrate in the soil between 60 and 120 cm, and produced the greatest yield and highest protein. In a Torripsamment, NH_3 application 2.5 to 4 months before seeding produced the highest protein grain and highest grain yields. Deep placement significantly increased yields of wheat and grain protein content in field studies in the state of Washington (Cochran et al., 1978).

Black and Siddoway (1976) studied the effects of three levels of N fertilization (0, 23, and 67 kg ha^{-1}) on the yield of different crops grown under various crop sequences. They found that each crop responded differently to fertilization within crop sequences and years, but that crops on continuous cropping systems always responded more.

In evaluating the effect of fertilizers on crop yield in drier regions, Viets (1966) concluded that, "most evidence shows that fertilizers can be used in the drier areas to correct nutrient deficiencies and to make more efficient use of a limited supply of water. Yield reductions from use of fertilizers are rare". Viets also stated that there can be "no doubt that fertilizers on dryland often stimulate growth and increase the water use rate. The significance of this depends on the balance between need and supply at critical phases of growth and whether rains come at the right time". To evaluate these effects on yield, he suggested that "the best approach is to examine the results of a large number of field trials conducted over a number of years in the geographical area of interest". This is precisely the kind of evaluation provided by a well-adapted crop growth model with inputs of long-term weather data and soils information.

CONCLUSIONS

In the semiarid Great Plains, management practices must be fine tuned for most efficient use of the limited water supply. Variations in soils and in weather require management practices to be site, soil, and crop specific. Because of the many possible combinations, the specific recommendations can best be quantified through a mechanistic modeling approach followed by estimations of yield frequencies and confidence bounds.

The general computer simulation methods used in this study can be applied to a range of models, site locations, and management techniques.

The steps involved can be summarized as follows:

(1) Make a series of seasonal, annual, or crop rotation sequences over the desired range of precipitation (or climate) and management techniques, or parameter values representing those techniques. The objective is to only cover the ranges in detail sufficient to define the important trends.

(2) Using the simulated values in (1) above, develop functional relationships for yield vs precipitation (site specific). Use regression or other curve fitting techniques.

(3) Use a stochastic climate generator or various time sequences from the historical relationships in item (2) above to develop the yield frequency relationship for each management practice.

This approach has the distinct advantage that long-term yield frequency information can be obtained from the application of a variety of relatively complex and comprehensive computer models without expending large amounts of resources. Skill is needed in selecting the data ranges and densities which are used. Some situations may require a continuous or semicontinuous simulation over an extended time period. The decision concerning the type of analysis required is critical. The examples presented in this paper illustrate types of analyses which do not require continuous simulations.

REFERENCES

Aase, J.K., Siddoway, F.H. and Black, A.L., 1976. Perennial grass barriers for wind erosion control, snow management and crop production. In: R.W. Tinus (Editor), Shelterbelts on the Great Plains. Proc. Symp., 20—22 April 1976, Denver, CO. Great Plains Agric. Counc. Publ. 78, pp. 69—78.

Allmaras, R.R., 1967. Soil water storage as affected by infiltration and evaporation in relation to tillage-induced soil structure. In: Conf. Proc. Tillage for Greater Crop Production, 12—17 December 1967, Detroit, MI. PROC-168, American Society of Agricultural Engineers, St. Joseph, MI, pp. 37—43.

Allmaras, R.R., 1983. Soil conservation: using climate, soils, topography, and adapted crops information to select conservation practices. In: H.E. Dregne and W.O. Willis (Editors), Dryland Agriculture in North America. Am. Soc. Agron. Monogr., 23: 139—153.

Amemiya, M., 1968. Tillage-soil water relations of corn as influenced by weather. Agron. J., 60: 534—537.

Bauer, A.Young, R.A. and Ozbun, J.L., 1965. Effects of moisture and fertilizer on yields of spring wheat and barley. Agron. J., 57: 354—356.

Black, A.L. and Siddoway, F.H., 1971. Tall wheatgrass barriers for soil erosion control and water conservation. J. Soil Water Conserv., 26: 107—111.

Black, A.L. and Siddoway, F.H., 1976. Dryland cropping sequences within a tall wheatgrass barrier system. J. Soil Water Conserv., 31: 101—105.

Black, A.L., Soddoway, F.H. and Brown, P.L., 1974. Summer fallow in the Northern Great Plains (winter wheat). In: Summer Fallow in the Western United States., Conserv. Res. Rep. 17, Agricultural Research Service, U.S. Department of Agriculture, Washington, DC, pp. 36—50.

Black, C.A., 1966. Crop yields in relation to water supply and soil fertility. In: W.H. Pierre, D. Kirkham, J. Pesek and R. Shaw (Editors), Plant Environment and Efficient Water Use. American Society of Agronomy/Soil Science Society of America, Madison, WI, pp. 177—206.

Burnett, E. and Hauser, V.L., 1967. Deep tillage and soil-plant-water relationships. In: Conf. Proc. Tillage for Greater Crop Production, 11—12 December 1967, Detroit, MI. PROC-168, American Society of Agricultural Engineers, St. Joseph, MI, pp. 47—57.

Burwell, R.E. and Larson, W.E., 1969. Infiltration as influenced by tillage-induced random roughness and pore space. Soil Sci. Soc. Am. Proc., 33: 449—452.

Campbell, C.S., 1977. An Introduction to Environmental Biophysics. Springer-Verlag, New York, NY, 159 pp.

CIA, 1974. USSR Agricultural Atlas. Central Intelligence Agency, U.S. Government Printing Office, Washington, DC, 59 pp.

Cochran, V.L., Warner, R.L. and Papendick, R.I., 1978. Effect of N depth and application rate on yield, protein content, and quality of winter wheat. Agron. J., 70: 964—968.

Compton, L.L., 1942. Moisture conservation practices and the relationship of conserved water to crop yields. Soil Sci. Soc. Am. Proc., 7: 368—373.

De Jong, R. and Best, K.F., 1979. The effect of soil water potential, temperature, and seedling depth on seedling emergence of wheat. Can. J. Soil Sci., 59: 259—264.

Dutt, G.R., Shaffer, M.J. and Moore, W.J., 1972. Computer simulation model of dynamic bio-physicochemical processes in soils. Univ. Ariz. Agric. Exp. Stn. Tech. Bull. 196, 101 pp.

Greb, B.W., 1979a. Technology and wheat yields in the Central Great Plains: Commercial advances. J. Soil Water Conserv., 34: 269—273.

Greb, B.W., 1979b. Reducing drought effects on croplands in the West-Central Great Plains. U.S. Dep. Agric. Inf. Bull. 420, 31 pp.

Greb, B.W., Smika, D.E. and Black, A.L., 1970. Water conservation with stubble mulch fallow. J. Soil Water Conserv., 25: 58—62.

Greb, B.W., Smika, D.E., Woodruff, N.P. and Whitfield, C.J., 1974. Summer fallow in the Central Great Plains. In: Summer Fallow in the Western United States. Conserv. Res. Rep. 17, Agricultural Research Service, U.S. Department of Agriculture Washington, DC, pp. 51—85.

Greb, B.W., Smika, D.E. and Welsh, J.R., 1979. Technology and wheat yields in the Central Great Plains: Experiment Station advances. J. Soil Water Conserv., 34: 264—268.

Gul, A. and Allan, R.E., 1976. Stand establishment of wheat lines under different levels of water potential. Crop Sci., 16: 661—615.

Hadas, A. and Russo, D., 1974. Water uptake by seeds as affected by water stress, capillary conductivity and seed-soil water contact. II. Analysis of experimental data. Agron. J., 66: 647—652.

Hanks, R.J. and Hill, R.W., 1980. Modeling Crop Responses to Irrigation in Relation to Soils, Climate, and Salinity. Pergamon Press, Oxford, 66 pp.

Hart, J., 1964. Wide choice in wheat and barley planting depth. Queensl. Agric. J., 90: 137—138, Abstr. Biol. Abstr. 47(1), No. 4054.

Herkelrath, W.N., Miller, E.E. and Gardner, W.R., 1977. Water uptake by plants: II. The root contact model. Soil Sci. Soc. Am. J., 41: 1039—1043.

Hillel, D., 1972. Soil moisture and seed germination. In: T.T. Kozlowski (Editor), Water Deficits and Plant Growth, Vol. III. Academic Press, New York, NY, pp. 65—89.

Hillel, D. and Hadas, A., 1972. Isothermal drying of structurally layered soil columns. Soil Sci., 113: 30—35.

Hillel, D. and Rawitz, E., 1972. Soil water conservation. In: T.T. Kozlowski (Editor), Water Deficits and Plant Growth, Vol. III. Academic Press, New York, NY, pp. 307—388.

Hobbs, J.A., 1957. The winter wheat and grazing region. In: SOIL, the 1957 Yearbook of Agriculture. U.S. Department of Agriculture, Washington, DC, pp. 505—515.

Holmes, J.W., Greacen, E.L. and Gurr, C.G., 1960. The evaporation of water from bare soils with different tilths. Trans. Int. Congr. Soil Sci., 7(1): 188—194.

Holt, R.F., 1979. Crop Residue, soil erosion and plant nutrient relationships. J. Soil Water Conserv., 34: 96—98.

Horner, G.M., 1960. Effect of cropping systems on runoff, erosion, and wheat yields. Agron. J., 52: 342—344.

Horner, G.M., Oveson, M.M., Baker, G.O. and Pawson, W.W., 1960. Effect of cropping practices on yield, soil organic matter and erosion in the Pacific Northwest wheat region. Bull. 1, Agricultural Experiment Stations, Idaho, Oregon, Washington/Agricultural Research Service, U.S. Department of Agriculture, Washington, D.C.

Johnson, W.C. and Unger, P.W., 1976. Mechanical fallow in Southern Great Plains. In: Conservation Tillage. Proc. Great Plains Workshop, 10—12 August 1976, Fort Collins, CO. Great Plains Agric. Counc. Publ. 77, pp. 93—123.

Johnson, W.C., Van Doren, C.E. and Burnett, E., 1974. Summer Fallow in the Southern Great Plains. In: Summer Fallow in the Western United States, Conserv. Res. Rep. 17, Agricultural Research Service, U.S. Department of Agriculture, Washington, DC, pp. 86—109.

Karkanis, P.G. and Cairns, R.R., 1981. Notes: Some effects of deep plowing and fertilizing a Solonetzic soil under irrigation. Can. J. Soil Sci., 61: 157—160.

Knisel, W.C. (Editor), 1981. CREAMS: a field-scale model for chemicals, runoff, and erosion from agricultural management systems. Conserv. Res. Rep. 26, U.S. Department of Agriculture, Washington, DC, 640 pp.

Korgman, K.K. and MacKay, D.C., 1980. Horizon mixing on Solonetzic and associated soils: effect on drought-stressed barley and wheat. Can. J. Soil Sci., 60: 721—729.

Linden, D.R., 1982. Predicting tillage effects on evaporation from the soil. In: P.W. Unger and D.M. van Doren (Editors), Predicting Tillage Effects on Soil Physical Properties and Processes, Proc. Symp. 30 November—5 December 1980, Detroit, MI. Spec. Publ., 44, American Society of Agronomy, Madison, WI, pp. 117—132.

Lindstrom, M.J., Papendick, R.I. and Koehler, F.E., 1976. A model to predict winter wheat emergence as affected by soil temperature, water potential, and depth of planting. Agron. J., 68: 137—141.

Lindwall, C.W. and Anderson, D.T., 1977. Effects of different seeding machines on spring wheat production under various conditions of stubble residue and soil compaction in no till rotations. Can. J. Soil Sci., 57: 81—91.

Mannering, J.V. and Meyer, L.D., 1961. The effects of different methods of cornstalk residue management on runoff and erosion as evaluated by simulated rainfall. Soil Sci. Soc. Am. Proc., 25: 506—510.

Mannering, J.V. and Meyer, L.D., 1963. The effects of various rates of surface mulch on infiltration and erosion. Soil Sci. Soc. Am. Proc., 27: 84—86.

McKinion, J. and Baker, D., in press. Dynamic crop modeling: A synergism of computers, experimental research, and scientific method. In: Proc. 3rd Int. Conf. State of the Art of Ecological Modeling, 24—28 May 1982, Fort Collins, CO.

Musick, J.T. and Dusek, D.A., 1971. Grain sorghum response to number, timing, and size of irrigations in the Southern Great Plains. Trans. ASAE, 14: 401—404, 410.

Musick, J.T., Wiese, A.F. and Allen, R.R., 1977. Management of bedfurrow irrigated soil with limited- and no-tillage systems. Trans. ASAE, 20: 666—672.

Nicks, A.O., 1974. Stochastic generation of the occurrence, pattern, and location of maximum amount of daily rainfall. In: Proc. Symp. Statistical Hydrology, August—September 1971, Tucson, AZ. Misc. Publ. No. 1275, pp. 154—171.

Olsen, S.R. and Barber, S.A., 1977. Effect of waste application on soil phosphorus and potassium. In: L.F. Elliott and F.J. Stevenson (Editors), Soils For Management of Organic Wastes and Waste Waters. American Society of Agronomy, Madison, WI, pp. 197—215.

Olsen, S.R. and Flowerday, A.D., 1971. Fertilizer phosphorus interactions in alkaline soils. In: R.A. Olson, T.J. Army, J.J. Hanway, and V.J. Kilmer (Editors), Fertilizer Technology and Use, 2nd Edition. Soil Science Society of America, Madison, WI, pp. 153—185.

Papendick, R.I. and Miller, D.E., 1977. Conservation Tillage in the Pacific Northwest. J. Soil Water Conserv., 32: 49—56.

Papendick, R.I., Swan, J.B. and Robacker, R.C., 1980a. Dryland agriculture in continental regions. In: Proc. Int. Congr. Dryland Farming, 25 August—5 September 1980, Adelaide, Australia (In press).

Papendick, R.I., Boersma, L.L., Colacicco, D., Kla, J.M., Kraenzle, C.A., Marsh, P.B., Newman, A.S., Parr, J.F., Swan, J.B. and Youngberg, I.G., 1980b. Report and Recommendations on Organic Farming. SEA, U.S. Department of Agriculture, Washington, DC.

Prihar, S.S., 1982. Management of physical environment of soil for increasing crops production. In: Whither Soil Research. Trans. Int. Congr. Soil Sci., 11: 1—24.

Ramig, R.E. and Rhoades, H.F., 1963. Interrelationships of soil moisture level at planting time and nitrogen fertilization on winter wheat production. Agron. J., 55: 123—127.

Ramig, R.E., Allmaras, R.R. and Papendick, R.I., 1983. Water conservation: Pacific Northwest. In: H.E. Dregne and W.O. Willis (Editors), Dryland Agriculture in North America. Am. Soc. Agron. Monogr., 23: 105—124.

Richardson, C.W., 1981. Stochastic simulation of daily precipitation, temperature, and solar radiation. Water Resour. Res., 17(1): 182—190.

Rogers, T.H. and Giddens, J.E., 1957. Green manure and cover crops. In: SOIL, the 1957 Yearbook of Agriculture. U.S. Department of Agriculture, Washington, DC, pp. 252—257.

Russell, E.W., 1961. Soil Conditions and Plant Growth. John Wiley, New York, NY, 688 pp.

Sepaskhah, A.R. and Ardekani, E.R., 1978. Effect of soil matric potential and seeding depth on emergence of barley. Agron. J., 70: 728—731.

Shaffer, M.J. and Gupta, S.C., 1981. Hydrosalinity models and field validation. In: T.K. Iskander (Editor), Modeling Wastewater Renovation (Chapter 7). John Wiley, New York, NY, pp. 136—181.

Shaffer, M.J. and Larson, W.E. (Editors), 1982. Nitrogen Tillage-Residue Management (NTRM) Model. Vol. 3, Technical Documentation. A.R.S. U.S. Dep. Agric. Res. Rep., St. Paul, MN, 450 pp.

Shaffer, M.J., Ribbens, R.W. and Huntley, C.W., 1977. Prediction of Mineral Quality of Irrigation Return Flow. Vol. V, Detailed Return Flow Salinity and Nutrient Simulation Model, U.S. EPA, EPA-6001, 2-77e, 226 pp.

Shaffer, M.J., Gupta, S.C., Molina, J.A.E., Linden, D.R. and Larson, W.E., in press. Simulation of nitrogen, tillage, and residue management effects on soil fertility. In: Proc. 3rd Int. Conf. State of the Art in Ecological Modeling, 24—28 May 1982, Fort Collins, CO.

Shrader, W.D. and Voss, R.D., 1980. Crop rotation vs. monoculture-soil fertility. Crops Soils Mag., 32: 15—18.

Skidmore, E.L., Kumar, M. and Larson, W.E., 1979. Crop residue management for wind erosion control in the Great Plains. J. Soil Water Conserv., 34: 90—94.

Sloan, R.F., 1976. Legume rotations-crop yield and nitrogen fixation. Kansas Agricultural Experiment Station, 5 pp. Keeping up with Research, No. 24.
Smika, D.E., 1976. Mechanical tillage for conservation fallow in the semiarid Central Great Plains. In: Conservation Tillage. Proc. Great Plains Workshop, 10—12 August 1976, Fort Collins, CO. Great Plains Agric. Counc. Publ. 77, pp. 78—92.
Smika, D.E. and Grabouski, P.H., 1976. Anhydrous ammonia applications during fallow for winter wheat production. Agron. J., 68: 919—922.
Smika, D.E., Black, A.L. and Greb, B.W., 1969. Soil nitrate, soil water, and grain yields in a wheat-fallow rotation in the Great Plains as influenced by straw mulch. Agron. J., 61: 785—787.
Snyder, J.R., Skold, M.D. and Willis, W.O., 1980. Economics of snow management for agriculture in the Great Plains. J. Soil Water Conserv., 35: 21—24.
Stewart, B.A. and Musick, J.T., 1982. Conjunctive use of rainfall and irrigation in semi-arid regions. In: D. Hillel (Editor), Advances in Irrigation, Vol. I. Academic Press, New York, NY, pp. 1—24.
Unger, P.W., 1978. Straw-mulch rate effect on soil water storage and sorghum yield. Soil Sci. Soc. Am. J., 42: 486—491.
Unger, P.W., 1981. Tillage effects on wheat and sunflower grown in rotation. Soil Sci. Soc. Am. J., 45: 941—945.
Unger, P.W. and McCalla, T.M., 1980. Conservation tillage systems. Adv. Agron., 33: 1—58.
Unger, P.W. and Wiese, A.F., 1979. Managing irrigated winter wheat residues for water storage and subsequent dryland grain sorghum production. Soil Sci. Soc. Am. J., 43: 582—588.
USDA, 1981. RCA Appraisal, Part I. U.S. Department of Agriculture, Washington, DC.
Van Doren, D.M., Jr. and Allmaras, R.R., 1978. Effect of residue management practices on the soil physical environment, microclimate, and plant growth. In: W.R. Oschwald (Editor), Crop Residue Management Systems. Proc. Symp. 28 November—3 December 1976, Houston, TX. Am. Soc. Agron. Spec. Publ. 31, pp. 49—83.
Viets, F.G., Jr., 1962. Fertilizers and the efficient use of water. Adv. Agron., 14: 223—264.
Viets, F.G., Jr., 1966. Increasing water use efficiency by soil management. In: W.H. Pierre, D. Kirkham, J. Pesek, R. Shaw (Editors), Plant Environment and Efficient Water Use. American Society of Agronomy/Soil Science Society of America, Madison, WI, pp. 259—274.
Voorhees, W.B. and Holt, R.F., 1969. Management of alfalfa to conserve soil moisture. Minn. Agric. Exp. Stn. Bull. 494, 10 pp.
Voss, R.D. and Shrader, W.D., 1979. Crop rotations effect on yields and response to nitrogen. PM-905, Cooperative Extension Service, Iowa State University, Ames, IA, 5 pp.
Williams, J., Dyck, P. and Jones, A., in press. EPIC - a model for assessing the effects of erosion on soil productivity. In: Proc. 3rd Int. Conf. State of the Art of Ecological Modeling, 24—28 May 1982, Fort Collins, CO.
Willis, W.O., 1976. Annual cropped versus cropped-fallow spring wheat in the Northern Great Plains. In: Conservation Tillage. Proc. Great Plains Workshop, 10—12 August 1976, Fort Collins, CO. Great Plains Agric. Counc. Publ. 77, pp. 124—131.
Young, R.A., Ozbun, J.L., Bauer, A. and Vasey, E.H., 1967. Yield response of spring wheat and barley to nitrogen fertilizer in relation to soil and climatic factors. Soil Sci. Soc. Am. Proc., 31: 407—410.

Agricultural Water Management, 7 (1983) 115—141
Elsevier Science Publishers B.V., Amsterdam — Printed in The Netherlands

FARM MANAGEMENT AND THE FUNCTION OF FIELD CROP ROOT SYSTEMS

BETTY KLEPPER[1], R.W. RICKMAN[1] and H.M. TAYLOR[2]

[1] *Columbia Plateau Conservation Research Center, Agricultural Research Service, U.S. Department of Agriculture, Pendleton, OR 97801 (U.S.A.)*
[2] *Plant and Soil Science Department, Texas Tech University, Lubbock, TX 79409 (U.S.A.)*

(Accepted 28 January 1983)

ABSTRACT

Klepper, B., Rickman, R.W. and Taylor, H.M., 1983. Farm management and the function of field crop root systems. *Agric. Water Manage.*, 7: 115—141.

Because most farmer-controlled factors influence the growth and function of plant roots, we need to develop root-centered models to describe below-ground plant behavior. A root-centered water extraction model is described. This model estimates water extraction rates from a soil layer in terms of the length of root present, the permeability of the soil-root system, and the difference in water potential between bulk soil and the root xylem in the mid-plane of the soil layer. Changes in xylem water potential resulting from frictional losses as water passes through small vessels is estimated using the Poiseuille-Hagen equation. These potential changes are used to calculate the root xylem water potential at any depth in the profile from the xylem water potential at the land surface. Field data for cotton (*Gossypium hirsutum* L.), soybean (*Glycine max* L.) and winter wheat (*Triticum aestivum* L. em. Thell.) were used to test the model. Measured values of root length density and soil water content over depth and time, field estimates of root-soil system permeability, measured diurnal changes in plant water potential and assumed values of numbers and radii of root xylem vessels were used to estimate water extraction patterns for these three crops for one time in the growing season. Estimates of the total water extraction and its distribution in the profile were in fair agreement with measured values. Principle uncertainties were absence of measurements for numbers and radii of root xylem vessels, questions of lack of homogeneity in field soils for application of a single water release curve to all sampling points in a large plot, and absence of confirming measurements for assessing changes in xylem water potential brought about by friction.

When farm managers make decisions on dryland farms, they impact the root system by changing soil physical and chemical properties, the activities of microbes and other organisms, water and nutrient concentrations and distributions. These may influence root growth and function and have an impact on yield. Before models can be used to help farmers make cost-effective decisions, we must learn more about crop root system phenology, environmental effects on root growth and function, and relationships between roots and shoots and between crop roots and their neighboring crop or weed roots. We must also develop models which deal properly with heterogeneity in the landscape, in the crop, and in the soil profile.

0378-3774/83/$03.00

INTRODUCTION

Models of dryland crop production often contain elaborate information about the size and shape of the canopy, changes in canopy composition and structure during the growing season, production and partitioning of above-ground dry matter, and evapotranspiration. However, below-ground models and data bases are usually an order of magnitude less specific and less sophisticated than their above-ground counterparts (Hillel and Talpaz, 1976; Van Bavel and Ahmed, 1976; Taylor and Klepper, 1978). This short fall is a necessary result of the fact that our research information on root growth, phenology, and function is about two decades behind such information for shoots. Compounding this problem is the fact that most root water extraction models have been developed by soil scientists rather than plant scientists, primarily because the complex soil physical problems were handled poorly by most botanists. This division was especially true prior to the development of unified terminology which permitted soil physicists and plant physiologists to describe root and soil water potentials and flows consistently (Kramer et al., 1966).

As a result of this history, most below-ground models are soil-centered, that is, they describe water movement in the soil using Darcy's Law and also provide for a 'root sink' which is a quantitative expression of the water which left the profile via roots because its disappearance cannot be accounted for by fluxes within the soil profile. Although the 'root sink' approach has elegance, it does take an indirect approach to roots. It says that roots or root activities are present in proportion to the amount of the water that has disappeared. Now that unified terminology is well-entrenched, however, more explicit account has been taken of roots and the physics of root-soil relations. Progress is also being made in the techniques available for measuring roots, for describing and investigating the physics and chemistry of the rhizoplane and the rhizosphere, and for monitoring fluxes of materials across the soil-root interface.

With better integration between the research of plant and soil scientists and better measurement capability, we also need to develop modelling ideas which are plant-centered, rather than soil-centered. This development is needed because it permits explicit expression of impacts of soil environment on root growth and morphology and the resultant effects on root function. It is especially important that below-ground agronomic models be plant-centered in dryland production areas. Practically all of the farmer-controlled factors affect the root-soil environment either directly or indirectly. All of the decisions made during the fallow year and many of those made during the crop year have their impact through effects on root growth or function. If we are to build farm-level management models, we must have sufficient sensitivity in the models to generate specific effects of farmer choices on the plants and on their stature, phenology and yield. Root-centered models can be more sensitive than soil-based ones because they deal in a direct way with the most dynamic part of the system. Not only do root systems change on a weekly basis in response to shifts in soil and shoot environment, they also change on an hour-by-hour basis during the day in the uptake process. Roots are the point of the steepest gradients of the driving forces which bring water and ions into the plant. Only by adopting root-centered approaches for on-the-farm management models can we hope to have sufficient sensitivity to predict plant responses to management. Root-centered models also are more definitive than soil-centered ones about root properties so that they may be useful in selecting root criteria for screening crop varieties for better utilization of soil resources. Finally, the root system integrates over space the heterogeneity of many local root-soil interfaces to determine the total supplying power of the soil-root system.

This paper presents a crude form of a root-centered model to describe water extraction by a crop root system and uses that model to point out areas of research needed before a farm manager can use a model to provide insight into how management choices will influence plant behavior and yield. Water uptake principles are stressed more than ion uptake ones because water is usually the crucial limiting factor in dryland production and is more difficult to change by management than is ion supply.

Model derivation

Consider a field planted at a seeding density of H (number per m^2) and a spacing between rows of W (m). Assume that the fraction of planted seed producing successful seedlings is F. Then there will be FHW plants per m of row. Each plant will have available a land surface area of $F^{-1}H^{-1}$ (m^2) and can be considered to occupy a rectangular area which measures $F^{-1}H^{-1}W^{-1}$ (m) within the row and W (m) across the row. The assumption is made that each plant has exclusive use of the resources in a volume generated as the product of these rectangular dimensions and the maximum rooting depth at a given time. This rooted soil volume is divided into units selected to contain soil at uniform water content and root length density. The units can be of any specifiable dimensions so long as the units are connected by roots capable of transporting water out of the more distal units into the more proximal ones. However, units of horizontal layers of uniform thickness, Y, are convenient, and the rest of this article will assume such an arrangement. Root length densities within and between rows are assumed to be uniform within a layer. However, soil beneath the interrow could be treated separately if necessary.

The root water extraction model of Taylor and Klepper (1978) has been used. Their model does not concern exchanges of matter and energy between the canopy and air. These exchanges are important because they determine rates of transpiration which ultimately control total water extracted from soil. Instead, the model concerns extraction patterns to be expected for a given crop when distributions of roots and water in the profile are known.

Plant xylem potential at the land surface (Ψ_p) is used to calculate potential gradients which move water into root xylem from soil and along xylem vessels to the shoot at the land surface. Using an Ohm's Law Analogue, Taylor and Klepper (1978) calculated U_i, the rate of water uptake ($m^3\ s^{-1}$) from the ith soil unit as:

$$U_i = V_i D_i K_i(\Psi_{si} - \Psi_p + \Psi_{zi} + \sum_{j=1}^{i} \Delta\Psi_{fj}) \qquad (1)$$

where V_i = volume of soil in the unit (m^3); D_i = root length density in the unit (m (root) m^{-3}); K_i = root-soil system permeability ($m^3\ s^{-1}$ m $(root)^{-1}$ MPa^{-1}); Ψ_{si} = soil water potential in the unit (MPa); Ψ_p = xylem water potential at the land surface (MPa); $\Delta\Psi_{zi}$ = loss in potential due to elevation (MPa); $\Delta\Psi_{fj}$ = water potential losses between units due to friction in the moving water column (MPa).

This equation permits estimates to be made of water uptake rates from units of soil as long as these units are connected by root xylem tissues. The units interact with one another in the uptake process because of the last term in Eq. (1), the frictional potential losses in moving water columns in interconnecting root xylem. Because of this term, uptake will decrease progressively in distal units, especially if proximal units have already supplied large amounts of water.

The root system is assumed to act as though all of the water extracted from the soil unit moves to the midplane of the unit for transport to other units. The water movement is approximated by one-dimensional flow with the K-term incorporating non-linearities associated with the real process of water movement through soil, across the soil-root interface, and through the root tissues. A value of $8.3 \times 10^{-12}\ m^3\ s^{-1}\ m^{-1}\ MPa^{-1}$ (0.03 $\mu l\ h^{-1}\ cm^{-1}\ bar^{-1}$) has been used for K (Taylor and Klepper, 1975) for theoretical calculations in this paper. There are N_1 xylem vessels available to transport water from the midplane (C_1) of the top layer to the land surface (Fig. 1A). These are assumed to be evenly distributed in the land area and all vessels are assumed to have the same structure and to behave identically. Likewise there are N_2 vessels transporting water from the midplane of the second to the first layer and so on down the profile. Through the midplane of a layer must pass all the water absorbed in distal layers as well as that absorbed in the layer itself. For example, through the midplane of the fifth layer passes water from the fifth layer as well as all water from the sixth layer (Fig. 1A). Fig. 1B shows the generalized water extraction pattern expected with flow-dependent frictional potential loss. During the field season, root growth and atrophy cause marked shifts in the root length density profile. Water use causes the soil water potential profile to change with time. Because Eq. (1) includes explicit values for these terms, such shifts can be accommodated on a daily basis.

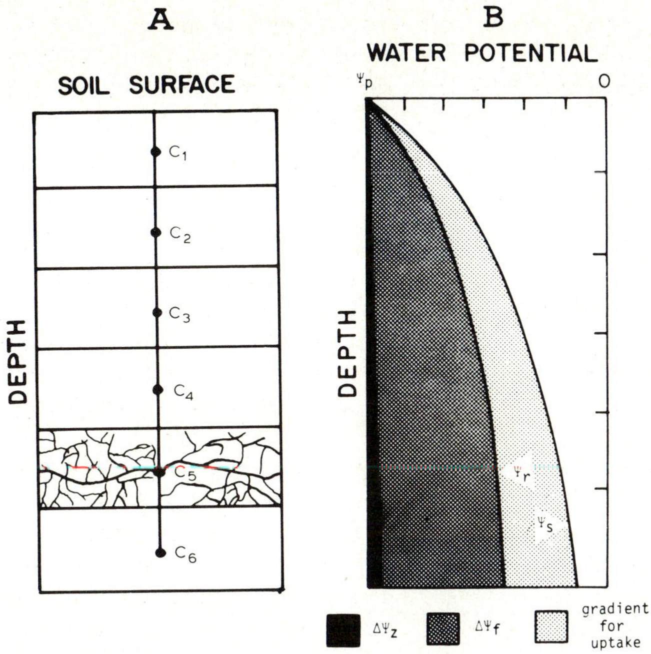

Fig. 1. Diagrams of depth relationships in model. (A) Relationships between absorbing and transporting roots. Water is extracted and transmitted to the mid-plane of a layer by absorbing roots; it is moved to the surface by transporting roots. (B) Potential relationships with flow-dependent axial resistance in transport roots.

Calculation of flow-dependent resistance

An expression for $\Delta\Psi_f$ can be derived using the Poiseuille-Hagen equation which relates flow through small pipes to pressure gradients. Use of this equation implies that all of the water potential change in xylem vessels is the result of changes in the pressure component with no change in osmotic component. This equation can be written:

$$\Delta E = \frac{8\, L\eta J}{\pi B^4} \tag{2}$$

where ΔE = pressure difference (MPa); L = vessel length (m); η = coefficient of viscosity of water (MPa s); J = flow rate ($m^3\ s^{-1}$); B = vessel radius (m). The J term is the flow rate through one vessel, or:

$$J = \frac{Q_i}{N_i} \tag{3}$$

where Q_i ($m^3\ s^{-1}$) is the total flow rate from the ith to the $(i-1)$th layer and N_i the number of vessels available for transporting water from the ith to the $(i - 1)$th layer. The value of B can be calculated from the arithmetic average of measured radii or from a weighted mean, such as the harmonic mean (Ponsana, 1975). The flow rate J can be entered into Eq. (2), to give the following:

$$\Delta\Psi_{fi} = \frac{8\,\eta L Q_i}{\pi N_i B^4} \tag{4}$$

where $\Delta\Psi_{fi}$ is the frictional potential loss (MPa) between midplanes of the ith and $(i - 1)$th layers. We have used 0.001 002 MPa s, the viscosity of pure water at 20°C, for η in all calculations and thus have implied that the entire root system is at 20°C.

Assuming by an Ohm's Law analogue that:

$$\Delta\Psi_{fi} = R_i Q_i \tag{5}$$

where R is the resistance to water flow through xylem between midplanes of the ith and $(i - 1)$th layers (MPa s m^{-3}), we can substitute Eq. (4) to obtain:

$$R_i = \frac{8\,\eta L}{\pi N_i B^4} \tag{6}$$

Assuming that flow rate between any two layers equals the uptake rates of all distal layers, we can substitute Eq. (5) into Eq. (1) to obtain i sets of equations of the form:

$$\begin{aligned} U_i = V_i D_i K_i \{ &\Psi_{si} - \Psi_p + \Delta\Psi_{zi} + R_1(U_1 + U_2 + \ldots + U_i) \\ &+ R_2(U_2 + \ldots + U_i) + \ldots + R_i(U_i) \} \end{aligned} \tag{7}$$

The value for R_i given in Eq. (6) can be substituted for all values of R_i in Eq. (7). These i equations in i unknowns can be solved by iteration on a computer.

Relationship of number of transporting xylem vessels and rooting density

The values of N_i and D_i are not independent because a certain percentage of the root length of any system is involved in transport. By assuming that 22% of the roots crossing a plane separating any two layers are transport roots and that each transport root contains on the average six conducting vessels, we calculated that $N_i = 0.66\ D_i$ by using the relationship (Melhuish and Lang, 1969) that the number of roots crossing the face of a unit cube is one-half of the rooting density in the cube.

Changing values of K

Although we have dealt with K as a constant, it changes both with the soil-crop system and with soil water content. Fortunately the relationship between K and soil water content is often linear over the range of interest (Taylor and Klepper, 1975; Herkelrath et al., 1977). We have chosen to use empirical linear relationships of K with soil water content for the model. Using hourly values of Ψ_p, diurnal uptake curves for each soil layer can be calculated with Eq. (7) and summed to give daily water extraction for each layer. This value can be compared to measured extraction rates, if evaporation and drainage are accounted for. In practice, we have systematically varied B, the least certain of the parameters, to determine which values of B yield water extraction rates for all soil layers, which most closely match those measured in the field.

Sensitivity analysis of flow-dependent model

Table I shows values for parameters in Eq. (7) under afternoon field conditions. Estimated errors of input terms, calculated as the ratio of one standard deviation to the mean (or coefficient of variation), range from 5% for factors like depth and soil volume to 25% for root-soil system permeability and 65% for losses in xylem potential caused by friction. The table also shows how much the estimated error of U_i is reduced by halving the error in the input value. This value was determined by combining estimates of error for the terms in Eq. (7), according to standard methods (Michels, 1958) for determining variance of sums and products to find an initial error estimate for U_i. Then one-half of the error for each term was combined with the original error of the other terms to provide the new error estimate of U_i. Column 4 in Table I is the ratio of the change in error estimate of U_i to the original error estimate for U_i. Unfortunately, D_i, K, B

TABLE I

Error contributions of terms in Eq. (7) to errors in calculated value of uptake (U_i); errors are given as the ratio of one standard deviation to the mean

Term	Mean value assumed for error analysis	Estimated error of input values	Change in error estimates of U_i from reduction of error in a term by 0.5
V_i	0.03 m^3	0.05	0.01
D_i	10^4 m m^{-3}	0.20	0.06
K	8.3×10^{-12} m^3 s^{-1} m^{-1} MPa^{-1}	0.25	0.08
Ψ_{si}	−0.5 MPa	0.10	0.01
Ψ_p	−2.0 MPa	0.10	0.01
$\Delta\Psi_{zi}$	−0.001 MPa m^{-1}	0.05	0.01
L_i	0.15 m for $i = 1$ 0.30 m for $i > 1$	0.05	0.01
N_i	660 for $D_i = 10^4$	0.10	0.01
B	3×10^{-5} m	0.10	0.10
$\Sigma\Delta\Psi_{fi}$	−0.5 MPa	0.65	0.25

and $\Sigma\Delta\Psi_{fi}$, the parameters most difficult to measure, contribute much of the error in flow estimates.

Mathematically, terms in Eq. (7) fall into four categories: (a) V_i and K are proportional to U_i; (b) the water potential terms, Ψ_{si}, Ψ_p, Ψ_{zi} and $\Sigma\Delta\Psi_{fi}$ influence U_i depending on their magnitudes relative to one another, but U_i is directly proportional to the summed potential term; (c) D_i enters the equation twice, once in direct proportion to U_i and once (as N_i) in the denominator of the $\Delta\Psi_{fi}$ term; (d) L_i and B_i enter only into the $\Delta\Psi_{fi}$ term. To illustrate the influence of terms in these categories on U_i, we used the input values shown in Table I for six 0.30 m layers (1.8 m total profile) and solved six equations in six unknowns by iteration on a computer. Then we systematically varied input values to determine effects on uptake from individual layers and on total uptake from the profile.

Fig. 2 shows water uptake rates in each layer and total uptake rate for four values of K and two values of B_i. Plant water potential was set at −2.0 MPa, root length density at all depths at 10^4 m m^{-3}, and Ψ_s at −0.01 MPa. A 1% change in K gives a 1% difference in U_i at low axial resistance (large xylem vessel radius) and about 0.7% difference in U_i at high axial resistance (smaller xylem vessel radius). A large axial resistance changes the shape of the water extraction profile so that water is taken preferentially from shallow layers.

Values in Fig. 2 were calculated for a case where the difference between the plant and soil water potentials was large. The water potential difference term in the equation consists of four components each of which influences U_i. Relative to the other three terms, the elevation term is always small

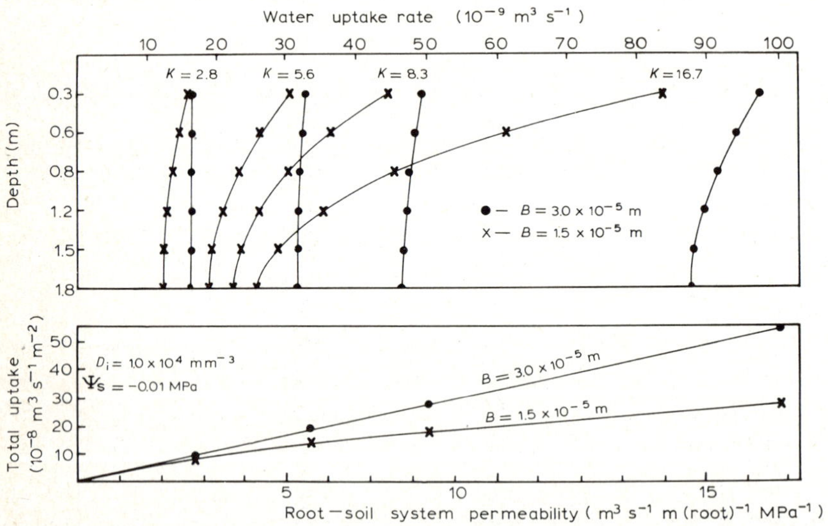

Fig. 2. Water uptake rate with depth for different values of K and B with Ψ_p at −2.0 MPa, D_i at 10^4 m m^{-3}, and Ψ_s at −0.01 MPa at all depths.

and a 1% difference in $\Delta\Psi_{zi}$ causes less than a 0.01% change in U_i for most cases. The plant water potential term (Ψ_p) can range from a few tenths MPa or less at night to —3.0 MPa in some crops, and soil water potential (Ψ_s) ranges from near —0.01 to around —2.0 MPa. Plant water potential changes dramatically during the day but soil water potential changes significantly only on a day-to-day basis. The $\Delta\Psi_{fi}$ term is also dynamic. It can range from a value as small and insignificant as the elevation term to a maximum value less in magnitude than ($\Psi_p - \Psi_s - \Delta\Psi_z$). The magnitude of the $\Delta\Psi_{fi}$ term is closely related to the Ψ_p term on any one day. Since the $\Delta\Psi_{fi}$ term depends on U_i, there is a non-linear relationship between U_i and the potential term. As flow increases in response to a change in potential, the new flow produces a new $\Delta\Psi_{fi}$ value which partially offsets the potential change and establishes a new flow rate.

Fig. 3 shows the effect of changing the root length density (D_i) on uptake at two values of xylem vessel diameter. Root length density determines the root length available to extract water from a layer as well as the number of xylem vessels available to carry water to the next layer. Root length density under field conditions generally ranges from nil to 3×10^4 m m^{-3} for most crops. For a 1% change in D_i there is a 1% change in U_i for all values of xylem vessel radius.

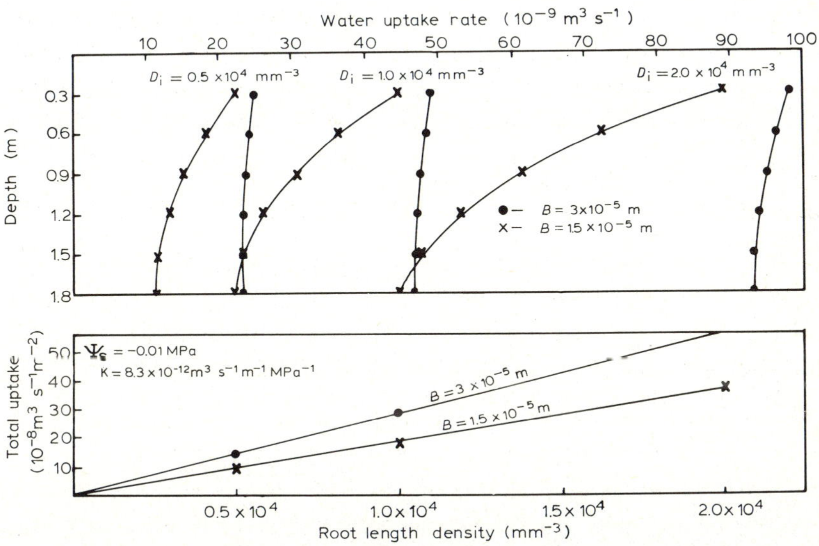

Fig. 3. Water uptake rate with depth for different values of D and B with Ψ_p at —2.0 MPa, K = 8.3 × 10^{-12} m^3 s^{-1} m^{-1} MPa^{-1}, and Ψ_s at —0.01 MPa at all depths.

Radii of xylem vessels are of the order of 10^{-5} m and may differ depending on species and on the soil environment in which the root grew and differentiated (Meyer and Alston, 1978). This value may differ from axis to axis

(Passioura, 1980) and may not be the same throughout the length of a particular axis (Meyer, 1976). The selection of a value for B is critical to the calculations of frictional potential loss, which is dependent on the 4th power of B. A 1% change in B causes less than 0.01% change in flow if U_i is small or if B is large. However, if B is small, if U_i is large, or both, then a 1% change in B is magnified into a change of several percent in U_i. Both Figs. 2 and 3 illustrate effects of different B values. The most noticeable influence of B is on the shape of the uptake profile, with small B restricting uptake from deep soil.

For Figs. 2 and 3 soil water potential was —0.01 MPa at all depths. This condition rarely occurs. Fig. 4 shows water uptake profiles when the water potential in surface layers changes and Ψ_p = —2.0 MPa. Where the top four layers decrease from —0.01 to —0.5 MPa, uptake from these layers decreases by 25% and uptake from the lower layers increases by 12%. If soil water potential in the top four layers drops from —0.01 to —1.5 MPa, uptake from the top four layers drops at least 75% and increases from the lower layers by almost 40%. In one of the profiles of Fig. 4, the top two layers have Ψ_{si} = 1.5 MPa, the center two, —0.5 MPa, and the lowest two —0.01 MPa. Here again, when uptake is small in surface layers, more water is taken from lower layers even though Ψ_{si} remains at —0.01 MPa in these lower layers. This effect is specifically the result of having axial resistance flow-dependent because it only occurs if B is small enough to restrict flow.

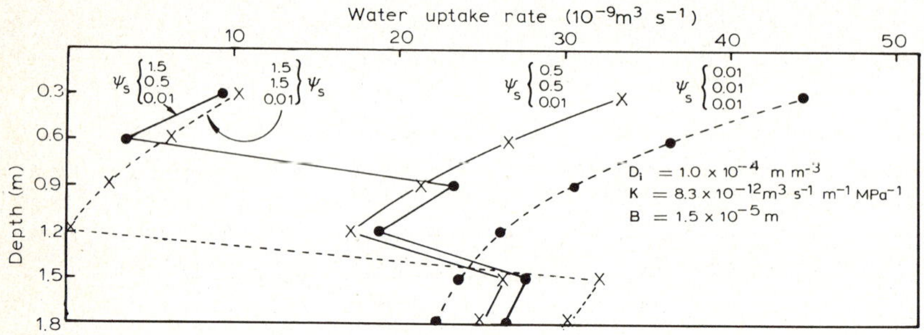

Fig. 4. Water uptake rate with depth for several values of Ψ_{si}.

Availability of model inputs. The present model requires the following information: plant xylem water potential at the land surface either continuously or for time intervals of the order of 1 to 2 h (Huck and Klepper, 1977; Turner, 1981); estimates of soil water potential as it changes through space and time so that units of soil may be readjusted every day or two; estimates of rooting depth and of root length density as it changes through space and time so that root growth and decay can be accommodated and the rooted soil volume can be adjusted; empirical estimates of the change in K_i with soil water content for the particular crop and soil under study, including any effects of change in texture with profile depth that may apply (Herkelrath et al., 1977); specific information on the vascular structure of the root system including the number and size of xylem vessels involved in transporting water to the shoot.

The last of these inputs is the most glaring failure in our present knowledge. We lack any real understanding of the root system parts and their relationship to one another and to different parts of the shoot. We do not yet know the detailed phenology of a crop root system. For example, cotton puts down a tap root which grows at a rate of several centimeters per day; this tap root branches and these branches explore the profile laterally with their elongation and with tertiary branches. This apparently happens randomly with no set spacing for branches. We do not know whether there are set angles for branch roots to grow relative to their parent root in cotton; in soybean, temperature is important in determining this angle (Kaspar et al., 1981). Both seminal and nodal roots of wheat appear to be generated during a specific time window, branch after a given period, and grow in a predictable direction while producing branches at definite angles to the parent root (R.K. Belford, Letcombe Laboratories, personal communication, 1982). Thus, root systems may indeed be more orderly and predictable than previously thought. Models which predict the amount of root material present in the profile and the history and relationships of those pieces of root to one another are needed to develop root-centered functional models.

Field data test. The model was tested with field data from three crops grown in three separate locations. Data for cotton (*Gossypium hirsutum* L.) were taken from experiments performed using Cahaba loamy fine sand in the Auburn rhizotron (Browning et al., 1972). Soybean data were obtained from experiments performed by Willatt and Taylor (1978) on Ida silt loam in western Iowa; numerical data for the specific time period selected were furnished by S. Willatt (La Trobe University, personal communication, 1979). Data for winter wheat were taken from datasets collected in 1977 by Rickman and Klepper (1979). Table II shows information and input values for the three datasets.

Fig. 5 compares measured water extraction profiles to water extraction rates calculated with the model. Agreement is very good for cotton, moderately good for soybean, and poor for wheat.

Model summary. The axial resistance model presented here is a modification of the one used by Taylor and Klepper (1978). It has been modified to permit more specific inputs of experimental data, including values for the number of xylem vessels involved in axial transport and the average vessel radius. It has been developed to permit the calculation of flow-dependent axial resistance by iterative techniques. The model predicts measured extraction profiles moderately well for taprooted crops (cotton and soybean) and less well for wheat. The model has the advantage of being cen-

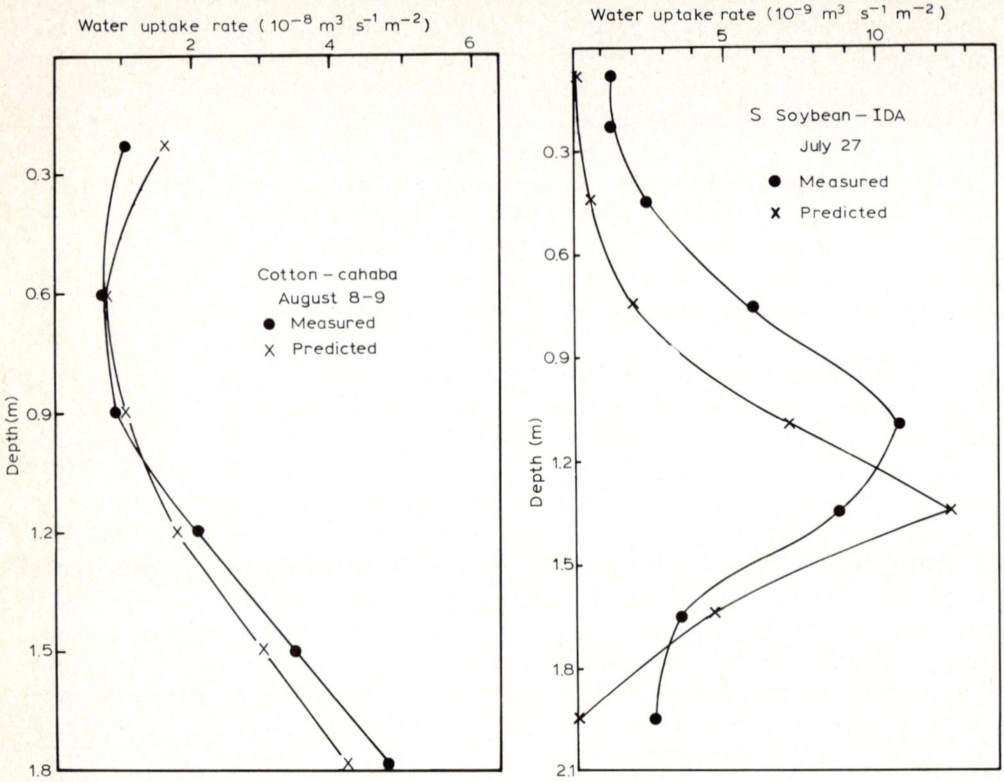

tered on root behavior, and thus would be useful for applications where impacts of different rooting patterns are under study.

IMPACTS OF MANAGEMENT CHOICES ON ROOTS

Table III shows some of the important choices made in managing a farm in a wheat-fallow rotation. Impacts of these decisions are made primarily through effects on plant roots. These impacts can be classed into 5 groups according to effects of nutrients, water, pests, pesticides and herbicides, and space (planting geometry). This section deals briefly with these impacts of management on root growth and function.

Nutrient effects

Soil fertility, whether it results from preplant amendments, from post-plant application or from mineralization of old (humus) or recent (crop residues) organic matter, affects both root growth and root function. Generally, roots proliferate in fertile bands by increasing growth rates and

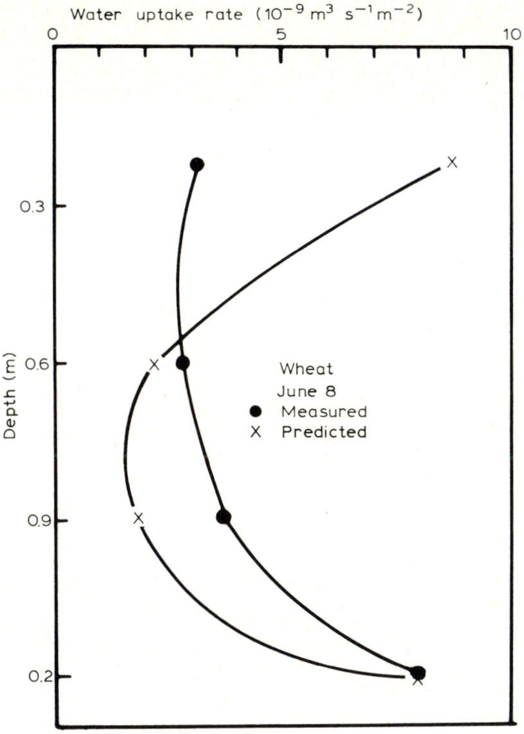

Fig. 5. Water uptake rates with depth for cotton, soybean, and wheat. See Table II for input values used for each crop.

sometimes frequency of branches whether the nutrient supplied is phosphorus (Christie, 1975; Anghinoni and Barber, 1980), nitrogen (Hackett, 1972; Maizlish et al., 1980) or potassium (Tennant, 1976). In grasses, effects on seminal and nodal root systems are different, with nodal roots being generally more responsive than seminals (Christie, 1975; Tennant, 1976). Fertilization can also increase overall rooting depth in some species (Holechek, 1982). These effects on root growth are coupled with effects on root function. The total nutrient uptake by root systems for supply to shoots depends on both the length of root and the activity of the material in the soil solution which contacts the various parts of the root (Brouwer and De Wit, 1969). Localized pockets of fertilizers are sites of high uptake rates, both because the root surface area in and near the pocket is high and because the concentration of fertilizer is high. Feedback control mechanisms for integrating uptake of nutrients over different parts of a heterogeneous root-soil system may or may not exist in the same way that water uptake at one depth impacts uptake at other depths through xylem water potential (See Eq. 1). The concentration of minerals in the xylem of transpiring plants is rarely sufficient to cause back diffusion (Fiscus, 1977).

TABLE II

Inputs used in testing the model with field data for three crops in three different soils

Crop and soil type	Cultivar	Location and soil classification	Date of data collection
Cotton — Cahaba loamy fine sand	*Gossypium hirsutum* L. Auburn 6238	Auburn, AL (Rhizotron) Fine-loamy siliceous, thermic Typic Hapludults	8 August 1972
Soybean — Ida silt loam	*Glycine max* [L.] Merr. 'Wayne'	Castana, IA Fine, silty, mixed (calcareous) mesic family of Typic Udorthents	23 July 1976
Winter wheat — Walla Walla silt loam	*Triticum aestivum* L. em. Thell 'Luke'	Pendleton, OR Mixed, mesic family of Typic Haploxerolls	7 June 1977

Water effects

Soil water content affects both the growth pattern and the function of a root system. Water held in the soil at high potential enhances root growth both by providing low penetration resistance for root elongation and by providing a steep water potential gradient for rapid water flow into the cells of the enlarging roots. Thus roots tend to proliferate in moist zones of soil. Water uptake by roots can be described by Eq. (1) in the model section and depends on the length of root present, the permeability of the root-soil system, and the water potential difference available for causing flow.

The soil water profile depends on a complex history of climate, tillage and residue management, weed control, crop water use, and such soil properties as water-holding-capacity, infiltration and internal water transmission rates, and profile depth. The following description was taken from Ramig et al. (1982) who described the tillage sequences and their impacts for drylands in the Pacific Northwest, where precipitation primarily falls in the winter. Primary tillage sets surface roughness and residue cover which, in

Soil input values			
Depth increment (m)	Ψ_s (MPa)	K (10^{-12} m^3 s^{-1} m^{-1} MPa^{-1})	D_i (10^4 m m^{-3})
0—0.45	—0.040	2.70	1.66
0.45—0.75	—0.04	2.70	1.25
0.75—1.05	—0.40	2.70	1.95
1.05—1.35	—0.025	5.75	1.54
1.35—1.65	—0.023	8.23	1.93
1.65—1.88	—0.020	13.76	2.13
0—0.15	—2.50	0.03	0.15
0.15—0.30	—1.27	5.56	0.17
0.30—0.60	—1.20	7.78	0.16
0.60—0.90	—0.74	22.22	0.11
0.90—1.20	—0.43	44.44	0.13
1.20—1.50	—0.24	66.67	0.12
1.50—1.80	—0.18	77.78	0.04
1.80—2.10	—0.14	88.89	0.001
0—0.45	—0.160	0.78	2.77
0.45—0.75	—0.320	0.42	1.75
0.75—1.05	—0.350	0.36	2.00
1.05—1.35	—0.080	1.25	2.12

Continued overleaf

turn, control fallow year water infiltration and surface evaporation. Depth and degree of compaction of the soil profile also usually accompanies the primary tillage operation. Secondary tillages and herbicides control weed growth and are used to control evaporative water loss. Soil temperature, water content, and strength profiles are determined by the specific set of tillage practices chosen in any field. These profiles as they develop by the time of seeding and crop establishment determine the relative success of seedlings. Compacted layers will slow root penetration. Short, thick, twisted roots with above average branching are common with soil compaction. Not only shape but root function is affected. Roots restricted to limited volumes of the soil profile will not be able to extract materials or may extract them inefficiently (Bowen, 1981). Surface residues tend to increase infiltration of water and to decrease surface evaporation to give a net gain of water; this extra water can produce differences in root distribution and yield in a dry year or a dry location (Unger, 1979). In very wet locations and on clay or impermeable soils, the added water in the profile can cause waterlogging and detrimental effects on root proliferation (Ellis and Barnes, 1980), soil nitrogen (Rickman and Klepper, 1979) and yield (Ellis et al., 1979).

TABLE II Continued

Input values for diurnal Ψ_p (MPa)											
Time (h)											
12.00	02.00	04.00	06.00	08.00	10.00	12.00	14.00	16.00	18.00	20.00	22.00
−0.47	−0.36	−0.30	−0.30	−0.30	−1.26	−1.58	−1.73	−1.60	−1.30	−0.92	−0.63
−0.30	−0.30	−0.30	−0.30	−0.45	−0.79	−1.01	−1.32	−1.45	−1.42	−1.16	−0.76
−1.05	−0.89	−0.75	−0.86	−1.74	−2.28	−2.40	−2.24	−2.26	−2.36	−2.03	−1.16

Pests

Below-ground pests and pathogens include plant parasitic nematodes bacterial rots, fungal parasites, and arthropods. These organisms impact root growth and function by changing root morphology, anatomy and growth rate (Schenck, 1977; Walters and Ayres, 1981), by loss of roots to ingestion and decay (Detling et al., 1980), by competition for nutrients (Elliott et al., 1981), by plugging vascular tissues, by phytotoxic effects (Katouli and Marchant, 1981) and by suppressing activities of beneficial organisms (Baker and Cook, 1974). Many organisms are specifically attracted to and stimulated by root exudates, and there is some possibility that resistant varieties differ in amount and composition of exudates released into the rhizosphere (Schenck, 1977). Management to avoid pest damage includes use of resistant varieties, use of pesticides, and careful management of rotations and surface residues to maintain low inoculum potentials of pests (Cook, 1981).

Pesticides and herbicides

Spray treatments for pest control and weed control may cause temporary impacts in non-target micro-organisms (Katan and Eshel, 1973; Stanton

TABLE III

Management options for production of dryland winter wheat on fallowed rotations with impacts of those choices on soil and root properties

Farmer choice	Principal soil property impacted	Model parameters available to describe impact of choices	Principal root property impacted
During fallow year			
Primary tillage			
Degree of mixing	Soil strength	D, N, B	Rooting patterns
Timing	Residue composition	D	Net nutrient availability
Depth	Residue placement	D	Timing of release of nutrients and phytotoxins
Surface roughness	Infiltration	Ψ_s, K	Water availability
Evapotranspiration control			
Herbicides	Profile water content	Ψ_s, K	Water availability
Tillage	Surface moisture and residue cover	Ψ_s, K	Temperature and surface water availability
Pre-planting			
Fertilizer application			
Timing	Nutrient location and form	D	Branching and uptake rates
Placement	Nutrient location	D	Branching and uptake rates
Composition	Nutrient balance and form in soil solution	—	Nutrient balance
Amount	Nutrient concentration	—	Rate of uptake
Crop and variety	Long-term changes in soil organic matter composition	Ψ_p, D, N	Rooting pattern
At planting			
Seed quality and treatments	Pest levels	D, N, B	Root number and growth rate
Legume inoculation	Inoculum potential	—	Nodulation
Depth of seeding	—	—	
Timing	—	Ψ_s, K	Temperature and moisture availability
Drill design	Seed-soil contact and overburden strength	D, N, B	Rate of emergence and radicle growth rate
Seed-seed distances	—	—	Root-root competition
Post-plant			
Fertilization			
Timing	Nutrient location	D	Branching and uptake location
Placement	Nutrient location	D	Branching and uptake location
Composition	Nutrient balance in soil tion	—	Nutrient balance
Formulation	Nutrient form	—	Nutrient availability
Amount	Nutrient concentration	—	Rate of uptake
Pest Control			
Weeds	Water and nutrient concentrations	Ψ_s, K	Water and nutrient availability
Fungi	Inoculum potential	D, N	Root decay and permeability
Bacteria	Inoculum potential	D, N	Root decay and permeability
Animals	Pest populations	D, N	Root loss

et al., 1981) and on crop roots (Currey and Teem, 1977). Of 16 classes of herbicidal chemicals, ten classes have been shown to have direct effects on crop roots (Ashton and Crafts, 1973). Effects include malformation of roots and root cells and cell components, changes in cell division, elongation rates, and geotropic responses of roots (Currey and Teem, 1977) and changes in room permeability (Reid and Hurtt, 1970). These effects can indirectly limit the functional ability of the root system. For example, Tottman and Davies (1978) found that commercial rates of two herbicides (MCPA and Mecoprop) produced short, swollen lateral branches ('tassels') on wheat roots. They suggested that such deformities could decrease the ability of the crop to withstand nutritional or water stresses.

Planting geometry

The planting density of dryland crops is generally higher, with more narrow rows and less distance between seedlings within the row, in moister parts of the country. The main effect of increasing plant density within the row is to cause roots of neighboring plants to compete sooner for water and minerals. In the same way, wider rows will delay the time when roots explore the inter-row soil and compete for below-ground resources.

Optimum row spacing and within-row plant distances are specific to site and season (Myers and Foale, 1980) and to crop and variety (Faris and DePauw, 1981). Generalized principles for selecting planting densities for commercial crops could be developed if we had better information on below-ground competition. In the future we may have the ability to seed with much greater precision than is now possible. Uniformity of spacing within the row has been shown to increase yield in cereal grains (Mülle and Heege, 1981).

Farmer inputs are related to yield through the interactive growth and function of plant roots and shoots. A tillage operation or soil amendment may change a soil property (strength, fertility, pH, etc.). This change in soil property provides an environment which is either more or less suitable for root growth. Root growth is also influenced by the supplies of carbohydrates and other nutrients produced by the functioning shoot. The end result of root growth is the amount and distribution pattern of root material in the soil profile or the root length density profile. This root profile determines how roots extract material from the soil; the rate of extraction depends on such soil properties as the concentration or activity of the material being extracted in the soil solution and soil temperature. Roots then supply shoots with raw materials, which, along with materials and energy from the aerial environment, provide for additional shoot growth and partitioning to yield. Management can also affect shoots directly. For example, many pest control measures extend the life span of green leaf area and enhance yield by prolonging the operation of shoot function.

RESEARCH NEEDS

Although the past two decades have seen great progress in understanding processes in the soil-root system, a number of areas are in need of special attention. More specific information is needed on: phenology of root system development relative to shoot development under field conditions; above- and below-ground environmental effects on root growth and death; functional relationships among roots, between roots and soil, and between roots and shoots under field conditions; genetic variation in root anatomy and rooting patterns; effect of variability in the landscape, soil profile, and plant populations on operation of models of the soil-root system.

Additional information on the above topics is needed to assess impacts of farm management on root system function. If only one of the above topics is studied and the others neglected, application of our knowledge to practical situations will not be possible. For example, the resistance of the root-soil interface to fluxes of materials is a crucial topic for study, but that study is not very useful if we do not know how long the roots are and how they are connected to one another. It would be like knowing the resistivity of wire with great precision and trying to predict the resistance of an unknown length of the wire. In order to get information on the length of root for use in a truly functional model, we must understand the first two topics listed above. Then, if we are to change root parameters and design root morphology to improve crop function in a particular environment, we need to understand how much genetic variation there is in rooting habit for different crops. Finally, we must concern ourselves with the sorts of variability seen in fields and with the effect of variability on the operation of plant yield models. Dynamic models produce different results when variances are given proper consideration. Consequently, we must find efficient ways to include field and plant heterogeneity in models.

Studies undertaken on roots will be more meaningful if they are done with careful reference to plant developmental stage, so that shifts in sinks, sources, and factors limiting to production can be taken into account. For example, a recent Japanese study on sorghum (Mirhadi and Kobayashi, 1980) provides a highly specific plant-centered framework for describing the sequence of events in shoot and root morphological development for both above- and below-ground phenology. Using phyllochrons to describe vegetative development, these authors related sizes and development of sheaths, blades, internodes and nodal roots to the history of development of successive nodes. Similar developmental sequences have been worked out for rice (Yamazki and Harada, 1982) and for small grains (Klepper et al., 1982a, b).

Most work on the developmental history of dicotyledons has not yet included specific information on root phenology. The root system of peas (*Pisum sativum* L.) during the period 6 to 20 days after germination has been examined (Yorke and Sager, 1970) but not under field conditions and not for mature plants. This connection between root and shoot development needs to be made because models which include detailed morphological schemes permit easier determination of shifts in sinks and sources for fluxes of materials and better definition of the time when particular transfer processes become limiting to growth (Jung, 1980; Yang, 1981; Bennie, 1982; Mason et al., 1982; Reicosky et al., 1982b; Taylor et al., 1982). Work on dicots will require more careful examination of the xylem and phloem connections between roots and their branches and between roots and shoots. Anatomical approaches such as those of Feldman (1977) and Byrne et al. (1982), are useful but tedious. Perhaps methods can be developed to trace xylem vessels with polymerizing liquids such as polystyrene (Taneda et al., 1979) or polyesters (Stieber, 1981). These liquids could be used either alone or in conjunction with radioactive or fluorescent tracers to permit rapid and specific detection. The technique would have the advantage of stabilizing a tracer, so that it would not diffuse with time into tissues adjacent to active xylem.

A recent technique (Belford et al., 1982) permits preservation of field-grown root axes and allows their spatial arrangement to be displayed. This preservation is accomplished by inserting sharpened square metal tubes in the ground after the crop has emerged. Tubes, containing soil and the complete root system of field-grown plants, are removed periodically during the season and examined by washing the soil away from intact axes thus preserving the integrity of the system. To develop three-dimensional display of roots, soil can be washed in alternating 2 to 5 cm horizontal layers, leaving exposed root axes, held in place by adjacent layers of soil. Then the cavities are filled with clear casting resin. Once set, these layers of casting resin hold the roots in place while the remaining soil is washed away and replaced with more casting resin. Thus the axes of field-grown roots can be preserved more or less in situ for observation and measurement.

In contrast root segments from cores inserted at the time of sampling are too fragmented to be of use in tracing the life history of individual axes. We need to improve techniques for recovering entire field-grown root systems and for non-destructive root observation under field conditions. Field lysimeter installations (Belford et al., 1982), monoliths, and trenching techniques need to be made more cost-effective (Köpke, 1981).

We also need better methods for determining when individual roots cease to function in absorption and in translocation of water and minerals. Vital staining techniques (Ward et al., 1978) will be of some use, especially if they can be combined with automated counting techniques (Voorhees et al., 1980) which are color-sensitive so as to count living roots and ignore debris and non-functional root material. Of even more potential use are functional approaches such as that used by Troughton (1981) who determined how long a grass root axis would supply water and minerals to support shoot growth when it was the sole supplier. Although the treatments (removal of all new root axes) were drastic, the important difference in this approach is that *function* is used as the measure of 'life'. The activity of the entire root axis complex (both transporting roots and absorbing roots) contributes to determining function in this test.

Another use of functionally-related parameters in measurement is illustrated by the work of Fiscus (1981; Fiscus and Markhart, 1979) on beans (*Phaseolus vulgarus* L.). Roots were divided into size classes and measurements of both leaf areas and root areas were made as bean plants aged (Fiscus, 1981). This meticulous work permitted Fiscus to study changes in root system conductance with plant age and to show effects of recent plant history on root system conductance. Apparently root conductance depends, in part, on recent root growth rates both in beans (Fiscus and Markhart, 1979) and in soybean and cotton (Bunce, 1978). These are only two of the recent papers that relate root and shoot growth and function. Such studies should now be done under field conditions to determine the relative importance of management factors on both roots and shoots and their function.

It is clear that roots offer resistance to water movement to shoots, but no one has successfully apportioned that resistance into radial and axial components for crop plants grown under field conditions. Theoretical analyses of relationships between the two resistances are available (Landsberg and Fowkes, 1978) as are techniques for measuring root water potentials (Fiscus, 1972; Adeoye and Rawlins, 1981). Potometer methods developed by Aston and Lawlor (1979) and by Rowse and Goodman (1981) could perhaps be adapted for measurements of fluxes. Also a careful analysis of xylem anatomy of field-grown crop plants such as was done by Jeje and Zimmerman (1979) would be useful in assessing the growing conditions, which might lead to increases in root resistance to water extractions.

Root-shoot relationships show how interactions occur between organs on the same plant. Also needed is work to relate plants within the row to one another and to competing weeds. Effects on roots of allelopathic substances from living plants (Nielsen et al., 1981) or from dead root and shoot materials left by previous plants (Elliott et al., 1978) are only poorly characterized. They represent one mechanism by which plants reserve space (soil volume) for the exclusive use of individuals of a given species. Other mechanisms of below-ground competition include differential rates of root exploration of soil (Harris, 1977; Berendse, 1979), differential sources of materials (Martin and Snaydon, 1982), different rates of extraction of the same resources, and different resource requirements for growth and production. We are used to thinking of below-ground competition of weeds for crop resources, but rarely do we apply quantitative approaches to this competition. Nor do we often consider mechanisms of plant-plant competition of the crop within the row and between rows. By shifting row spacing and seeding density within the row, we can manage competition of the crop with itself to our advantage, only if we understand the dynamics of the process.

Assuming that techniques are available for monitoring the developmental history of roots and shoots, further screening of genetic material must be done. Morphological traits such as branching habit can be used as they were for studies of alfalfa (McIntosh and Miller, 1980, 1981) and cotton (Quiesenberry et al., 1981) where branching habit was found to be positively correlated with ability to grow under limited water supplies. Wheat studies in Australia (Richards and Passioura, 1981a, b) have related genetic material with different seminal root xylem vessel diameters to water use history. Soybean studies in Iowa have related taproot elongation rates to the pattern of water extraction for several sources of genetic material (Kaspar, 1982). These studies are illustrations of the use of *functional* criteria for screening genetic material and show the necessary interaction of physiologist and breeders in integrating physiological traits into breeding programs. This type of team work is much more productive than is work where genetic material is screened for some morphological trait with no consideration of field conditions and physiological stress.

As better techniques become available for studying the phenology of root systems and the functional relationships between roots and shoots, we need to develop the capability to study the interactions of numerous factors all operating at once on plants. Almost all studies are limited to one or two factors. Even if both roots and shoots are studied, they are usually studied with respect to one process such as uptake and translocation of water, minerals, nitrogen, or carbon. Rarely are fluxes of these important materials studied together. Thus, there are subunit models for nitrogen flows, carbon flow, water flow, mineral uptake, root growth, and shoot growth. Rarely are the data available to integrate these subunits properly.

Multi-factor experiments are sometimes done. For example, a group of researchers at Ames, IA (Jung, 1980; Yang, 1981; Bennie et al., 1982; Mason et al., 1982; Reicosky et al., 1982a, b; Taylor et al., 1982) investigated the reason(s) that narrow row (0.25 m) soybeans outyielded wide row (1.0 m) ones by 18% during 1979. They measured, throughout the growing season, biomass accumulation, leaf area development, water potentials of leaves and soil, root lengths and masses, soil water contents, accumulation of twelve ions in various plant parts, stages of reproductive and vegetative growth, radiation interception, and evapotranspiration rates. Cushman (1980) modelled both root growth and fluxes of minerals; Bosatta et al. (1980) studied nitrogen mineralization, immobilization, leaching, and uptake by roots as related to soil water content, temperatures, microbial activity, and climate for Scots pine (*Pinus sylvestris* L.). Teskey and Hinckley (1981) studied root growth of white oak (*Quercus alba* L.) as related to temperature, soil water potential, and leaf water potential. Use of irreversible thermodynamics has permitted development of models which interrelate solute and water flow in root systems (Fiscus, 1975; Dalton et al., 1975). These developments have permitted new insights into the functioning root, but have not as yet been adapted to a field-level model. Nor have they been integrated into ion uptake models based on electrochemical potential theory or into water extraction models (Molz, 1981). The same general criticism can be made for recent work on the properties of the soil-root interface (Harley and Russell, 1979); the results are of great potential importance but will only realize that potential when they can be related to mature plants functioning under field conditions. Likewise, one of the important features of the soil-root interface is the presence of microbes which alter root function. Work in this area will be more meaningful if it is related to the complex of environmental variables present in the field and to fluxes of materials in crop plants. An attempt to integrate fluxes of matter and energy by Christian and Milthorpe (1981) was moderately successful for short-term processes in a lucerne (alfalfa) crop.

Assuming that we have better understanding of plants and of factors which limit plant processes, we must expand our modelling ideas to include heterogeneity as an integral part of models. We use the word 'heterogeneity' to include spatial variability or 'patchiness' of plant populations within the row (Chesson, 1981; Weiner and Conte, 1981), for time distributions of plants with different emergence times (Rickman et al., 1981), for relative growth rates of individuals and of clones within a population (Burdon and Harper, 1980), and for differences in soil properties on both a horizontal (landscape) scale (Bresler et al., 1981; Vieira et al., 1981) or vertically within the profile. It is important to incorporate variability into models because operation of environmental factors may favor one subgroup of a population more than another (Ford and Diggle, 1981). Also operation of a model, which includes variance, permits real data to be installed for testing and allows use of statistical decision theory (English, 1981) in optimization of resource use.

ACKNOWLEDGEMENTS

The authors wish to thank Steve Willatt for permission to use unpublished data on soybeans, Sue Waldman for help in computer programming, and Tami Toll, T.W. Tucker, and G.R. Fischbacher for technical assistance.

REFERENCES

Adeoye, K.B. and Rawlins, S.L., 1981. A split-root technique for measuring root water potential. Plant Physiol., 68: 44—47.

Anghinoni, I. and Barber, S.A., 1980. Phosphorus application and distribution in the soil and phosphorus uptake by corn. Soil Sci. Soc. Am. J., 44: 1041—1044.

Ashton, F.M. and Crafts, A.S., 1973. Mode of Action of Herbicides. John Wiley, New York, NY, 504 pp.

Aston, M.J. and Lawlor, D.W., 1979. The relationship between transpiration, root water uptake, and leaf water potential. J. Exp. Bot., 30: 169—181.

Baker, K.F. and Cook, R.J., 1974. Biological Control of Plant Pathogens. Freeman, San Francisco, CA, 433 pp.

Belford, R.K., Rickman, R.W. Klepper, B. and Allmaras, R.R., 1982. A new technique for collecting and evaluating intact root systems of cereal plants grown in the field. 1982 Agron. Abstr.

Bennie, A.T.P., Mason, W.K. and Taylor, H.M., 1982. Responses of soybeans to two row spacings and two soil water levels. III. Concentration, accumulation, and translocation of 12 elements. Field Crop Res., 5: 31—43.

Berendse, F., 1979. Competition between plant populations with different rooting depths. I. Theoretical considerations. Oecologia, 43: 19—26.

Bosatta, E., Bringmark, L. and Staaf, H., 1980. Nitrogen transformations in a Scots pine forest mor-model analysis of mineralization, uptake by roots and leaching. Ecol. Bull., 32: 565—589.

Bowen, H.D., 1981. Alleviating mechanical impedance. In: G.F. Arkin and H.M. Taylor (Editors), Modifying the Root Environment to Reduce Crop Stress. American Society of Agricultural Engineers, St. Joseph, MI, pp. 21—57.

Bresler, E., Dasberg, S., Russo, D. and Dagan, G., 1981. Spatial variability of crop yield as a stochastic process. Soil Sci. Soc. Am. J., 45: 600—605.

Brouwer, R. and De Wit, C.T., 1969. A simulation model of plant growth with special attention to root growth and its consequences. In: W.J. Whittington (Editor), Root Growth. Butterworths, London, pp. 224—244.

Browning, V.D., Taylor, H.M., Huck, M.G. and Klepper, B., 1975. Water relations of cotton: a rhizotron study. Ala. Agric. Exp. Stn. Bull. 467, 70 pp.

Byrne, J.M., Byrne, J.M. and Emmitt, D.P., 1982. Development and structure of the vascular connection between the primary and lateral root of *Lycopersicon esculentum*. Am. J. Bot., 69: 287—297.

Bunce, J.A., 1978. Effects of shoot environment on apparent root resistance to water flow in whole soybean and cotton plants. J. Exp. Bot., 29: 595—601.

Burdon, J.J. and Harper, J.L., 1980. Relative growth rates of individual members of a plant population. J. Ecol., 68: 953—957.

Chesson, P.L., 1981. Models for spatially distributed populations: the effect of within-patch variability. Theor. Popul. Biol., 19: 288—325.

Christian, K.R. and Milthorpe, F.L., 1981. A systematic approach to the simulation of short-term processes in the plant-environment complex. Plant Cell Environ., 4: 275—284.

Christie, E.K., 1975. Physiological responses of semiarid grasses. II. The pattern of root growth in relation to external phosphorus concentration. Aust. J. Agric. Res., 26: 437—446.

Cook, R.J., 1981. The influence of rotation crops on take-all decline phenomenon. Phytopathology, 71: 189—192.

Currey, W.L. and Teem, D.H., 1977. Herbicides and root development. Proc. Soil Crop Fla., 36: 23—28.

Cushman, J.H., 1980. Comparison of convective and diffusive absorption of nutrients by growing roots. Plant Soil, 57: 407—421.

Dalton, F.N., Raats, P.A.C. and Gardner, W.R., 1975. Simultaneous uptake of water and solutes by plant roots. Agron. J., 67: 334—339.

Detling, J.K., Winn, D.T., Proctor-Gregg, C. and Painter, E.L., 1980. Effects of simulated grazing by below-ground herbivores on growth, CO_2 exchange, and carbon allocation patterns of *Bouteloua gracilis*. J. Appl. Ecol., 17: 771—778.

Elliott, L.F., McCalla, T.M. and Waiss, A., Jr., 1978. Phytotoxicity associated with residue management. In: Crop Residue Management Systems. Proc. Symp., 28 November—3 December 1976, Houston, TX. Am. Soc. Agron. Spec. Publ. 31, pp. 131—146.

Elliott, L.F., Cochran, V.L. and Papendick, R.I., 1981. Wheat residue and nitrogen placement effects on wheat growth in greenhouse. Soil Sci., 113: 48—52.

Ellis, F.B. and Barnes, B.T., 1980. Growth and development of root systems of winter cereals grown after different tillage methods including direct drilling. Plant Soil, 55: 283—295.

Ellis, F.B., Elliott, J.G., Pollard, F., Cannell, R.Q. and Barnes, B.T., 1979. Comparison of direct drilling, reduced cultivation and loughing on the growth of cereals. 3. Winter wheat and spring barley on a calcerous clay. J. Agric. Sci., Camb., 93: 391—401.
English, M.J., 1981. The uncertainty of crop models in irrigation optimization. Trans. ASAE, 24: 917—921, 928.
Faris, D.G. and DePauw, R.M., 1981. Effect of seeding rate on growth and yield of three spring wheat cultivars. Field Crops Res., 3: 289—301.
Feldman, L.J., 1977. The generation and elaboration of primary vascular tissue patterns on roots of *Zea.* Bot. Gaz., 138: 393—401.
Fiscus, E.L., 1972. In situ measurement of root water potentials. Plant Physiol., 50: 191—193.
Fiscus, E.L., 1975. The interaction between osmotic- and pressure-induced water flow in plant roots. Plant Physiol., 55: 917—922.
Fiscus, E.L., 1977. Determination of hydraulic and osmotic properties of soybean root systems. Plant Physiol., 59: 1013—1020.
Fiscus, E.L., 1981. Analyses of the components of the area growth of bean root systems. Crop Sci., 909—913.
Fiscus, E.L. and Markhart, A.H., 1979. Relationships between root system water transport properties and plant size in *Phaseolus.* Plant Physiol., 64: 770—773.
Ford, E.D. and Diggle, P.J., 1981. Competition for light in a plant monoculture modelled as a spatial stochastic process. Ann. Bot., 48: 481—500.
Hackett, C., 1972. A method of applying nutrients locally to roots under controlled conditions, and some morphological effects of locally applied nitrate on the branching of wheat roots. Aust. J. Biol. Sci., 25: 1169—1180.
Harley, J.L. and Russell, R.S. (Editors), 1979. The Soil-Root Interface. Academic Press, London, 448 pp.
Harris, G.A., 1977. Root phenology as a factor of competition among grass seedlings. J. Range Manage., 30: 172—177.
Herkelrath, W.N., Miller, E.E. and Gardner, W.R., 1977. Water uptake by plants: II. The root contact model. Soil Sci. Soc. Am. J., 41: 1039—1043.
Hillel, D. and Talpaz, H., 1976. Simulation of root growth and its effect on the pattern of soil water uptake by a nonuniform root system. Soil Sci., 121: 307—312.
Holechek, J.L., 1982. Fertilizer effects on above- and below-ground biomass of four species. J. Range Manage., 35: 39—42.
Huck, M.G. and Klepper, B., 1977. Water relations of cotton. II. Continuous estimates of plant water potential from stem diameter measurements. Agron. J., 69: 593—597.
Jeje, A.A. and Zimmermann, M.H., 1979. Resistance to water flow in xylem vessels. J. Exp. Bot., 30: 817—827.
Jung, Y.S., 1980. Water uptake and transport of soybeans as a function of rooting patterns. Ph.D. Diss., Iowa State University, Ames, IA, 189 pp.
Kaspar, T.C., 1982. Evaluation of the taproot elongation rates of soybean cultivars. Ph.D. Diss., Iowa State University, Ames, IA, 106 pp.
Kaspar, T.C., Wooley, D.G. and Taylor, H.M., 1981. Temperature effect on the inclination of lateral roots of soybeans. Agron. J., 73: 383—385.
Katan, J. and Eshel, Y., 1973. Interactions between herbicides and plant pathogens. Residue Rev., 45: 145—177.
Katouli, M. and Marchant, R., 1981. Effect of phytotoxic metabolites of *Fusarium culmorum* on barley root and root-hair development. Plant Soil, 60: 385—397.
Klepper, B., Rickman, R.W. and Peterson, C.M., 1982a. Quantitative characterization of vegetative development in small cereal grains. Agron. J., 74: 789—792.
Klepper, B., Belford, R.K. and Rickman, R.W., 1982b. The phenology of root development in winter wheat. 1982 Agron. Abstr., p. 16.
Köpke, U., 1981. A comparison of methods for measuring root growth of field crops. Z. Acker Pflanzenbau, 150: 39—49.

Kramer, P.J., Knipling, E.B. and Miller, L.N., 1966. Terminology of cell-water relations. Science, 153: 889—890.
Landsberg, J.J. and Fowkes, N.D., 1978. Water movement through plant roots. Ann. Bot., 42: 493—508.
Maizlish, N.A., Fritton, D.D. and Kendall, W.A., 1980. Root morphology and early development of maize at varying levels of nitrogen. Agron. J., 72: 25—31.
Martin, M.P.L.D. and Snaydon, R.W., 1982. Root and shoot interactions between barley and field beans when intercropped. J. Appl. Ecol., 19: 263—272.
Mason, W., Rowse, H.R., Bennie, A.T.P., Kaspar, T.C. and Taylor, H.M., 1982. Responses of soybeans to two row spacings and two soil water levels. II. Water use, root growth and plant water status. Field Crops Res., 5: 15—29.
McIntosh, M.S. and Miller, D.A., 1980. Development of root-branching in three alfalfa cultivars. Crop Sci., 20: 807—809.
McIntosh, M.S. and Miller, D.A., 1982. Genetic and soil moisture effects on the branching-root trait in alfalfa. Crop Sci., 21: 15—18.
Melhuish, F.M. and Lang, A.R.G., 1969. A new technique for estimating diameter, total length, and surface area of roots grown in soil. In: W.J. Whittington (Editor), Root Growth. Butterworths, London.
Meyer, W.S., 1976. Seminal roots of wheat: manipulation of their geometry to increase the availability of soil water and to improve the efficiency of water use. Ph.D. Thesis, University of Adelaide, Australia,
Meyer, W.S. and Alston, A.M., 1978. Wheat responses to seminal root geometry and subsoil water. Agron. J., 70: 981—986.
Michels, W.C., 1958. Electrical Measurements and their Applications. Van Nostrand, New York, NY.
Mirhadi, M.J. and Kobayashi, Y., 1980. The relationship between the growth of different plant organs of grain sorghum hybrid H-726. Jpn. J. Crop Sci., 49: 420—427.
Molz, F.J., 1981. Models of water transport in the soil-plant system: a review. Water Resour. Res., 17: 1245—1260.
Mülle, G. and Heege, H.J., 1981. Kornverteilung über die Fläche and Ertrag bei Getreide. Z. Acker Pflanzenbau, 150: 97—112.
Myers, R.J.K. and Foale, M.A., 1980. Row spacing and population density in Australian grain sorghum production. J. Aust. Inst. Agric. Sci., 46: 214—220.
Nielsen, R.L., Stuthman, D.D. and Barnes, D.K., 1981. Interference between oats and alfalfa in mixed seedings. Agron. J., 73: 635—638.
Passioura, J.B., 1980. The transport of water from soil to shoot in wheat seedlings. J. Exp. Bot., 31: 333—345.
Ponsana, P., 1975. Drainage and water uptake terms in the water balance. Ph.D. Thesis, University of Adelaide, S.A.
Quisenberry, J.E., Jordan, W.R., Roark, B.A. and Fryear, D.W., 1981. Exotic cottons as genetic sources for drought resistance. Crop Sci., 21: 889—895.
Ramig, R.E., Allmaras, R.R. and Papendick, R.I., 1982. Water conservation: the Pacific Northwest. In: Dryland Agriculture. Am. Soc. Agron. Monogr. (In press).
Reicosky, D.C., Kaspar, T.C. and Taylor, H.M., 1982a. The diurnal relationship between evapotranspiration and leaf water potential of field grown soybeans. Agron. J., 74: 667—673.
Reicosky, D.C., Rowse, H.R., Mason, W.K. and Taylor, H.M., 1982b. Effect of irrigation and row spacing on soybean water use. Agron. J., 74: 958—964.
Reid, C.P.P. and Hurtt, W., 1970. Root permeability as affected by picloram and other chemicals. Physiologia Plant., 23: 124—130.
Richards, R.A. and Passioura, J.B., 1981a. Seminal root morphology and water use of wheat. I. Environmental effects. Crop Sci., 21: 249—252.
Richards, R.A. and Passioura, J.B., 1981b. Seminal root morphology and water use of wheat. II. Genetic variation. Crop Sci., 21: 253—255.

Rickman, R.W. and Klepper, B.L., 1979. Wet season aeration problems beneath surface mulches in dryland winter wheat production. Agron. J., 72: 733—736.

Rickman, R.W., Klepper, B. and Peterson, C.M., 1981. Time distributions for describing appearance of specific culms of wheat. 1981 Agron. Abstr., p. 113.

Rowse, H.R. and Goodman, D., 1981. Axial resistance to water movement in broad bean (*Vicia faba*) roots. J. Exp. Bot., 32: 591—598.

Schenck, N.C., 1977. Microorganisms and root development and function. Proc. Soil Crop Sci. Soc. Fla., 36: 32—36.

Stanton, N.L., Allen, M. and Campion, M., 1981. The effect of the pesticide carbofuran on soil organisms and root and shoot production in short grass prairie. J. Appl. Ecol., 18: 417—431.

Stieber, J., 1981. A new method of examining vessels. Ann. Bot., 48: 411—414.

Taneda, K., Kawakami, H., Ishida, S. and Oktani, J., 1979. Observation of the polymer in wood-polymer composite. Mokuzai Gakkaish (J. Jpn. Wood Res. Soc.), 25: 209—215.

Taylor, H.M. and Klepper, B., 1975. Water uptake by cotton root systems: an examination of assumptions in the single root model. Soil Sci., 120: 57—67.

Taylor, H.M. and Klepper, B., 1978. The role of rooting characteristics in the supply of water to plants. Adv. Agron., 30: 99—128.

Taylor, H.M., Mason, W.K., Bennie, A.T.P. and Rowse, H.R., 1982. Responses of soybeans to two row spacings and two soil water levels I. an analysis of biomass accumulation, canopy development, solar radiation interception and components of seed yield. Field Crops Res., 5: 1—14.

Tennant, D., 1976. Root growth of wheat. I. Early patterns of multiplication and extension of wheat roots including effects of levels of nitrogen, phosphorus, and potassium. Aust. J. Agric. Res., 27: 183—196.

Teskey, R.O. and Hinckley, T.M., 1981. Influence of temperature and water potential on root growth of white oak. Physiol. Plant., 52: 363—369.

Tottman, D.R. and Davies, E.L.P., 1978. The effect of herbicides on the root system of wheat plants. Ann. Appl. Biol., 90: 93—99.

Troughton, A., 1981. Length of life of grass roots. Grass Forage Sci., 36: 117—120.

Turner, N.C., 1981. Correction of flow resistances of plants measured from covered and exposed leaves. Plant Physiol., 68: 1090—1092.

Unger, P., 1979. Effects of deep tillage and profile modification on soil properties, root growth, and crop yields in the United States and Canada. Geoderma, 22: 275—295.

Van Bavel, C.H.M. and Ahmed, J., 1976. Dynamic simulation of water depletion in the root zone. Ecol. Modelling, 2: 189—212.

Vieira, S.R., Nielsen, D.R. and Biggar, J.W., 1981. Spatial variability of field-measured infiltration rate. Soil Sci. Soc. Am. J., 45: 1040—1048.

Voorhees, W.B., Carlson, V.A. and Hallauer, E.A., 1980. Root length measurement with a computer-controlled digital scanning microdensitometer. Agron. J., 72: 847—850.

Walters, D.R. and Ayres, P.G., 1981. Growth and branching pattern of roots of barley infected with powdery mildew. Ann. Bot., 47: 159—162.

Ward, K.J., Klepper, B., Rickman, R.W. and Allmaras, R.R., 1978. Quantitative estimation of living wheat-root lengths in soil cores. Agron. J., 70: 675—677.

Weiner, J. and Conte, P.T., 1981. Dispersal and neighborhood effects in an annual plant competition model. Ecol. Modelling, 13: 131—147.

Willatt, S.T. and Taylor, H.M., 1978. Water uptake by soya-bean roots as affected by their depth and by soil water content. J. Agric. Sci., 90: 205—213.

Yamazki, K. and Harada, J., 1982. The root system formation and its possible bearings on grain yield in rice plants. Jpn. Agric. Res. Q., 15: 153—161.

Yang, R.L., 1981. The relationship between growth and various water relation parameters in irrigated and nonirrigated soybean. MS Thesis, Botany Department, Iowa State University, Ames, IA.

Yorke, J.S. and Sagar, G.R., 1970. Distribution of secondary root growth potential in the root system of *Pisum sativum*. Can. J. Bot., 48: 699—704.

Agricultural Water Management, 7 (1983) 143—155
Elsevier Science Publishers B.V., Amsterdam — Printed in The Netherlands

PRINCIPLES OF WATER MANAGEMENT UNDER DROUGHT CONDITIONS

W.R. GARDNER[1] and H.R. GARDNER[2]

[1] *Soils, Water and Engineering Department, University of Arizona, Tucson, AZ 85721 (U.S.A.)*
[2] *Agricultural Research Service, U.S. Department of Agriculture, Fort Collins, CO 80522 (U.S.A.)*

Arizona Agricultural Experiment Station Paper No. 436, University of Arizona, AZ 85721 (U.S.A.)

(Accepted 15 March 1983)

ABSTRACT

Gardner, W.R. and Gardner, H.R., 1983. Principles of water management under drought conditions. *Agric. Water Manage.*, 7: 143—155.

In rain-fed agriculture the options for water management are much more limited than for irrigated agriculture and the management rools available to the farmer are few. The major variable at his disposal is the selection of the crop to be grown. The economic return per unit of water is the major factor to be considered. This usually implies the maximum possible ratio of grain or other harvested plant part to total dry matter. When the rainfall is just below amount needed for maximum dry matter production, this maximum can be achieved through appropriate manipulation of the plant population. The optimum population depends both upon the characteristics of the plant and the amount and distribution of the precipitation. When the amount of precipitation is significantly less than about 25% of the optimum, then other management approaches are required. A widespread solution to this problem has been summer-fallowing. However, in certain circumstances some variation on surface water management or water harvesting may be more advantageous. Criteria for decision making for optimum production under dry-land conditions are given.

INTRODUCTION

As Paleg and Aspinall point out in the preface to the monograph, *Drought Resistance in Plants* (Paleg and Aspinall, 1981), the general response towards meeting the water requirements of various crops has been via an engineering response. The history of water management in arid regions is the history of moving water from one place to another. Modern construction techniques and low-cost energy have allowed this process to proceed to the point where plans to turn Canadian and Siberian rivers completely around are given serious consideration. As energy costs escalate, the engineering solutions to water shortages are found to be economically and politically less acceptable. The recent defeat of the proposal for the Peripheral Canal in California does not augur well for future grand water projects in the U.S.A. After a brief flurry of interest in chemical solutions to agricultural water shortages (such as anti-transpirants), we must come to the full realization that agronomic solutions are generally the best long term solutions for drought.

WATER MANAGEMENT: CROP RESPONSE

The principles of infiltration, redistribution, and water uptake by plants are now quite well understood, though there is undoubtedly much yet to be learned. Such topics have been the subject of innumerable reviews and will not be rehashed here. However, to discuss the potential for agronomic solutions to drought and the role soil water management contributes to those solutions, it is useful to review some of the relations between water use and crop response.

Fig. 1 shows a graph of total corn dry matter produced as a function of evapotranspiration, as reported by Sorenson et al., (1980) for northern Utah. The linear relation between dry matter and evapotranspiration is characteristic of most of the data now reported for uniform field plots. Linearity is also to be expected from theoretical considerations (Tanner and Sinclair, 1983). The intercept along the evapotranspiration axis presumably represents evaporation from the soil surface. There is also undoubtedly a modest component of evaporation in the water use at higher levels of evapotranspiration though most of this water is consumed in transpiration. In general this evaporation of soil water is not beneficial. The possibility

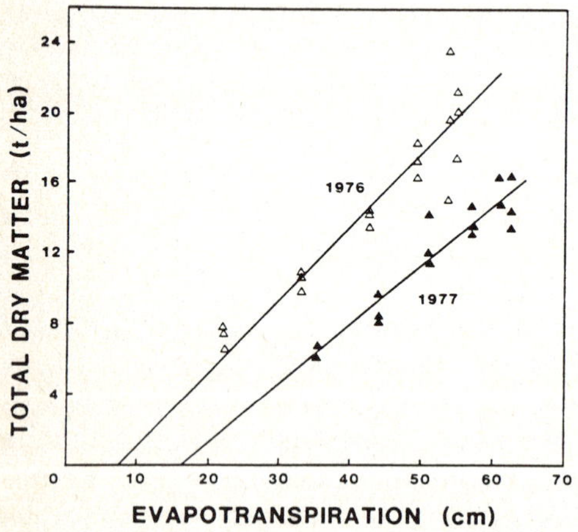

Fig. 1. Total dry matter as a function of evapotranspiration for corn (Sorenson et al., 1980).

of reducing it will be considered in a later section. We are more concerned at this point about the linear relation between biomass production and transpiration. Although not apparent from Fig. 1, when water is applied to the field in amounts greater than the maximum possible evapotranspiration, dry matter production is not increased further. In fact, dry matter may be reduced if aeration is inadequate or nutrient leaching is excessive. Actual crop yield-water use curves for large fields often do not exhibit this linear behavior. The observed relation tends to be parabolic in shape. This departure from linearity is almost certainly due to non-uniformity in the distribution of available water in the soil profile associated with both soil and irrigation non-uniformity. This non-uniformity can be characterized by a coefficient of variability (CV). Warrick and Gardner (1983) discuss this in more detail and show that, to a good degree of approximation, the two variabilities can be lumped to get the total coefficient.

Fig. 2 shows the average crop yield versus water applied response curves for CV equal to zero and to one. A log-normal water distribution is assumed and two different values of the intercept on the water use axis for use axis for curves such as Fig. 1 are shown. The existence of this intercept is due to both evaporation from the soil surface and the fact that some crops must achieve a minimum biomass before there is harvestable yield. The parameters in Fig. 2 are normalized so that when $\bar{w}$ = 1, the amount of available water is equal to the maximum possible evapotranspiration, and $\bar{y}$ = 1 corresponds to the maximum possible yield. This curve allows us to draw some conclusions about the best strategy for soil water management when water is limiting. If the intercept on the evapotranspiration axis is small (solid lines in Fig. 2), then a reduction in the value of CV always

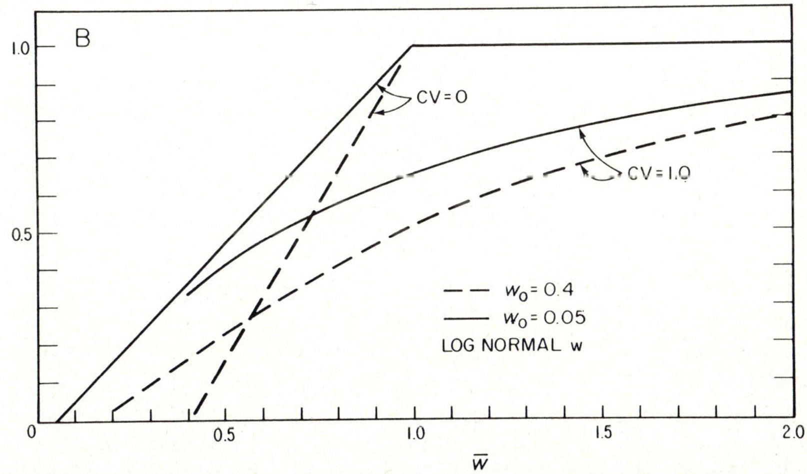

Fig. 2. Average relative yield, $\bar{y}$, plotted as a function of average relative rate of water applied, $\bar{w}$, for two different values for the coefficient of variability, CV. w_0 represents the rate of water application necessary before any yield is achieved relative to $\bar{w}$.

results in increased yield. If, however, this intercept occur at a value relatively large compared to $\bar{w}$ (w_0 = 0.4 $\bar{w}$ in our example) and CV is relatively large, we find that when $\bar{w}$ is significantly less than one we get the highest yield with the larger CV value one. Physically speaking, in the first case when the amount of water is relatively high, non-uniformity results in some parts of the field receiving more water than can be used beneficially to produce dry matter. Thus, a more uniform distribution improves the yield of the dry spots with little cost to the wet spots. However, when the amount of available water is significantly less than the maximum useful amount, it is better to distribute it non uniformly. Non uniformity is, in a sense, a useful characteristic of desert ecosystems and many arid zone plant species have adaptations which actually enhance it. The agriculturalist has long since recognized this principle and employs it in his selection of plant spacing or in his employment of summer fallow.

HARVEST INDEX

If we were interested only in total dry matter production we could end the discussion here since more water means more yield. Our only problem is the engineering problem of how to distribute the water as uniformly as possible. However, most farmers must be concerned with the harvest index, i.e. that fraction of dry matter which is of direct economic value. In order to examine this factor adequately, we must consider the relation between the grain yield and the total dry matter production. Fig. 3 shows the total grain produced per plant plotted against total dry weight per plant for three varieties of corn and one sorghum variety. The corn data are from Sorenson et al. (1980) and the sorghum data from Brown and Shrader (1959).

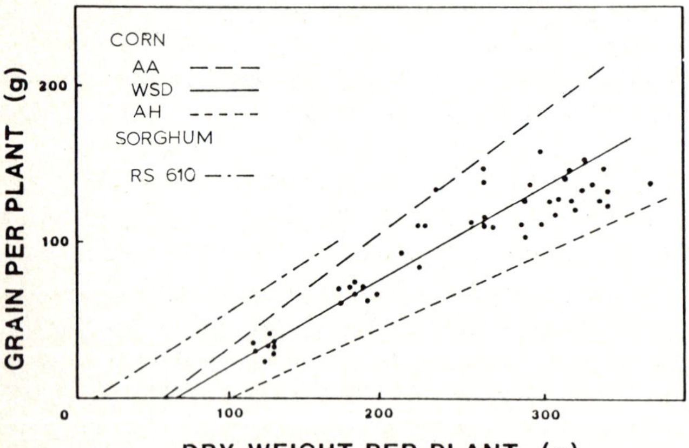

Fig. 3. Grain yield per plant plotted as a function of total dry weight per plant. Corn data are by Sorenson et al. (1980) and sorghum data from Brown and Shrader (1959). Data points shown are for the hybrid variety WSD.

Three aspects of these data are of immediate interest.

First, the relation between grain and total dry matter tends to be very nearly linear. This fact is apparently not well known, but seems to hold true for many different species of grain, for cotton seed and lint, for beans and other legumes and a number of other crops. Mathematically, these linear relations can be expressed by:

$$g = a(d - d_0) \tag{1}$$

where g is the amount of grain per plant, and d is the total dry matter per plant. The constant d_0 is the intercept on the abscissa and represents the hypothetical minimum plant size which just produces grain. The actual minimum may be somewhat larger than this hypothetical value since the line is extrapolated beyond the limit of the data. The constant, a, represents the increase in grain per each unit increase in dry weight per plant.

Secondly, the relation varies with different cultivars of the same species, both in slope and intercept.

Thirdly, sorghum has a much smaller intercept, d_0, than corn. Small grains have a very small value of d_0. In the case of wheat and barley, the line virtually goes through the origin. This is extremely important with respect to drought tolerance. When the line goes through the origin, the fraction of dry matter which is harvested as grain is independent of plant size. A plant weighing only a few grams will still produce grain, though the amount may be too small to harvest profitably. However, when d_0 is relatively large, as is the case for corn, the fraction of grain drops off much more rapidly and a plant smaller than 100 g produces very little grain.

The two parameters in Eq. (1), slope a and intercept d_0 are extremely important in characterizing the drought adaptation of crops. It is not to be presumed that these parameters are fixed and unchangeable. It is well known that water stress at a critical period, such as flowering, can dramatically reduce grain yields. In predicting yields, it is common to reduce the projected yield according to the degree of stress experienced at various stages of growth. Unfortunately, the data on which such estimates are based usually consist of yield responses measured at the end of the season. We need more quantitative models of crop response to stress based upon quantitative measurements of both soil and atmospheric conditions and the internal plant response at critical stages throughout the season. For example, salinity stress appears to result mainly in smaller plants, with a consequent reduction in grain yield, leaving the slope and intercept in Eqn. (1) unchanged. The data in Figs. 1 and 3 were obtained with the line-gradient sprinkler irrigation technique in which water stress is introduced more or less uniformly during the growing season. Many other factors, such as nitrogen fertilization, may affect the relation for cotton. However, examination of many sets of data shows these relations to be remarkably robust. Although they may shift modestly from year to year with the weather, they provide an effective means of distinguishing between cultivars with respect to many aspects of drought adaptation.

The conclusion to be drawn from relations such as those shown for corn in Fig. 3 is that the larger the plant, the more grain obtained from a given amount of dry matter. Because water limits the total dry matter production under drought conditions, the agronomist should, therefore, be concerned with achieving the largest plants possible, which requires that plant population be taken into account. Because of competition, fewer plants mean larger plants. However, there is a trade-off between size and number since the total dry matter production tends to increase with increasing plant population. The challenge is to find the optimum population. This subject has been discussed in detail by Gardner (1983). The relation between plant size and plant population is well studied and tends to be of the same mathematical form as the well known Langmuir equation. If the inverse of plant size is plotted against plant population the result is a straight line, with an intercept which corresponds to the hypothetical plant size at zero population. The total dry matter is the product of the population times the plant size, which also is the Langmuir form.

The relation between the dry matter per plant, d, and the plant population, P, is given by the equation:

$$d = d_m P_h / (P + P_h) \quad (2)$$

where d_m is the maximum plant size and P_h is that plant population which results in one-half of the maximum dry matter production. The total dry matter production is given by the product of the dry matter per plant and the plant population, i.e., $D = d\,P$. Multiplying Eq. (2) by P gives:

$$D = D_m T / (P + P_h) \quad (3)$$

Equations (2) and (3) appear to be obeyed quite well when maximum growth is limited by light or by soil fertility. However, when water or nitrogen become limiting, one finds the curve followed only up to a limiting dry matter yield. At higher populations the yield is constant. This point is also illustrated in Fig. 4. The straight horizontal lines represent the maximum dry matter production when dry matter is limited by water to various indicated fractions of the total (no stress) production.

For plant species for which the grain is a constant fraction of the total dry weight, these curves also can be used to represent the grain yield as a function of population since grain is a constant fraction of total dry matter. However, for a species for which the intercept d_0 is not neglible, e.g. corn, the situation is quite different.

Fig. 5 shows the effect of the magnitude of the ratio d_0/d_m on the optimum population. Here the grain yield is plotted as a function of population for values of d_0/d_m equal to 0.05, 0.02, 0.01, 0.005 and 0.001. For values of d_0/d_m less than about 0.01 the intercept is not critical and the harvest index is virtually independent of plant size.

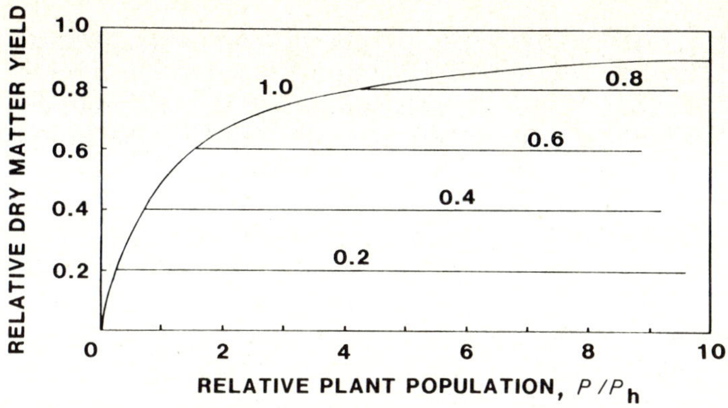

Fig. 4. Relative dry matter yield as a function of relative plant population for crop obeying Langmuir type equation. Upper curve represents case when water is adequate. Horizontal lines represent total dry matter yield when available water is given as a fraction of that just required for maximum yield.

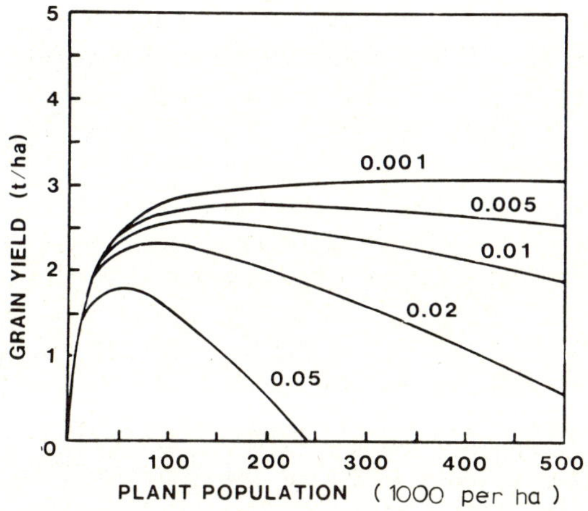

Fig. 5. Grain yield in tons per ha for hypothetical crop as a function of total plant population for different ratios of minimum to maximum plant size, d_0/d_m.

It is found that d_0/d_m varies from zero for barley and wheat and about 0.05 for sorghum to as high as 0.4 for corn grown under conditions of poor fertility. The result of increasing the plant population when $d_0 \neq 0$ is quite different from the case when $d_0 = 0$. The higher plant populations result in more total dry matter, but as the plants decrease in size there is a resulting decrease in the harvest index. It turns out, therefore, that the grain yield goes through a maximum. In this particular plot, P_h is about 25 000 plants per ha. The maximum yield occurs at populations ranging from 50 000 to 200 000 plants per ha. The location of the maximum depends very strongly upon the intercept on the grain/dry matter plot. For a zero intercept, the maximum occurs at infinite population. For plants stressed at an inopportune time, resulting in a large intercept, this maximum could come at a relatively low population. Examination of the data in the literature verifies that plant breeders have developed increased drought adaptation both by increasing the slope of the line a, and reducing the intercept d_0. This is particularly true of cotton, where large, rank plants are now seldom found in US fields.

When available water is limiting, there are important consequences insofar as optimum plant population is concerned. As seen in Fig. 4, once dry matter yield is limited by water, increased plant population does not result in increased total dry matter production. However, as the plant population increases, the average plant size must decrease proportionately. When d_0, (the minimum plant size which results in any harvestable yield) is other than zero, the harvest index must decrease also as the population increases. The exact equation for this is given by Gardner (1983). This is illustrated in Fig. 6 where the grain yield in arbitrary units is plotted as a function of the plant population relative to the plant population at half yield, P/P_h. The straight lines represent different levels of water availability, relative to the amount just necessary to produce maximum dry matter. It can be seen that the optimum plant population decrease with decreasing available water. This means that in order to truly maximize yield a farmer would have to know the amount of water which will be available during the growing season before selecting the proper plant population.

Figs. 4 and 5 help illustrate a common weakness of many studies of plant population. It is unusual in such studies to find more than three or

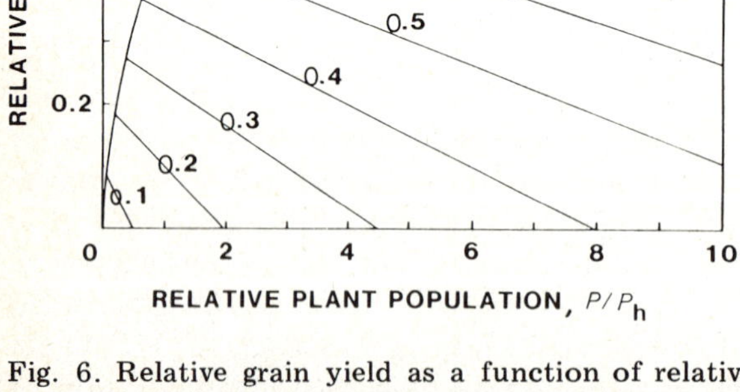

Fig. 6. Relative grain yield as a function of relative plant population, P/P_h, for several levels of available water. The available water is calculated relative to the amount just necessary to achieve maximum dry matter yield.

four populations represented in any experiment on the effect of plant population on yield. Often these are chosen so they are grouped around the maximum yield. Thus, one could find, in any given year, yields increasing with population, depending upon the precipitation. Three successive years might give three different results, making interpretation difficult, if not impossible.

The best strategy for optimizing yield depends upon the probability distribution of the precipitation. Without an analysis such as that represented here, it seems unlikely that the effect of population on yield could be unambiguously revealed through conventional field experiments and statistical analyses. In order to understand the interaction between plant population and water use, it is helpful to plot contours of plant population against water available. Such a plot is shown in Fig. 7. The relative population P/P_h is plotted against relative precipitation for several values of relative yield. As in Fig. 6, $d_0/d_m = 0.1$. To the left of the zero curve, there is no yield at all for any combination of population and water. For precipitation or available water above about 80% of that needed for maximum growth there is a relatively wide range of populations which will give at least 95% of the maximum possible yield. Therefore, if rainfall is

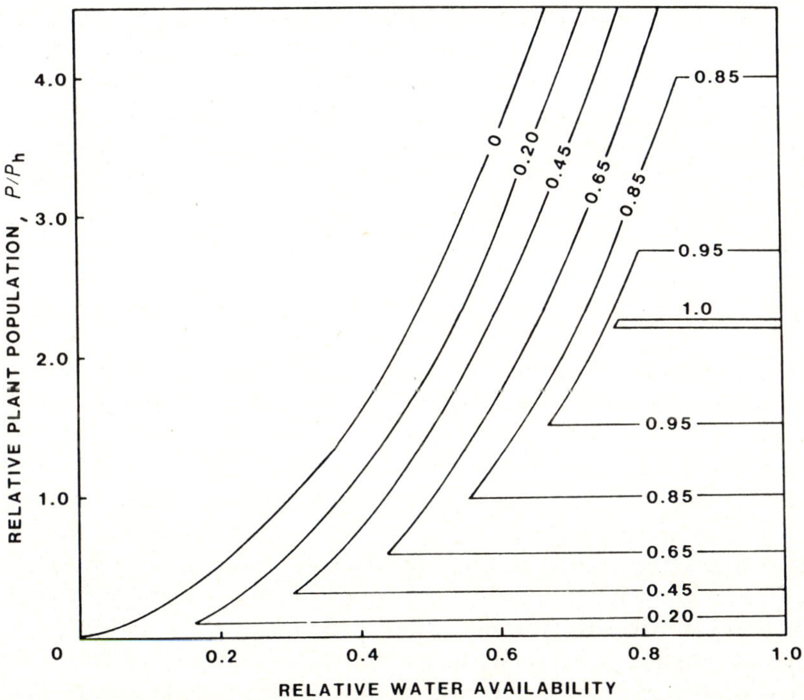

Fig. 7. Contour plot of relative plant population to relative water availability for different levels of relative yield. P_h is the plant population which gives one half the maximum dry weight yield.

relatively dependable, one need not guess too accurately what the precipitation will be in order to determine the best plant population. Because of the shape of the contours, it is better to be somewhat pessimistic and use a lower population than optimum, since the penalty for guessing wrong is not then so severe as that for being too optimistic. This is true at lower precipitation rates also. If one is willing to settle for 85% of the maximum yield, this can be obtained at only 60% of the optimum precipitation if one chooses a sufficiently low population. Conversely, if one plants a population which will give maximum yield only under good rainfall conditions, the 60% level of precipitation could result in a yield of only 50% of the maximum. One can also use such curves to estimate the potential value of irrigation. The uniformity of the irrigation technique must be taken into account (Hart et al., 1980; Gardner et al., 1982; Warrick and Gardner, 1983) in order to be quantitative.

As noted, Fig. 7 was calculated for the ratio $d_0/d_m = 0.1$. For a ratio as low as 0.02, which one might find for sorghum, the optimum population occurs at much higher ratios of P/P_h. The principle, however, is still the same. The best strategy is to be pessimistic because the loss in yield resulting from a low population in a wet year is not so severe as the loss resulting from too high a population in a dry year. A crop with a small d_0/d_m ratio can tolerate a low precipitation and still produce a yield since the minimum plant size is small. It comes as no surprise, therefore, that sorghum is better than corn under drought conditions. This response is in addition to other physiological differences. The advantage of this kind of an analysis is that it allows one to make quantitative comparisons. At the other end of the scale from sorghum, or small grains e.g., for fruit crops and other sensitive crops for which d_0/d_m is large, even modest reductions in available water can result in substantial reductions in yield. For these, low plant population alone will not solve the drought problem for such crops and more drastic measures must be adopted if such crops are to be successful.

As the total amount of available seasonal water decreases, a point is eventually reached where reduction of population alone no longer gives the maximum, or even adequate, pay-off. Plant size is no longer significantly limited by competition for light or nutrients, and water is all important. For our particular example this situation begins when the precipitation is less than about 20% of the optimum. In this range a given increase in available water results in a disproportionately large increase in yield. It is in this range, therefore, that manipulation of the soil water as well as plant population becomes agronomically advantageous. One form of manipulation is to go to a crop-fallow system. Precipitation from the fallow years is saved in the soil profile to be used during the crop years. We will not go into the physics of water storage and loss during fallow since this is covered in many articles and has been long researched. However, a careful examination of curves such as Fig. 5 through 7, or their mathematical formulations, can give one a quantitative idea of the potential benefits of a fallow system.

For example, if the precipitation is 20% or more of that which is optimum, it appears that a judicious choice of plant population and continuous cropping would result in greater yields than alternate or fallow cropping, all other factors being equal. Only under conditions of very erratic or very low precipitation does fallowing become attractive.

However, plant population does not tell the whole story. Evaporation losses from the soil have already been mentioned. As plant population decreases, the evaporation loss becomes proportionately greater relative to transpiration since more soil surface is exposed longer at the lower plant populations. Fig. 8 shows the ratio of transpiration to evapotranspiration as a function of plant population for surghum. These data were obtained at Akron, CO, with different row spacings giving different populations.

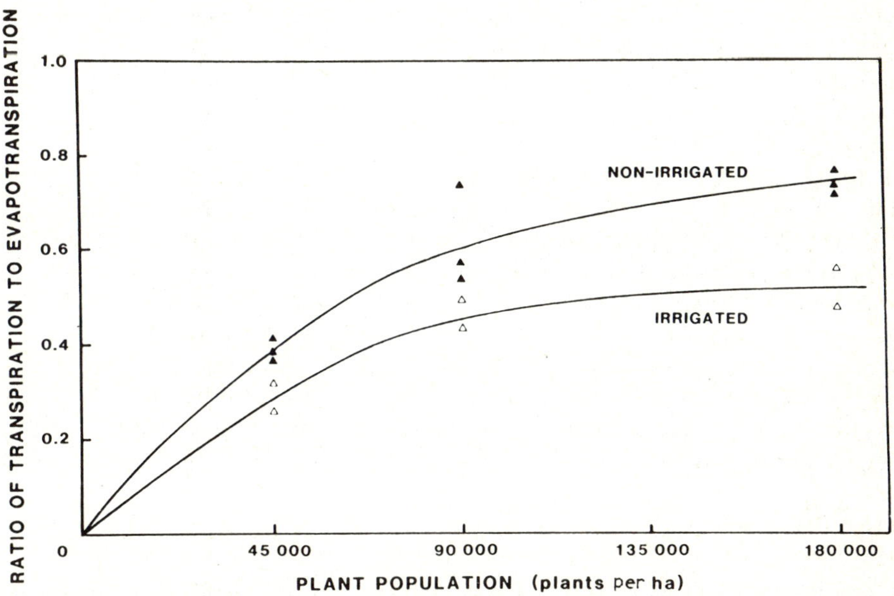

Fig. 8. Ratio of transpiration to evapotranspiration as a function of plant population for sorghum at Akron, CO.

As the plant population increases it can be seen that the fraction of water which is transpired increased, though at a decreasing rate. At the lowest population of 45 000 plants per ha only about 40% of the ET was transpired, while at four times that population, this ratio increased to almost 80% for the non-irrigated treatment. It is interesting that for the treatment which was irrigated by sprinkling, the efficiency actually decreased, with a greater proportion of the irrigation water going to evaporation than in the case of the natural precipitation. Clearly, when irrigating low populations or small plants, the less soil surface that is wetted, the better. Drip irrigation and alternate furrow irrigation achieve significant advantage in this way.

If, however, one is dependent entirely upon natural precipitation, then there are only two recourses available under summer drought. One is to cover the surface to reduce surface evaporation. Dust and other mulches are of only limited effectiveness, because they retard but do not eliminate evaporation. Only a relatively vapor tight cover offers hope for water conservation but unfortunately, this is thus far too costly for any but the most valuable crops. The other alternative is some variation of run-off farming. In this technique, origins of which are lost in antiquity, the soil surface is shaped and treated to enhance runoff from some areas so as to increase water accumulation and infiltration in adjacent areas. This subject brings us full circle back to an engineering solution. It is a solution which, in modern times has failed to catch the attention of very many agriculturalists and is appreciated more for its novelty value than anything else. It is true that terraces, benches, microcatchments, etc. have been studied by many scientists, but a convincing economic case for runoff farming has yet to be made so far as US farmers are concerned. At the Page Ranch experimental station of the University of Arizona, grapes have been successfully grown for the past 10 years on an annual precipitation of 300 mm, using sodium-chloride treated microcatchments. It is not yet known how long these catchments will survive without rehabilitation, but if they last at least ten years, it is estimated that the water costs will be competitive with alternative sources of water. There is little doubt that biomass production can be increased and soil water evaporation reduced in any of our drought prone areas by more intensive management of the surface water. Whether this management can be done economically without serious wind and water erosion problems still remains to be determined.

CONCLUSIONS

The principles of water use by crops are sufficiently well understood to better optimize the system under drought conditions. How well a given plant species is adapted to drought depends not only upon its ability to produce a harvestable yield under stress, but also upon the relation between plant size and grain yield (or other harvestable plant parts). Plant breeders have already gone far towards improving drought adaptation, but more can undoubtedly be accomplished. Plant population is one of the major management variables available to the farmer. We need better data relating evaporation and transpiration to plant size and plant population. When these relations are adequately understood, the best strategy for adapting to or mitigating drought can more readily be worked out. We particularly need a more quantitative understanding of the relation between the duration and intensity of water stress upon the grain/dry matter relation. More economic studies of water harvesting, and other techniques of surface water manipulation are needed.

REFERENCES

Brown, P.L. and Shrader, W.D., 1959. Grain yields, evapotranspiration and water use efficiency of grain sorghum under different cultural practices. Agron. J., 51: 339—343.

Gardner, H.R., 1983. Optimizing plant population under limited water. Agron. J. (In press).

Gardner, W.R., Warrick, A.W. and Halderman, A.D., 1982. Soil variability and measures of irrigation efficiency. Am. Soc. Agric. Eng. Pap. 82-2105.

Hart, W.E., Norum, D.I. and Peri, G., 1980. Optimal seasonal irrigation application analysis. J. Irrig. Drain. Div. ASCE, 106: 221—235.

Paleg, L.G. and Aspinall, D.D., 1981. The Physiology and Biochemistry of Drought Resistance in Plants. Academic Press, New York, NY, 492 pp.

Sorensen, V.M., Hanks, R.J. and Cartree, R.L., 1980. Cultivation during early season and irrigation influences on corn production. Agron. J., 72: 266—270.

Tanner, C.B. and Sinclair, T.R., 1983. Efficient water use in crop production. In: T.R. Sinclair, W.R. Jordan and H.M. Taylor (Editors), Limitations to Efficient Water Use in Crop Production. American Society of Agronomy, Madison, WI (In press).

Warrick, A.W. and Gardner, W.R., 1983. Crop production as affected by spatial variations of soil and irrigation. Water Resour. Res., 19: 181—196.

Agricultural Water Management, 7 (1983) 157—178
Elsevier Science Publishers B.V., Amsterdam — Printed in The Netherlands

IRRIGATION IN THE GREAT PLAINS

E.T. KANEMASU[1], J.L. STEINER[1], A.W. BIERE[2], F.D. WORMAN[2] and J.F. STONE[3]

[1]*Agronomy Department* and [2]*Economics Department, Kansas State University, Manhattan, KS 66506 (U.S.A.)*
[3]*Agronomy Department, Oklahoma State University, Stillwater, OK 74078 (U.S.A.)*

Contribution No. 82-532-J, Agronomy Department, and Economics Department, Kansas Agricultural Experiment Station, Kansas State University, Manhattan, KS 66506

(Accepted 4 February 1983)

ABSTRACT

Kanemasu, E.T., Steiner, J.L., Biere, A.W., Worman, F.D. and Stone, J.F., 1983. Irrigation in the Great Plains. *Agric. Water Manage.*, 7: 157—178.

Irrigation scheduling answers the question of when to irrigate and how much. The techniques used for scheduling include the monitoring of soil moisture, physiological indicators and water balance models. The areas of major concern are: (a) a soil moisture sensor which is inexpensive, rapid and accurate; (b) a means of assessing the upper and lower limits of soil water content in the changing root zone; (c) development of a rapid technique for estimating leaf area; and (d) a means of measuring canopy transpiration.

A major research thrust that can be identified is the separation of evaporation from the soil surface and transpiration by the canopy. As one manipulates the canopy geometry to assess the cultural practices and irrigation systems, the need to clearly identify the contribution of evaporation and transpiration becomes increasingly important.

The objective for irrigation for the farmer is to maximize his net returns. There are three components of an economic model for determining that decision: (a) a water balance; (b) a growth response function; and (c) an economic optimization function. A few of the problem areas in such a scheme are: (a) adaptation by the plant to water stress; (b) dynamics of the root system; (c) interaction between fertility and water; and (d) risk analysis.

INTRODUCTION

Dryland crop production in the Great Plains is primarily water limited with cropping patterns determined by average rainfall, rainfall variability, and distribution of rainfall within the year. Extensive irrigation development has increased the productivity and stability of agriculture in the Great Plains. The irrigated area in the Great Plains states are given in Table I. The variability of rainfall from year to year has promoted irrigation development, with irrigation buffering against crop losses during periods of low rainfall, as well as increasing average crop yields (Rosenberg, 1978).

0378-3774/83/$03.00

TABLE I

Irrigated cropland in 1944 and 1978 and total cropland in 1978 in the Great Plains states (USDA, 1981, Agricultural Statistics, U.S. Government Printing Office, Washington, DC)

State	Irrigated cropland (1000 ha)		Total cropland (1000 ha)
	1944	1978	1978
Colorado	1 080	1 399	3 634
Kansas	39	1 087	10 873
Montana	629	844	5 913
Nebraska	256	2 306	8 231
New Mexico	217	366	736
North Dakota	9	57	11 285
Oklahoma	1	244	3 862
South Dakota	21	138	6 757
Texas	534	2 840	9 955
Wyoming	548	682	873

The major water supply for irrigation in the Great Plains is the Ogallala aquifer that extends from western Texas and eastern New Mexico through southern South Dakota, and underlies about 45 million ha of land (Luckey et al., 1981). In 1980 about 6.5 million ha of land were irrigated in the High Plains from about 170 000 wells. Withdrawal from the aquifer system exceeds recharge in most areas where extensive irrigation development has occurred, particularly in the southern portions of the aquifer, and groundwater levels are declining. Serious depletion of the aquifer has occurred in areas that were developed early, where large percentages of the regional area are irrigated, where annual application rates are high, or where smaller saturated thickness levels were available at the time of development (Luckey et al., 1981). Unrestricted withdrawals during drought periods may lead to accelerated depletion of groundwater, but restricted withdrawals during drought periods could lead to crop yield reductions (Riefler, 1978).

As the water level of the aquifer declines, irrigators face increased energy requirements and reduced water supplies. In addition, energy costs have risen dramatically since much of the irrigation development occurred, as have loan interest costs. Since irrigated production is both energy and capital intensive, increasing costs, with little increase in crop prices likely, force irrigators to re-evaluate their irrigation strategy. Stegman et al. (1981) pointed out that most irrigators cannot reduce the yield levels of their crops and maintain profitability. To reduce pumpage requirements and costs without reducing the water available to the crops, irrigators are looking for improvements in system design, management, and scheduling procedures to improve the efficiency of irrigation, to prolong the economic life of groundwater supplies and to maintain the profitability of irrigation.

Irrigation scheduling techniques have been developed to reduce irrigation applications while improving or maintaining crop yields. These techniques should avoid excessive applications as well as untimely or excessive soil moisture depletion. Irrigation scheduling requires knowledge of root zone water storage capacity, the soil moisture levels which exist at particular times during the season, and the relationship between soil moisture and yield. Heermann (1980) pointed out that a scheduling program should forecast dates when irrigation should be applied, as well as the amount of water required for an individual field, so that irrigation can be coordinated with other farming operations.

IRRIGATION SCHEDULING TECHNIQUES

Irrigation scheduling techniques fall into the general categories of hydrological and physiological. Hydrological techniques are those concerned with estimating soil moisture either directly or by water balance. Physiological techniques assess the need for water from plant measurements such as stomatal resistance, leaf water potential, growth stage, or canopy temperature.

Hydrological

The water balance of a field can be defined as:

$$\mathrm{SM} = \mathrm{SM} + P_\mathrm{e} + I - D - \mathrm{ET}$$

where SM is the soil moisture on a given day, SM is the initial soil moisture, P_e is the effective precipitation, I is irrigation, D is drainage below the root zone, and ET is evapotranspiration. Van Bavel and Wilson (1952) found good agreement in determining irrigation dates by the water balance method and by tensiometer monitoring of soil moisture. They found the greatest errors involved the determination of rainfall infiltration and ET losses. Jensen and Wright (1978) described the confidence intervals of predicted irrigation dates using a water balance model. They found that the largest errors involved determination of the amount of water applied during irrigation and flux across the lower boundary of the root zone. Additional error was involved in the determination of effective rainfall, ET estimate, and the soil moisture measurement. Rouse and Wilson (1972) found that spatial variability of soils and variability in the determination of effective rainfall or irrigation limited the use of water balance methods over short time periods.

Water balance models require initial soil moisture, water holding capacity of the root zone, and other crop and soil characteristics as initialization inputs. Daily climatic variables, irrigation dates and irrigation amounts are required throughout the growing season. The error involved in measuring initial soil moisture depends upon the spatial variability of the field and upon the method of measurement (Schmugge et al., 1980; Cary, 1981).

Available water holding capacity of a soil for irrigation scheduling purposes is generally determined from generalized soil type information. Field capacity is often referred to as the soil moisture which is held in the soil at 30 kPa of tension; wilting point as the soil moisture which is held at 1 500 kPa of tension; and plant available water use as the difference between the two. The available water holding capacity of an undisturbed profile is affected by soil structure, layering in the soil profile, and the root characteristics of the crop being grown. Ritchie (1981a, b) suggests that field determination of the upper and lower limits of soil moisture better describe the water storage capacity of the soil. Hearn and Constable (1981) developed a water balance model in which the depth of the root zone is deeper when there is infrequent rainfall early in the season than when there is frequent rainfall or irrigation. This response is reasonable in a case where the upper part of the profile is depleted of water, and moisture levels favorable for root growth exist at greater depths. Most irrigation scheduling models do not consider variable rooting depths, but instead use a fixed root zone for a given crop and soil.

Evapotranspiration (ET) estimates for water balance models are obtained by many different methods. Measurement of actual ET is difficult, so the correlation of ET to climatic conditions is often used to estimate ET losses under various conditions. Daily ET, or ET over a short term, must be estimated for use in an irrigation scheduling model. Detailed reviews of methods of measuring or estimating ET are given by Tanner (1967), Jensen (1973) and Kanemasu et al. (1979b).

Evapotranspiration includes all water lost by evaporation from the soil surface or by transpiration from plant surfaces. Ritchie (1972), Kanemasu et al. (1976) and Rosenthal et al. (1977) make separate calculations of evaporation from the soil surface and transpiration. Partitioning of energy to the soil surface and to the crop canopy is calculated as a function of leaf area index (LAI), which is the ratio of green leaf surface area to soil surface area. Soil evaporation occurs in two phases: a constant rate phase when the surface is wet, which occurs at the potential rate; and a falling rate phase, which depends upon the water transmitting properties of the soil and decreases with the square root of the number of days into the drying phase (Ritchie, 1972). Transpiration is related to the potential ET and the leaf area index of the crop.

Jury and Tanner (1975) described a modification of the proportionality constant, α, to account for advected energy in the Priestley-Taylor equation (Priestley and Taylor, 1972).

$$\alpha' = 1 + (\alpha - 1)\,(e_z^* - e_z)/\overline{(e_z^* - e_z)}$$

where α' is the modified proportionality constant, $(e_z^* - e_z)$ is the average vapor pressure deficit for the day, and $(e_z^* - e_z)$ is a long term mean vapor pressure deficit for periods when advection is low. This modification has the disadvantage of requiring a vapor pressure input. When air temperature exceeds canopy temperature, there is movement of sensible heat into the canopy, increasing transpiration losses. Kanemasu et al. (1976) includes a temperature based advection component which is added to the transpiration component on days when the temperature is high. There is some indication that the magnitude of the difference between the canopy and air temperature is dependent upon the vapor pressure deficit when water is not limiting (Idso et al., 1982).

Other components of the water balance are related to infiltration of rainfall into the soil and redistribution within the profile. Daily rainfall is usually included in water balance models, because intensity and duration of the precipitation are not known. Runoff of precipitation depends on the vegetative or mulch cover, soil texture and structure, slope, roughness, and soil moisture, as well as intensity and duration of the rainfall. Simple, empirical equations are used to determine effective rainfall in most irrigation scheduling models. Irrigation input is usually the effective irrigation depth, so the efficiency of the irrigation system, as well as the gross irrigation depth, must be known in order to use an irrigation scheduling model.

Flux across the lower boundary of the root zone must be estimated. Kanemasu et al. (1979a), Steiner et al. (1982) and Ritchie (USDA/ARS, Temple, TX, unpublished communication) used layered profiles that assume downward flux in the profile, from onle layer to the next, whenever the soil moisture in the higher layer exceeds the upper limit of soil moisture. Detailed soil hydraulic data are required to make a more complete analysis of flux at the lower boundary of the root zone (Nimah and Hanks, 1973). This may be important in assessing leaching requirements.

The major problems in hydrological type approaches are: (a) the need for a rapid and accurate sensor for assessing soil water content with depth; (b) a technique for estimating the upper and lower soil water limits in the actual root zone; (c) a technique for estimating or predicting leaf area; and (d) a method for measuring canopy transpiration rates. There is a need to estimate soil water contents rapidly and accurately. While the neutron probe is accurate when calibrated correctly, access tubes must be positioned and revisited. Thus, a significant investment must be made in terms of equipment and trained personnel. Because the rooting depth changes with time, a dynamic method of predicting the limits of soil water content must be developed. The growth and development of leaves is one of the most complex interactions that takes place in the canopy space. New leaves initiate and expand while old leaves senesce and die. The canopy expands upwards and horizontally intercepts radiation which is used in processes of photosynthesis, transpiration and sensible heat. There has been significant progress in estimating leaf area index by remotely sensed spectral data (Pollock and Kanemasu, 1979). Because of the close relationship between transpiration and photosynthesis, evaporation from the soil surface is water lost from the potential transpiration reservoir. In order to better understand the partitioning of energy between transpiration and evaporation, an improved technique for measuring transpiration is required. A possible technique is a heat pulse meter. While relatively successful on trees, it must be adapted to vascular plants such as cotton, soybeans and corn.

Soil moisture monitoring

Schmugge et al. (1980) and Cary (1981) review advantages and disadvantages of several methods of determining soil moisture. Gravimetric soil sampling is simple but can require considerable investment in labor. Spatial variability of soil water content requires a large number of samples (Warrick and Nielsen, 1980) since one does not return to the exact same location for resampling. Campbell and Campbell (1982) calculated a sample number of 20 for estimating the soil water content of a field to within 0.01 cm^3/cm^3 if the standard deviation was 0.03 cm^3/cm^3.

Neutron attenuation provides a convenient and accurate measurement of soil moisture. The neutron probe can be used to schedule irrigation. The neutron probe samples a relatively large volume of soil and allows for repeated sampling at a given location in a field. Because neutron probes are quite expensive and require a licensed operator, they are practical only for a service agency or large operators which can use the equipment on many fields in an area. Available soil moisture can then be calculated from the volumetric soil moisture measurements.

Kanemasu and Raney (1982) compared corn yields over a 5-year period in which irrigations were scheduled at 50% available soil water content using neutron probe and a computerized water balance. The results showed that there was no significant yield differences between the treatments (Table II). Thus, one would conclude that an adaptable water balance technique could be used to estimate soil moisture for irrigation management.

TABLE II

Comparison of corn grain yields (15.5% moisture) for irrigation scheduling treatments (kg/ha) using neutron probe and the computerized water balance

	1977	1978	1979	1980	1981
T1 (dryland)	1 650c	8 150b	7 340b	95b	7 970b
T2 (50% neutron probe)	10 170a	9 940a	8 980a	8 140a	11 280a
T3 (50% computer)	6 820b	10 230a	9 040a	8 040a	9 990a
T4 (35/65% computer)	8 925a	9 145ab	8 750a	8 120a	10 190a

Yields showing different letter within a column differ significantly (0.05) by Duncan's Multiple Range Test.

Physiological

Crop growth and development is affected by plant water potential and indirectly by soil water potential (Gerakis and Carolus, 1970). A plant undergoing water stress exhibits decreased plant water potential, increased diffusive resistance and increased leaf temperature. Often the plants exhibit responses such as leaf rolling or changes in leaf orientation. Thorough reviews of plant responses to water stress have been given by Vaadia and Waisel (1967), Hsiao (1973), Boyer and McPherson (1975), Begg and Turner (1976) and Turner (1979). Blum (1974), Bielorai and Hopmans (1975), Turner et al. (1978), and Meyer and Green (1980, 1981) reported leaf water potential responses of major crops plants to soil water deficits under field conditions. Ehrler and Van Bavel (1967), Blum (1974), Sumayo et al. (1977) and Sumayo and Kanemasu (1979) reported leaf diffusive resistance responses to soil water deficits. Ehrler and Van Bavel (1967), Sumayo and Kanemasu (1979) and Gardner et al. (1981) reported elevated crop temperatures under soil water stress. Gardner et al. (1981) also reported an increase in the standard deviation of crop temperature measurements as the stress levels increase in a field.

If plant indicators are to be useful for irrigation scheduling, they must be simple to measure and must show detectable changes before adverse affects on plant growth and development occur. Plant water potential and diffusive resistance changes are generally detectable only at fairly low soil moisture levels when onset of serious plant water stress follows quickly (Meyer and Green, 1980, 1981). In addition, the measurements are made on single plants or leaves and several measurements must be made to obtain a representative sample of the field. Frequent sampling requirements will probably limit the widespread use of plant water stress indicators for irrigation scheduling (Stegman et al., 1976).

Hiler and Clark (1971) introduced the stress day index (SDI) concept which has been developed for use in irrigation scheduling.

$$\mathrm{SDI} = \sum_{i=1}^{n} (\mathrm{CS}_i \times \mathrm{SD}_i)$$

where CS is the crop susceptibility and SD is the stress day factor. Crop susceptibility is dependent upon growth stage. The SD factor could be based on plant, soil, or climatic and soil factors. Bordovsky et al. (1974) compared soil water potential and fixed and variable plant water potential indices for irrigation scheduling and found higher water use efficiencies for the plant based indicators.

The canopy-air temperature differential is desirable for irrigation scheduling, because it integrates the effects of soil moisture and atmospheric demand (Jackson, 1982). Also, use of an infrared thermometer gives an average temperature for all of the canopy within the field of view of the instrument rather than single leaf temperatures. Techniques for using hand-held infrared thermometers are given by Jackson et al. (1980).

Jackson et al. (1979) developed a stress degree day index (SSD) which sums the daily positive values of canopy minus air temperatures ($T_c - T_a$). One would start the summation immediately after an irrigation and continue until the SSD summed to a predetermined level and the crop was irrigated. Geiser et al. (1982) developed critical ($T_c - T_a$) values based on net radiation and relative humidity. If the observed ($T_c - T_a$) are greater than the critical ($T_c - T_a$), an irrigation is required. Corn yields from the ($T_c - T_a$) treatments were not significantly different from treatments irrigated by the electrical resistance blocks and water balance method. However, the resistance block and the water balance treatments required more water than the ($T_c - T_a$) treatment.

Clawson and Blad (1982) suggested the use of the variability in canopy temperature as plants undergo stress. By measuring the canopy temperature in several locations within a field, they found that the field needed irrigation when the range in temperature of several readings exceeded 0.7°C. They also examined the difference in canopy temperature between a stressed and well-watered plant as a means of assessing irrigation. They concluded that irrigation applied when the stressed treatment was 1°C warmer than the well-watered treatment, yields were already being reduced.

Canopy temperature appears to offer potential for assessing stress and transpiration differences. However, its use in an operational program of irrigation scheduling is somewhat tenuous. Some of the problems are: (a) the infrared thermometer sees soil as well as vegetal surfaces; therefore a mixture of temperatures is seen especially when stand density is low; (b) frequent (daily) measurements are desirable; and (c) canopy temperatures are dependent upon environmental conditions (net radiation, wind speed, and relative humidity). Therefore, relationships developed in one area may not extend to another because of soil, climate and crop differences. Further research is required to address these issues.

Infrared thermometry does offer a means of signaling stress from water, nutrients, weeds, etc. For example, problem areas within a field due to non-uniformity in water application, soil type, fertilizer application or herbicide application may be detected by aerial infrared thermometry.

IRRIGATION SYSTEMS AND MANAGEMENT OPTIONS

Tanner and Sinclair (1982) reviewed research in crop and water relationships and found that the transpiration-dry matter relationships are relatively constant for a specific crop in a specific environment. The efficiency of water use can be improved by reducing evaporation from the soil surface or by manipulating the partitioning of dry matter into harvestable yield. In addition, improving the efficiency of irrigation systems reduces the irrigation requirement without reducing the water available for plant use. Several irrigation and management systems have been developed to improve the efficiency of water use. Fig. 1 shows that improved efficiency of irrigation can be achieved by minimizing evaporation from the soil surface and by reducing irrigation system losses (e.g., runoff, drainage, spray losses, disuniformity, distribution losses). Stegman et al. (1981) discussed strategies for irrigating with a limited water supply.

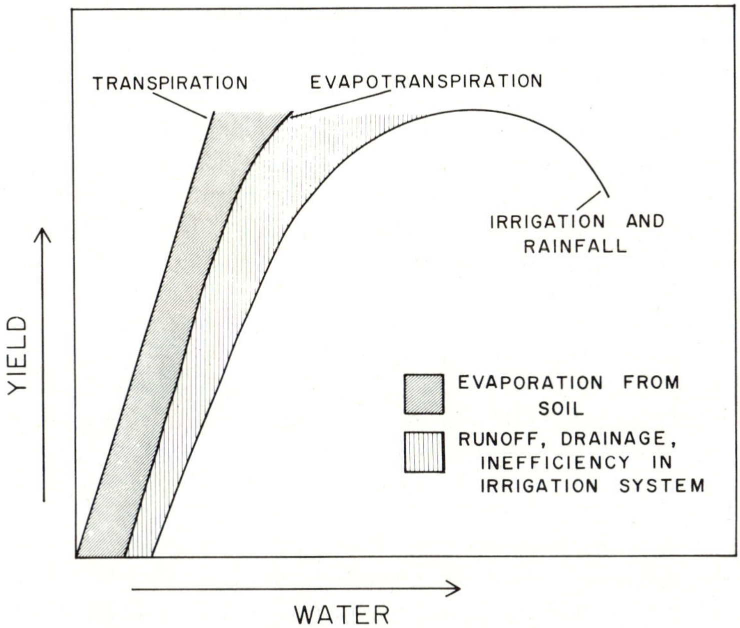

Fig. 1. Relative yield versus relative transpiration, evapotranspiration and irrigation + precipitation.

Sprinkler irrigation contains many inefficiencies. High evaporation losses may occur when water is pumped into the system at high pressures and sprayed into the air. Steiner et al. (1982) found about 15% spray loss on corn at Garden City, KS. Recent advances in research and in management techniques are pointing towards improved water efficiency and reduced energy requirements. Traveling or pivoting systems are being designed to release the water closer to the ground. Lyle and Bordovsky (1979, 1981) used a low pressure nozzle requiring only 7 kPa pressure drop. Water was applied at comparatively high flow rates and held in place by a system of dikes at the soil surface.

Deficit, high frequency irrigation has been proposed by Rawlins and Raats (1975) to enhance the plant growing conditions with limited irrigation. They proposed high frequency irrigation at rates less than ET, to maintain a high soil matric potential and a high osmotic potential near the source of water, yet maintain the capacity to absorb precipitation and enhance gradual depletion of stored soil moisture. Fereres et al. (1978) found that deficit, high frequency irrigation depressed yield and water use efficiency of sorghum, beans, and tomatoes at Davis, CA. Irrigation at the 100% ET rate resulted in similar yields with frequent or infrequent irrigation. Under deficit irrigation, full crop cover was not achieved, leading to increased soil evaporation losses under high frequency irrigation. The authors concluded that deficit irrigation is useful only if the moisture requirements of the crop are met throughout the season, either through depletion of stored moisture or through precipitation.

Reduction of plant population has long been used as a manner of coping with limited water availability in drought-prone areas. Each plant extracts water with less competition from a larger volume of soil, stretching the water supply through a season. Areal yields are commonly low but the risk of crop failure is reduced. A common irrigation practice is to omit selected rows from irrigation usually on a regular spacing basis — for example, two rows of crop and one row skipped. Other spacings have been reported. For example, a four row, two skip basis might provide irrigation to only the center furrow of the four rows. In general, the less water applied the lower the yield, but water use efficiency (as defined by yield over water applied) is highest for the wider irrigation skips (Newman, 1967; Musick and Dusek, 1982).

Stone et al. (1979, 1982) reported on wide-spaced furrow irrigation of uniformly spaced row crops (cotton, soybean, and grain sorghum). They believe that the potential exists for reducing evapotranspiration by reducing evaporation from the soil. Wide-spaced furrow irrigation presents a drier surface to the atmosphere and this may be one of the reasons for the maintenance of yield levels with less water applied. The method requires a fine text-textured soil which will wet laterally, as well as vertically. Alternating the irrigated furrow with each irrigation early eliminated the yield reduction in most of the years studied (Stone et al., 1982). This was true in furrows as long as 0.8 km.

Stewart et al. (1981) have combined the furrow diking practice with furrow irrigation as a means of eliminating tail-water from furrow irrigated fields. In essence, the lower portion of the field is adapted to rainfed conditions, with low planting and fertility rates, and the upper end of the field is planted for full irrigation. Irrigation rates are sufficient to maintain the dikes at the upper end of the field while the lower end of the field remains as dry-land. They were completely successful at eliminating tail water losses. They reported reasonable water use efficiencies in the upper end of the field but care must be taken that the furrow dikes do not cause greater deep percolation losses at the head end of the field. The method has application for water-limited irrigation situations since some of the field is a dry-land agriculture. Some producers are known to be using wide-spaced furrow irrigation in combination with furrow diking in the nonirrigated furrows to retain all rain water.

Particularly for a determinant crop, stress at one growth stage can affect growth during other stages by limiting both source and sink size. Much re-

TABLE III

Crop sensitivity to water stress at different growth stages

Crop	Sensitive growth stage	Reference	Comments
Corn (*Zea mays*)	Silking	Denmead and Shaw (1960)	Drought stress during the tasseling, silking, pollination, and early grain fill periods can limit sink size and greatly reduce yield
	Early silk	Stone et al. (1978)	
	12-leaf-to blister kernel	Stegman (1982)	
	Flowering	Doorenbos and Kassam (1979)	
		Salter and Goode (1967)	
		Herrero and Johnson (1981)	
Sorghum (*Sorghum bicolor*)	Boot to bloom	Lewis et al. (1976)	Sensitivity to stress is not great in any single stage
		Musick and Grimes (1961)	
	Boot stage	Salter and Goode (1967)	
	Flowering	Doorenbos and Kassem (1979)	
	No distinct sensitive period	Stone et al. (1978)	
Wheat (*Triticum aestivum*)	Jointing	Mohiuddin et al. (1981)	Sensitivity to stress is not great in any single growth stage
	Flowering	Marim and Valez (1979)	
		Doorenbos and Kassem (1979)	
Soybeans (*Glycine max*)	Flowering and pod fill	Hearn and Constable (1981)	
	Flowering	Salter and Goode (1967)	
		Doorenbos and Kassem (1979)	
Cotton (*Gossypium hirsutum*)	Flowering	Levin and Shmueli (1964)	Quality of the crop is affected by excessive water; crop development by drought stress
		Doorenbos and Kassem (1979)	
Crop	Sensitive growth stage	Reference	Comments
	Flowering and ripening (interaction)	Salter and Goode (1967)	
		Constable and Hearn (1981)	
		Cull et al. (1981)	
Alfalfa (*Medicago sativa*)	None	Salter and Goode (1967)	Irrigation scheduled to promote regrowth after cutting; allow drying prior to cutting
		Doorenbos and Kassem (1979)	

search has been conducted to identify the sensitivity of crops to water stress at various growth stages. A summary of growth stage sensitivity to drought stress for major Great Plains crops is presented in Table III. Stegman et al. (1980) shows yield responses of sensitive and insensitive crops to water stress.

Different plant functions show different sensitivities to soil moisture stress (Hsaio, 1973) but, in general, yield reducing plant stress does not occur until about 65% of the available soil moisture has been depleted (Ritchie, 1981b). Plant water potential, diffusive resistance, leaf temperature, and transpiration responses remain stable across a wide range of soil moisture conditions until 30—35% of available soil moisture remains then a rapid onset of stress occurs (Sumayo et al., 1977; Meyer and Green, 1980, 1981). The generalized crop response to soil moisture is illustrated in Fig. 2. Onset of crop stress is affected by evaporative demand, as well as soil moisture depletion (Ritchie, 1981b).

A prevalent approach to irrigation with limited water supplies is to begin the growing season with a full soil moisture supply and then to irrigate the crop at sensitive growth stages to ensure the maximum yield response. This approach is effective if the soil water storage capacity is large. Fall or early spring irrigation can be used to bring the soil profile to field capacity if winter and spring precipitation are not adequate to do so. One of the benefits of this approach is that it promotes growth of the roots deep within the profile. The most favorable conditions for root extension occur when there are moist, but not excessively wet conditions throughout the profile

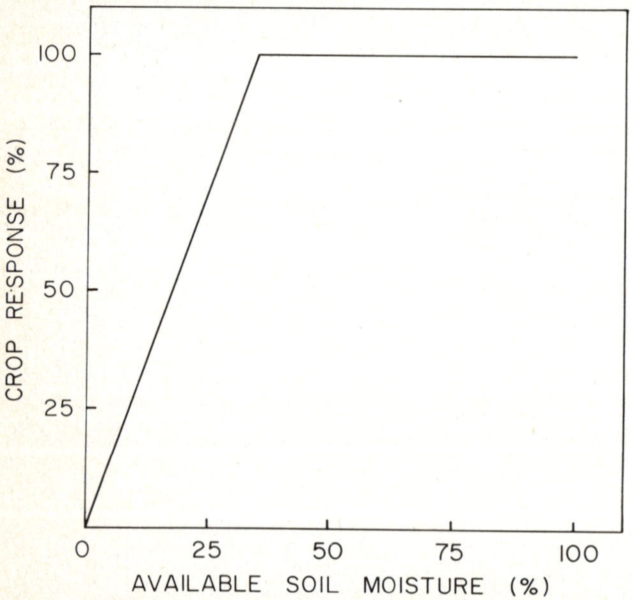

Fig. 2. Generalized crop response to available soil moisture.

and when there is gradual depletion of moisture of the surface layers (Pearson, 1965). Irrigation in the spring is preferable to fall irrigation. Redistribution of water within and below the root zone may continue over long periods of time after the soil reaches 'field capacity' (Hillel, 1971, pp. 162—164). Stone et al. (1981) found that large quantities of water could be lost over the winter period, particularly in moist soils. In addition, irrigation in the fall or early spring is likely to result in less effective storage of rainfall that occurs in the late spring.

There appears to be a potential for increased efficiency in water use by altering or modifying the irrigation system. However, there is no preferred system for all situations; each situation must be assessed separately. Similarly, management strategies may be different for each operator depending upon his irrigation system, soil, environment, and crop moisture. There have been some general rules of thumb developed such as the increased probability for wasting water by fall or winter irrigation (off-season irrigation) when the soil profile is above or about 50% available water content. Also, there are water savings when one can minimize the wetted surface soil area and, hence, reduce evaporation.

ECONOMIC OPTIMIZATION

Irrigators may have one of several objectives they wish to optimize (Fig. 3). While some irrigators may wish to maximize yield per acre or water use efficiency, it is assumed that most farmers wish to maximize total net farm income. This objective is too general for irrigation scheduling so a more limited management objective, maximization of net returns per hectare, is often used. Then, the rule for irrigation scheduling is to irrigate until marginal costs of irrigating equal the marginal returns from that irrigation.

Stegman et al. (1980, 1981) point out a number of reasons why the maximization of profit is not traditionally utilized in day-to-day irrigation scheduling even though it is a principal motivation in farming: (a) the variable costs of irrigation have been low; (b) there is a stochastic nature to the availability and impact of production inputs; (c) prices of inputs and outputs may not be known; (d) production of crop-water response functions suitable for day-to-day management decisions are not available. It has been difficult to integrate economic criterion with day-to-day irrigation management. Estimated total annual production costs for various cropping systems are compared to expected returns given local constraints. These estimates are used to determine a crop mix providing the maximum potential profit. Irrigations are then scheduled to achieve the optimum yield. In recent years, several approaches have been taken in efforts to integrate economic optimization criteria into day-to-day irrigation decisions. The needed components for an economic irrigation-scheduling model are: (a) a water balance equation and a evapotranspiration model to track available soil moisture; (b) a function to describe the plant's growth response to available soil moisture;

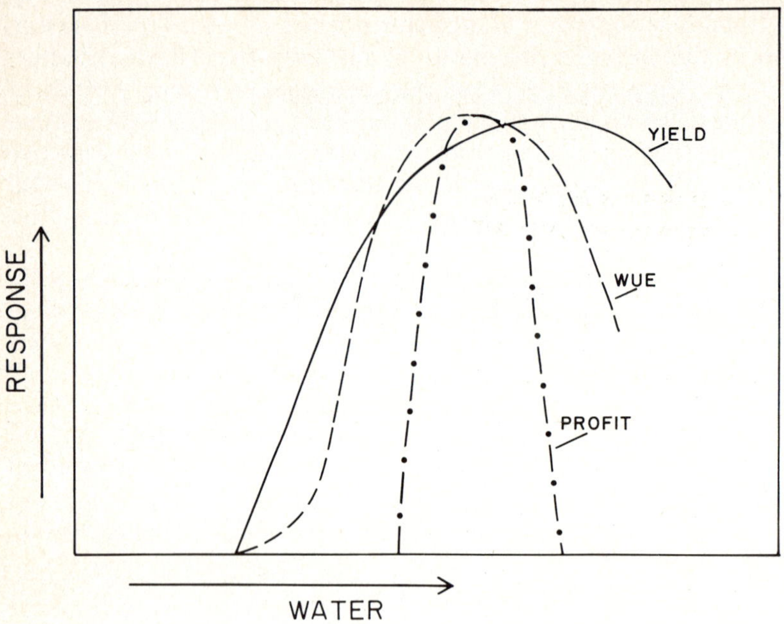

Fig. 3. Generalized response curves of water-use efficiency (WUE), yield, and profit to applied water.

(c) an economic optimization function. These components and their interrelationships are outlined in Fig. 4.

The first component is discussed elsewhere in this chapter. Plant responses to water (soil moisture) have been modeled by estimating the growth function statistically from experimental data or, more recently, by developing crop response functions from physiological relationships. There are significant questions concerning the definition of crop response functions, and a need exists to examine the response of stress-adapted plants to limited irrigation amounts. The statistically estimated functions or simulation models appear to have two general orientations. There are simulation models based on synthetic functions and data (Anderson and Maass, 1971; Mapp et al., 1975; Ahmed et al., 1976; Tscheschke et al., 1978). The other type of model simulates plant growth, based on experimental data (Mapp et al., 1975; Arkin et al., 1976; Childs et al., 1976).

Different approaches to the economic function to be used have also been taken. Moore (1961) utilized an annual marginal cost and marginal revenue forgone function, while Palacios (1981) developed a function involving total water applied and the ratios of shadow price of water to price of product. However, the most common function has been to equate marginal price of expected yield and marginal cost of irrigation. After the type of economic maximization function has been selected, there remains the question of how it should be applied. One of the major optimization techniques found in irrigation work is dynamic programming. A deterministic dynamic pro-

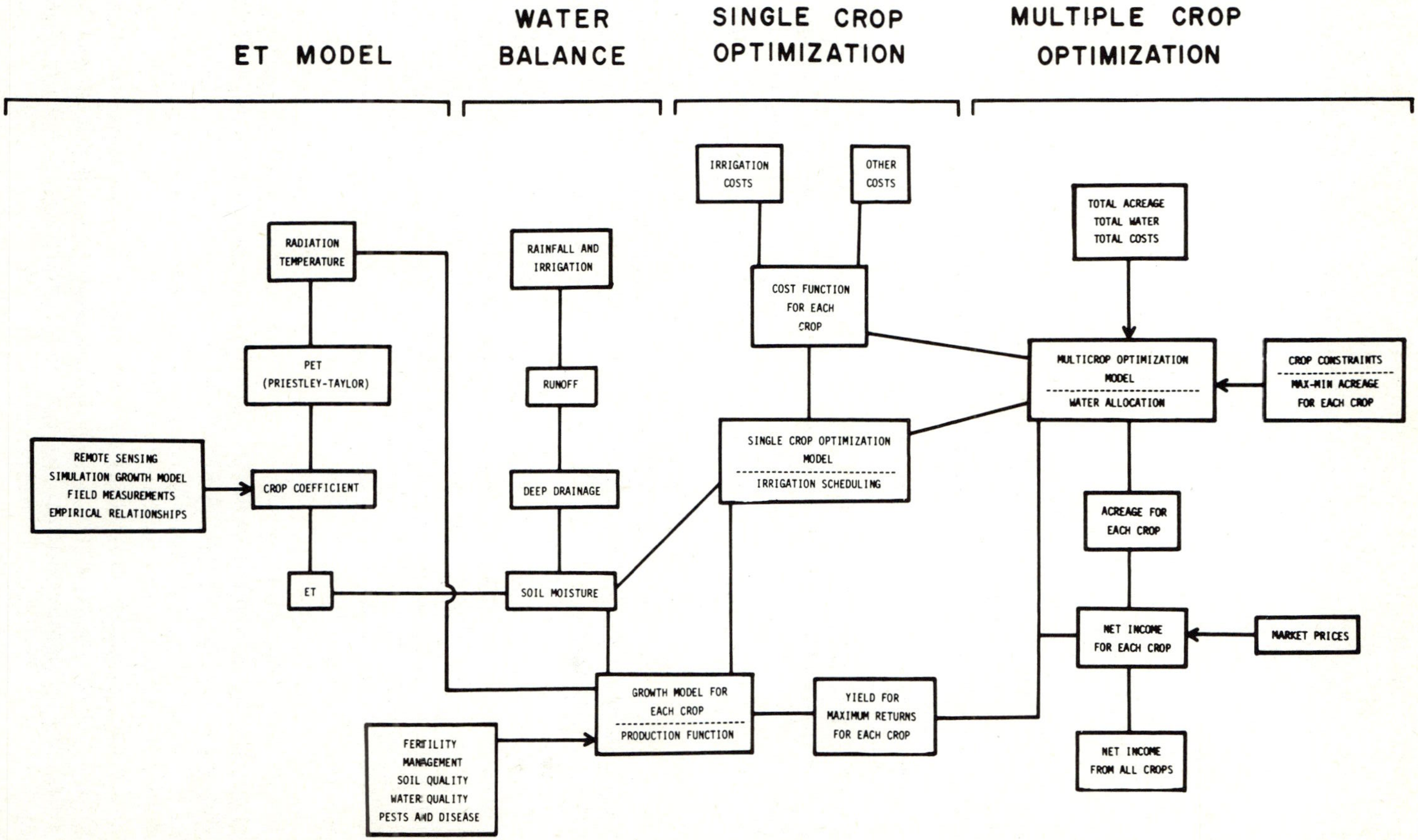

Fig. 4. Schematic of an irrigation scheduling model.

gramming model can allocate limited quantities of water to maximize profits when there are delivery system constraints. Butcher and Hall (1968) provided a theoretical framework for the application of dynamic programming in irrigation scheduling. Burt and Stauber (1971) utilized dynamic programming in a model that divided the season into five critical periods and provided for an interaction of water available in one period with the growing conditions in the other periods. Raju et al. (1983) have proposed an irrigation scheduling model, that incorporates a dynamic response model (Morgan et al., 1980). The scheduling criterion is to maximize net returns to irrigated production; optimal irrigation schedules are obtained using forward dynamic programming.

Several other techniques have been suggested for including the economic optimization criterion in irrigation scheduling. Harris (1981) utilized the optimal control theory to schedule irrigation while Ahmed et al. (1976) used a closed loop control system, and Zavaleta et al. (1980) used an open loop stochastic control system.

In a situation where there is insufficient water to irrigate for a maximum yield over an area, then single crop irrigation scheduling models may be useful in allocating available water within a growing season for a particular crop. However, multi-crop models may be more useful for planning cropping patterns, distributing limited water supplies, planning irrigation projects, and analyzing the potential impact of drought situations (Fig. 4). Several simulation models have been used in the multicropping context. Mapp et al. (1975) and Rydzewski and Nairizi (1979) used simulation models to determine allocation of water. Linear programming has been used by Blank (1975) and Matanga and Marino (1979), to determine optimal allocation of water between crops.

There are a number of critical aspects to irrigation scheduling models utilizing net returns as an economic criterion that deserve further consideration. One such critical aspect is the influence of prior stress on a crop's current response to irrigation (Ashton, 1956). Stewart et al. (1975) reported that corn's tolerance to stress during the pollination period is dependent on whether or not the plant was previously stressed. Sullivan and Bennett (1981) found that sorghum which had been preconditioned to stress maintained higher photosynthetic rates under stress, than did non-conditioned plants. Dynamic growth models offer a means of incorporating the effect of prior stress on current crop response. From the economic point of view, it is important to know if some irrigations have more impact than others and, thus, should be valued at a higher rate of yield.

Closely related is the fact that the impact of one irrigation on yield is not independent of whether or not other irrigations are applied, that is the effects of the irrigations are not additive, for three major reasons: (a) plant growth is cumulative - the current dry matter accumulation is dependent on the prior dry matter accumulation; (b) a plant's sensitivity to soil moisture stress may vary with the stage of development; (c) daily soil moisture levels are interrelated, so the impacts of irrigations on yield cannot be additive.

Another significant area to consider in modeling for drought conditions is the dynamics of root growth. Stewart et al. (1975) provided data on the extraction of water in unirrigated corn from successively deeper soil layers. The plants' ability to produce roots to effectively utilize available water requires a dynamic root growth model. Such a model must be sensitive to the type of soil in which the plant is growing and the effects of a nonhomogeneous soil profile, as both can affect the amount of soil moisture available to the plant and so have a direct impact on potential yield.

Most irrigation scheduling models assume an optimum input of fertilizer. However, there is the possibility of varying the amounts of fertilizer and water, substituting one for the other to some extent. This interaction between irrigation and soil fertility is an area of particular interest when water is limited because an improper water-fertility balance could greatly reduce yields and the net economic return. Similar arguments can be made about herbicides and pesticides.

A final critical area, which has not been treated extensively in irrigation scheduling models, is the question of risk and uncertainty. According to English (1979) there will be a great deal of uncertainty in crop production models due to the variability in model inputs and the uncertainties inherent in the models themselves. As a consequence, a farmer who uses crop production models to optimize irrigation practices will face some degree of risk. The amount of uncertainty may be great enough that the irrigator "may choose to follow irrigation practices which reduce expected net income in order to reduce risk". How an irrigation model handles risk and uncertainty can have an impact on the economic criterion used to optimize the model, particularly if, as English suggests, the risk factor causes a shift in the economic objectives.

REFERENCES

Ahmed, J., Van Bavel, C.H.M. and Hiler, E.A., 1976. Optimization of crop irrigation strategy under a stochastic weather regime: a simulation study. Water Resour. Res., 12: 1241—1247.

Anderson, R.L. and Maass, A., 1971. A simulation of irrigation systems: the effects of water supply and operating rules on production and income on irrigated farms Tech. Bull. 1431, Economic Research Service, U.S. Department of Agriculture, 57 pp.

Arkin, G.I., Vanderlip, R.L. and Ritchie, J.T., 1976. A dynamic grain sorghum growth model. Trans. ASAE, 19: 622—626, 630.

Ashton, F.M., 1956. Effects of a series of cycles of alternating low and high soil water contents on the rate of apparent photosynthesis in sugar cane. Plant Physiol., 31: 266—274.

Begg, J.E. and Turner, N.C., 1976. Crop water deficits. Adv. Agron., 28: 161—217.

Bielorai, H. and Hopmans, P.A., 1975. Recovery of leaf water potential, transpiration and photosynthesis of cotton during irrigation cycles. Agron. J., 67: 629—632.

Blank, H.G., 1975. Optimal irrigation decisions with limited water. Ph. D. Diss., Colorado State University, Fort Collins, CO, 241 pp.

Blum, A., 1974. Leaf water potential and stomatal activity in sorghum and influenced by soil moisture stress. Isr. J. Bot., 23: 14—19.
Bordorvsky, D.G., Jordan, W.R., Hiler, E.A. and Howell, T.A., 1974. Choice of irrigation timing indicator for narrow row of cotton. Agron. J., 66: 88—91.
Boyer, J.S. and McPherson, H.G., 1975. Physiology of water deficits in cereal crops. Adv. Agron., 27: 1—23.
Burt, O.R. and Stauber, M.S., 1971. Economic analysis of irrigation in subhumid climate. Am. J. Agric. Econ., 53: 33—46.
Butcher, W.S. and Hall, W.A., 1968. Optimal timing of irrigation. J. Irrig. Drain. Div. ASCE, 94: 267—275.
Campbell, G.S. and Campbell, M.D., 1982. Irrigation scheduling using soil moisture measurements: theory and practice. In: D. Hillel (Editor), Advances in Irrigation, Vol. I. Academic Press, New York, NY, pp. 25—42.
Cary, J.W., 1981. Irrigation scheduling with soil instruments: error levels and microprocessing design criteria. In: Irrigation Scheduling for Water and Energy Conservation in the 80's. Publ. 23-81, American Society of Agricultural Engineers, St. Joseph, MI.
Childs, S.W., Gilley, J.R. and Splinter, W.E., 1976. A simplified model of corn growth under moisture stress. Pap. 76-2528, American Society of Agricultural Engineers, St. Joseph, MI, 11 pp.
Clawson, K.L. and Blad, B.L., 1982. Infrared thermometry for scheduling irrigation of corn. Agron. J., 74: 311—316.
Constable, G.A. and Hearn, A.B., 1981. Irrigation for crops in a submitted environment. VI. Effect of irrigation and nitrogen fertilizer on growth, yield, and quality of cotton. Irrig. Sci., 3: 17—28.
Cull, P.D., Hearn, A.B. and Smith, R.C.G., 1981. Irrigation scheduling of cotton in a climate with uncertain rainfall. I. Crop water requirements and response to irrigation. Irrig. Sci., 2: 127—141.
Denmead, O.T. and Shaw, R.H., 1960. The effects of soil moisture stress at different stages of growth on the development and yield of corn. Agron. J., 52: 272—274.
Doorenbos, J. and Kassam, A.H., 1979. Yield response to water. FAO Irrig. Drain. Pap. 33, 193 pp.
Ehrler, W.L. and Van Bavel, C.H.M., 1967. Sorghum foliar response to changes in soil water content. Agron. J., 59: 243—246.
English, M.J., 1979. Use of crop models in irrigation optimization. ASAE Pap. 79-4521, American Society of Agricultural Engineers, St. Joseph, MI, 12 pp.
Fereres, E., Amry, B., Faci, J., Kamgar, A., Henderson, D.W. and Resende, M., 1978. A closer look at deficit high-frequency irrigation. Calif. Agric., 32 (8): 4—5.
Gardner, B.R., Blad, B.L. and Watts, D.G., 1981. Plant and air temperatures in differentially-irrigated corn. Agric. Meteorol., 25: 207—217.
Gear, R.D., Dransfield, A.S. and Campbell, M.D., 1977. Irrigation scheduling with neutron probe. J. Irrig. Drain. Div. ASCE, 103 (IR3): 291—298.
Geiser, K.M., Slack, D.C., Albred, E.R. and Stange, K.W., 1982. Irrigation scheduling using crop canopy-air temperature difference. Trans. ASAE, 25: 689—694.
Gerakis, P.A. and Carolus, R.L., 1970. Controlling internal plant water balance through microclimate manipulation. Agrochimica, 14: 441—452.
Harris, T.R., 1981. Analysis of irrigation scheduling for grain sorghum in the Oklahoma Panhandle. Ph. D. Thesis, Oklahoma State University Library, Stillwater, OK, 248 pp.
Hearn, A.B. and Constable, G.A., 1981. Irrigation for crops in a subhumid environment. V. Stress day analysis for soybeans and an economic analysis. Irrig. Sci., 3: 1—15.
Heermann, D.F., 1980. Irrigation scheduling. In: D. Yaron and C. Tapiaro (Editors), Operations Research in Agriculture and Water Resources. North-Holland Publishing Co., Amsterdam, pp. 501—516.
Herrero, M.P. and Johnson, R.R., 1981. Drought stress and its effects on maize reproductive systems. Crop Sci., 21: 105—110.

Hiler, E.A. and Clark, R.N., 1971. Stress day index to characterize effects of water stress on crop yields. Trans. ASAE, 14: 757—761.

Hillel, D., 1971. Soil and Water: Physical Principles and Processes. Academic Press, New York, NY, 288 pp.

Howell, T.A. and Hiler, E.A., 1975. Optimization of water use efficiency under high frequency irrigation: evapotranspiration and yield relationship. Trans. ASAE, 18: 873—878.

Hsiao, T.C., 1973. Plant responses to water stress. Annu. Rev. Plant Physiol., 24: 519—570.

Idso, S.B., Jackson, R.D. and Reginato, R.J., 1982. Soil- and atmosphere-induced plant water stress in cotton as inferred from foliate temperatures. Water Resour. Res., 18: 1143—1148.

Jackson, R.D., 1982. Canopy temperature and crop water stress. In: D. Hillel (Editor), Advances in Irrigation, Vol. I. Academic Press, New York, NY, pp. 43—85.

Jackson, R.D., Reginato, R.J. and Idso, S.B., 1979. Wheat canopy temperature: A practical tool for evaluating water requirements. Water Resour. Res., 13: 651—656.

Jackson, R.D., Pinter, R.J., Jr., Reginato, R.J. and Idso, S.B., 1980. Hand-held radiometry. Agricultural Reviews and Manuals, ARM-W-19, USDA-SEA, Phoenix, AZ.

Jensen, M.E. (Editor), 1973. Consumptive Use of Water and Irrigation Requirements. American Society of Civil Engineers, New York, NY, 215 pp.

Jensen, M.E. and Wright, J.L., 1978. The role of evapotranspiration models in irrigation scheduling. Trans. ASAE, 21: 82—87.

Jury, W.A. and Tanner, C.B., 1975. Advection modification of the Priestley and Taylor evapotranspiration formula. Agron. J., 67: 840—842.

Kanemasu, E.T. and Raney, R., 1982. Field comparisons of irrigation scheduling by neutron probe and a computerized water balance. Keeping up with Research, No. 60, Kansas Agricultural Experiment Station, Manhattan, KS, 3 pp.

Kanemasu, E.T., Stone, L.R. and Powers, W.L., 1976. Evapotranspiration model tested for soybean and sorghum. Agron. J., 68: 569—572.

Kanemasu, E.T., Steiner, J.L., Rasmussen, V.P. and Bagley, J., 1979a. Estimating water requirements for corn with a programmable calculator. Kans. Agric. Exp. Stn. Bull. 615 (Revised), 24 pp.

Kanemasu, E.T., Wesely, M.L., Hicks, B.B. and Heilman, J.L., 1979b. Techniques for calculating energy and mass fluxes. In: B.J. Barber and J.F. Gerber (Editors), Modification of the Aerial Environment of Plants. ASAE Monogr. 2, American Society of Agricultural Engineers, St. Joseph, MI, pp. 156—182.

Levin, I. and Shmueli, E., 1964. The response of cotton to various irrigation regimes in the Hula Valley. Isr. J. Agric. Res., 14: 211—225.

Lewis, R.B., Hiler, E.A. and Jordan, W.R., 1976. Susceptibility of grain sorghum to water deficit at three growth stages. Agron. J., 66: 589—591.

Luckey, R.R., Gutentag, E.D. and Weeks, J.B., 1981. Water-level and saturated thickness changes, pre-development to 1980, in the High Plains aquifer. USGS Hydrologic Investigations Atlas HA-652, USGS, Denver, CO.

Lyle, W.M. and Bordovsky, J.P., 1979. Traveling low energy precision irrigator. In: Irrigation and Drainage in the Nineteen-Eighties. Irrigation and Drainage Division, American Society of Civil Engineers, New York, NY, pp. 121—131.

Lyle, W.M. and Bordovsky, J.P., 1981. Low energy precision application (LEPA) irrigation system. Trans. ASAE, 24: 1241—1245.

Mapp, H.P., Eidman, V.R., Stone, J.F., and Davidson, J.M., 1975 Simulating soil water and atmospheric stress-crop yield relationships for economic analysis. Okla. Agric. Exp. Stn. Tech. Bull. 140, 63 pp.

Marim, R.M. and Velez, E.P., 1979. Repuesta del cultivo de tigro a variaciones de la humedad en el suelo en diferentes etapas de crecimiento. Agronciencia 38: 3—16.

Matanga, G.B. and Marino, M.A., 1979. Irrigation planning: 1. Cropping pattern. Water Resour. Res., 15: 672—678.

Meyer, W.S. and Green, G.C., 1980. Water use by wheat and plant indicators of available soil water. Agron. J., 72: 253—257.

Meyer, W.S. and Green, G.C., 1981. Plant indicators of wheat and soybean crop water stress. Irrig. Sci., 2: 167—176.

Mohiuddin, S.H., Hooker, M.L. and Kanemasu, E.T., 1981. The effect of irrigation timing on yield of winter wheat. Agron. Abstr., p. 109.

Moore, C.V., 1961. A general analytical framework for estimating the production function for crops using irrigation water. J. Farm Econ., 63: 876—888.

Morgan, T.H., Biere, A.W. and Kanemasu, E.T., 1980. A dynamic model of corn yield response to water. Water Resour. Res., 16: 59—64.

Musick, J.T. and Dusek, D.A., 1982. Skip-row planting and irrigation of graded furrows. Trans. ASAE, 25: 82—92.

Musick, J.T. and Grimes, D.W., 1961. Water management and consumptive use of irrigated grain sorghum in Western Kansas. Kans. Agric. Exp. Stn. Tech. Bull. 113, 20 pp.

Newman, J.S., 1967. Yield and fiber properties of cotton planted in solid and skip-row systems under minimal soil moisture levels. Texas High Plains, 1963—65. Bull. MP-843, South Plains Research and Extension Center, Texas Agricultural Experiment Station, Lubbock, TX, 16 pp.

Nimah, M.N. and Hanks, R.J., 1973. Model for estimating soil water and atmospheric interrelations. Soil Sci. Soc. Am. Proc., 37: 522—527.

Palacios, E.V., 1981. Response functions of crop yield to soil moisture stress. Water Resour. Bull., 17: 699—703.

Pearson, R.W., 1965. Soil environment and root development. In: W.H. Pierre, D. Kirkham, J. Pesek, and R. Shaw (Editors), Plant Environment and Efficient Water Use. American Society of Agronomy/Soil Science Society of America, Madison, WI, pp. 95—126.

Pollock, R.B. and Kanemasu, E.T., 1979. Estimating leaf area index of wheat with LANDSAT data. Remote Sensing Environ., 8: 307—312.

Priestley, C.H.B. and Taylor, R.J., 1972. On the assessment of surface heat flux and evaporation using large-scale parameters. Mon. Weather Rev., 100: 81—92.

Raju, K.S., Lee, E.S., Biere, A.W. and Kanemasu, E.T., 1983. Irrigation scheduling based on a dynamic crop response model. In: D. Hillel (Editor), Advances in Irrigation, Vol. II. Academic Press, New York, NY (In press).

Rawlins, S.L. and Raats, P.A.C., 1975. Prospects for high frequency irrigation. Soil Sci. Soc. Am. Proc., 37: 604—610.

Riefler, R.F., 1978. Drought: an economic perspective. In: N.J. Rosenberg (Editor), North American Droughts. AAAS Sel. Symp. Ser. 15, Westview Press, Boulder, CO, pp. 63—78.

Ritchie, J.T., 1972. Model for predicting evaporation from a row crop with incomplete cover. Water Resour. Res., 8: 1204—1213.

Ritchie, J.T., 1981a. Soil water availability. Plant Soil., 58: 327—338.

Ritchie, J.T., 1981b. Water dynamics in the soil-plant-atmosphere system. Plant Soil., 58: 81—96.

Rosenberg, N.J., 1978. Technological options for crop production in drought. In: N.J. Rosenberg (Editor), North American Droughts. AAAS Sel. Symp. Ser. 15, Westview Press, Boulder, CO, pp. 123—142.

Rosenthal, W.D., Kanemasu, E.T., Raney, R.J. and Stone, L.R., 1977. Evaluation of an evapotranspiration model for corn. Agron. J., 69: 461—464.

Rouse, W.R. and Wilson, R.G., 1972. A test of the potential accuracy of the water-budget approach to estimating evapotranspiration. Agric. Meteorol., 9: 421—446.

Rydzewski, J.R. and Nairizi, S., 1979. Irrigation planning based on water deficits. Water Resour. Bull., 15: 316—325.

Salter, P.J. and Goode, J.R., 1967. Crop responses to water at different stages of growth. Res. Rev. 2, Commonwealth Bureau of Horticultural and Plantation Crops, Commonwealth Agricultural Bureau, Farnham Royal, Berks., 246 pp.

Schmugge, T.J., Jackson, T.J. and McKim, H.L., 1980. Survey of methods for soil moisture determination. Water Resour. Res., 16: 961—979.

Stegman, E.C., 1982. Corn grain yields as influenced by timing of evapotranspiration deficits. Irrig. Sci., 3: 75—87.

Stegman, E.C., Schiele, L.J. and Bauer, A., 1976. Plant water stress criteria for irrigation scheduling. Trans. ASAE, 19: 850—855.

Stegman, E.C., Hanks, R.J., Musick, J.T. and Watts, D.G., 1981. Irrigation water management — adequate or limited water. In: Irrigation: Challenges for the 80's Proceedings of the Second National Irrigation Association, October 1980, Lincoln, NE. American Society of Agricultural Engineers, St. Joseph, MI, pp. 154—165.

Stegman, E.C., Musick, J.T. and Stewart, J.T., 1980. Irrigation water management. In: M.E. Jensen (Editor), Design and Operation of Farm Irrigation Systems. ASAE Monogr. 3, American Society of Agricultural Engineers, St. Joseph, MI, pp. 763—816.

Steiner, J.L., Kanemasu, E.T. and Clark, R.N., 1982. Spray losses and partitioning of water under a center pivot sprinkler system. Pap. 82-2006, American Society of Agricultural Engineers, St. Joseph, MI, 12 pp.

Stewart, B.A., Dusek, D.A. and Musick, J.T., 1981. A management system for the conjunctive use of rainfall and limited irrigation of graded furrows. Soil Sci. Soc. Am. J., 45: 413—419.

Stewart, J.I., Misra, R.D., Pruitt, W.O. and Hagan, R.M., 1975. Irrigating corn and grain sorghum with a deficit water supply. Trans. ASAE, 18: 270—280.

Stone, L.R., Gwin, R.E. and Dillon, M.A., 1978. Corn and grain sorghum yield response to limited irrigation. J. Soil Water Conserv., 33: 235—238.

Stone, J.F., Garton, J.E., Webb, B.B., Reeves, H.E. and Keflemariam, J., 1979. Irrigation water conservation using wide-spaced furrows. Soil Sci. Soc. Am. J., 43: 407—411.

Stone, L.R., Gwin, R.E. and Gallagher, P., 1981. Fall irrigation's net benefit in March. Kansas Water Newsl. 23 Kansas State Water Resources Board, Topeka, KS, 12—17.

Stone, J.F., Reeves, H.E. and Garton, J.E., 1982. Irrigation water conservation by using wide-spaced furrows. Agric. Water Manage., 5: 309—317.

Sullivan, C.Y. and Bennett, J.M., 1981. Effects of water stress preconditioning on net photosynthetic rate of grain sorghum. Photosynthetica, 15: 330—337.

Sumayo, C.R. and Kanemasu, E.T., 1979. Temperature and stomatal resistance of soybean leaves. Can. J. Plant Sci., 59: 153—162.

Sumayo, C., Kanemasu, E.T. and Hodges, T., 1977. Soil moisture effects on transpiration and net carbon dioxide exchange of sorghum. Agric. Meteorol., 18: 401—408.

Tanner, C.B., 1967. Measurement of evapotranspiration. In: R.M. Hagan, H.R. Haise and T.W. Edminster (Editors), Irrigation of Agricultural Lands. Agron. 11, American Society of Agronomy, Madison, WI, pp. 534—574.

Tanner, C.B. and Sinclair, T.R., 1983. Efficient water use in crop production: research or re-search. In: Limitations to Efficient Water Use in Crop Production. American Society of Agronomy, Madison, WI (In press).

Tscheschke, P., Gilley, J.R., Thompson, T. and Fishbach, P., 1978. IRRIGATE — a scheduling model. Agric. Eng., 59: 45—46.

Turner, N.C., 1979. Drought resistance and adaptation to water deficits in crop plants. In: H. Mussell and R.C. Staples (Editors), Stress Physiology in Crop Plants. John Wiley, New York, NY, pp. 344—372.

Turner, N.C., Begg, J.E., Rawson, H.M., English, S.D. and Hearn, A.B., 1978. Agronomic and physiological responses of soybean and sorghum crops to water deficits. III. Components of leaf water potential, leaf conductance, $^{14}CO_2$ photosynthesis and adaptation to water deficits. Aust. J. Plant. Physiol., 5: 179—194.

Vaadia, Y. and Waisel, Y., 1967. Physiological processes as affected by water balance. In: R.M. Hagan, H.R. Haise and T.W. Edminster (Editors), Irrigation of Agricultural Lands. Agron. 11, American Society of Agronomy, Madison, WI, pp. 354—372.
Van Bavel, C.H.M. and Wilson, T.V., 1952. Evapotranspiration estimates as criteria for determining time of irrigation. Agric. Eng., 33: 417—418, 420.
Warrick, A.W. and Nielsen, D.R., 1980. Spatial variability of soil physical properties in the field. In: D. Hillel (Editor), Applications of Soil Physics. Academic Press, New York, NY, pp. 319—344.
Zavaleta, L.R., Lacewell, R.D. and Taylor, C.R., 1980. Open-loop stochastic control of grain sorghum irrigation levels and timing. Am. J. Agric. Econ., 62: 785—792.

Introduction to Part 3

PLANT BREEDING AND GENETICS

This section deals with drought as it relates to plant breeding. Plant breeding is defined as the art and science of manipulating genetic material to achieve a more productive, stable genetic potential under specific or general environments. The current 'state of the art' is the point of departure in addressing variety and hybrid development for drought conditions.

Plant breeders are faced with two general challenges. First, they must respond to the specific clientele (producer) demands with an associated reward-penalty system. A successful breeder produces, on a frequent basis hybrids or varieties with improved characteristics resulting in higher yield potential. Breeder productivity is continually being evaluated by the growers. Breeder success or failure is generally measured by the frequency with which competitive and superior varieties or hybrids are developed. Second, the breeder must deal with a complex combination of scientific knowns and unknowns. These factors include the genetics of the species, the peculiarities of the environment, and the interaction of these two components. Breeders are often forced to use a combination of a confirmed data base and 'best guess' estimates in decision making.

The following papers relate to the question of breeding for response to drought conditions. It becomes immediately apparent that the two previously described conditions of program productivity and scientific unknowns play a major role in breeder response to this question. Breeding for positive genetic response to drought conditions has been uniformly difficult across all species. This difficulty is due to the complex environmental variables associated with genetic characteristics and the relatively small amount of scientific data available on the genetic control of plant characteristics responsible for positive reaction to water stress. However, significant gains are being made in addressing this issue and the breeders emphasize the areas that are either being used successfully or hold promise for the future.

JAMES R. WELSH
Montana Agricultural Experiment Station
Bozeman, MT, U.S.A.

Agricultural Water Management, 7 (1983) 181—194
Elsevier Science Publishers B.V., Amsterdam — Printed in The Netherlands

DROUGHT RESISTANCE AND WHEAT BREEDING

JOHN W. SCHMIDT

Department of Agronomy, East Campus, University of Nebraska, 322 Keim Hall, Lincoln, NE 68583 (U.S.A.)

Paper No. 6974, Journal Series, Nebraska Agricultural Experiment Station

(Accepted 20 October 1982)

ABSTRACT

Schmidt, J.W., 1983. Drought resistance and wheat breeding. *Agric. Water Manage.*, 7: 181—194.

The breeding of wheat (*Triticum* species) for drought resistance has been indirect for the most part. The most direct approach has been identifying an appropriate crop maturity that maximizes growth and the accompanying yield component development. Beyond that, the improvement of response to drought has been rare. This is due, in great part, to the complexity of the problem and the lack of techniques and facilities that can be used to screen thousands of genotypes in practical plant breeding programs. Yield improvement is the prime goal in most wheat breeding programs. Unfortunately, many of the favourable plant responses to moisture stress leave negative effects on grain yield.

An impressive number of research reports in the 1970's relating to the various aspects of drought resistance in wheat bodes well for directed selection for improved genotypic performance under moisture-stress conditions. Cooperative research between plant breeders and plant physiologist is the key to accomplishing this goal. This should begin with careful choices of parental and germplasm materials.

Regional cultivar-yield-testing was initiated by the U.S. Department of Agriculture and cooperating state experiment stations in the early 1930's. This program has been eminently successful in identifying cultivars that had the desired adaptation characteristics. This guideline for a successful testing program for advanced lines needs to be introduced early in the breeding program. Therefore, breeding procedures should include as much early generation multilocational testing and selection as feasible. This may include early generation bulks, F_2- or F_3-derived lines, and F_2 selection composites. High average-yield and stability of performance are still the best criteria for selection. The earlier in the breeding program that adaptive (high average) yield can be identified the better. Variability of selected phenotypes, within limits, may provide populational buffering in moisture-stress situations that are variable in their location, intensity, duration and seasonal appearance.

INTRODUCTION

Drought is an ever-present restraint to maximum wheat production somewhere in the Great Plains of the United States. It will vary in its location, intensity, duration and in seasonal appearance. It may interfere with stand

0378-3774/83/$03.00

establishment in the fall and this may predispose the winter wheat crop to winter injury. Sometimes the entire growing season is a continuous high moisture-stress situation but more often intermittent moisture stress occurs. Most reports indicate that, while damage may occur at any point in the crop developmental cycle, the most critical stage is from the last leaf visible (very early boot) to anthesis. Those of us from the northern Great Plains winter wheat area often relate the aborting of the basal spikelets to a hot, dry, windy day or two at a much earlier period in development. Grafius (1971) pointed out that the developmental process is sequential with one event influencing another, and that stress at any time during the developmental process can have considerable effect on future events but may be ameliorated by component compensation.

Breeding wheats for the drought situations encountered in the Great Plains is very complex and few examples of progress stand out. Boyer and McPherson (1975) stated that improvement in the response of plants to drought has been rare. Townley-Smith and Hurd (1979) reported that research into a plant's response to drought has intensified recently and such research has emphasized the complexity of the problem. Blum et al. (1981) stated, "The total drought resistance of a genotype cannot yet be defined physiologically and most probably it does not exist as a unique plant trait". A review of the rapidly expanding literature in this field supports the thesis that responses to drought are complex phenomena. Thus, research reports on various aspects of drought response are sometimes contradictory, difficult to integrate, often based on greenhouse studies, and not of practical usefulness in a plant breeding program where large numbers of lines must be screened. However, as a practical plant breeder, I am deeply appreciative of the tremendous effort by an increasing number of researchers to provide insights on the effects of moisture stress at various stages in the developmental processes of the wheat plant and to provide means for combatting one of the most serious restraints to maximum wheat production.

Development of drought-resistant wheat cultivars will require the integration of plant breeding and plant physiology. By necessity, the approaches of these two fields in efforts to solve the problem of moisture stress have been somewhat different. According to Specht and Williams (1978), plant physiological research on a fundamental plant process probably involved in stress responses often is quite specific, usually involving a limited number of genotypes. They term this a mechanistic approach. In contrast, plant breeding research, which they call the empirical approach, may involve the integration of many unidentified responses and is usually measured in terms of total yields for hundreds and often thousands of genotypes. The challenge, then, is for research in these two fields to become more complementary, with the observant plant breeder providing an array of genotypes for research and the plant physiologist providing evidence for traits correlated with moisture-stress response that can be screened for on a large scale.

RESPONSES TO DROUGHT STRESS

Plants respond to drought stress in a number of ways depending partly on the time of occurrence of the stress period. Thus, drought stress may lead to reduced germination, reduced seminal root growth and a delay in the establishment and development of a coronal root system. Foliar responses include stomatal closure, leaf rolling, wilting, and leaf senescence with their attendant effect on photosynthesis and shortening of the grain-filling period. Yield components are affected by failure to establish tillers, loss of tillers, and a reduction in spike yield-component values — reduced spikelet and floret number, and reduced kernel weight.

PERFORMANCE UNDER DROUGHT STRESS

The ability of a cultivar to produce satisfactory yields over a range of stress or nonstress environments is as important as its yield potential. Sojka et al. (1981) believe that "there is a high correlation between cultivar's seasonal plant water status and its yield, which is unique to each cultivar". The ability of a cultivar to perform satisfactorily in areas where moisture stress may be expected can be due to a number of mechanisms. Turner (1979) following the lead of May and Milthorpe identified three mechanisms: (a) drought escape; (b) drought tolerance with high tissue-water potentials — termed drought avoidance by many researchers; and (c) drought tolerance with low tissue-water potentials. Since some morphological and physiological attributes of the wheat plant relate to these mechanisms and may provide criteria for selection for drought resistance, they will be discussed briefly. This paper will use terminology provided by Turner (1979).

DROUGHT ESCAPE

Ray et al. (1974) suggest that a crop in semiarid regions, such as the Great Plains, has certain optimal seasonal growth patterns and is not always under drought stress. Therefore, breeders should select cultivars whose grain-producing periods coincide with the maximum seasonal precipitation pattern. Thus, the cultivar may often escape drought damage because of the timing of its grain-filling period. Mac Key (1966) points out that in semiarid regions, the best conditions for wheat growth usually will occur early in the growing season, and production will be correlated with those components of yield that develop early ontogenetically. Thus, tiller production would be a principal component of yield under those conditions. Reitz (1974) reported that for winter wheat in the Great Plains, earliness of cultivars (1 to 20 days earlier in anthesis than the late or moderately late 'Kharkof' cultivar) was worth 0.6 q/ha for each day earlier at Woodward, OK, 0.54 q/ha at Manhattan, KS, and from 0.6 to 1.2 q/ha at Lincoln, NE (q, quintal = 100 kg). Growers realized the importance of this value in the winter wheat area of the southern

Great Plains and were quick to adopt the earlier-maturing cultivars. However, once the optimum maturities have been established, no more progress is possible. Mac Key (1970) pointed out that earliness tends to reduce the total plant water requirement and may increase the grain to shoot ratio. Thus, while earliness may reduce the overall total plant production, harvestable grain yield may be increased. Derera et al. (1969) reported that in Australia, earliness in maturity accounted for 40 to 90% of the variation observed for drought tolerance.

Root development and rooting patterns also may provide escape mechanisms (classified as drought avoidance by many). Success of altered root systems will depend at least in part on seasonal precipitation patterns and location of soil moisture in the profile. Hurd (1976) and associates noted that 'Pelissier' durum had good early root growth and deep root development in western Canada and thus was in a better position to avoid late-season drought than other cultivars. This rooting pattern was an inherited trait which they transferred in crosses with 'Lakota' durum to lines later named 'Wascana' and 'Wakooma'. However, in regions where deep soil-moisture reserves are not present and the crop depends on precipitation received during the growing season, this trait would be of little value. Rate of root growth and timing is related to maturity (Atsmon, 1980). Thus, an early-maturing cultivar may exploit early season moisture while a late-maturing cultivar may better utilize deeper soil moisture (Asana and Williams, 1965). Tall cultivars tend to have deeper root systems than short cultivars, and tillering cultivars a more branched root system relying more on coronal roots than nontillering types but these correlations show considerable variation, (Mac Key, 1966, 1973; Welsh et al., 1975).

DROUGHT TOLERANCE

Drought tolerance with high tissue-water potential

Sensitivity of stomata to water stress has been investigated by many researchers. Adjei and Kirkham (1980) reported that stomatal resistance was higher in the drought-resistant 'Concho' hard red winter wheat cultivar than in 'Centurk' hard red winter wheat. They classified Centurk as drought sensitive. Shimshi and Ephrat (1975) reported that wheat cultivars with wider stomatal apertures were more productive without greater water use. Thus, while such cultivars were inclined to have lower stomatal resistance, they did not have high rates of water use even with high leaf permeability. Blum et al. (1981) observed that most of the avoidant genotypes maintained relatively open stomates. On the other hand, Jones (1979) reported that the values for leaf conductance during the critical period before anthesis had a strong negative correlation with yield. Teare et al. (1971) could not relate stomatal number to performance across various wheat species. Stomatal frequency was highest on the first leaf and decreased with each succeeding leaf through the fourth leaf. Jones (1979) points out that, while there are reports suggesting a relationship between lower stomatal frequency and drought resistance, this relationship would be effective only if stomatal size and leaf area are not allowed to increase. He found no evidence that stomatal frequency could be used to predict stomatal conductance (Jones, 1977). Finally, Begg and Turner (1976) point out that leaf transpiration is the major driving force in extracting water and nutrients from the soil by creating a lower water potential in the plant than in the soil.

Morphological features listed by Begg and Turner (1976) that may aid in drought tolerance with high water potential are accelerated leaf senescence and shedding, tiller death (Turner, 1979), reduction of leaf expansion and the increased reflectance of energy through leaf pubescence and epicuticular wax. Radiation interception may be reduced also by leaf orientation (Mac Key, 1968/69) and leaf rolling (Turner, 1979).

High water potential may sometimes be maintained by rooting depth and patterns (Turner, 1979). These were considered earlier as escape mechanisms because such rooting is part of optimum phenological development. Atsmon (1980) reported that the rate of root growth was directly related to earliness. Increased root weight due to either greater root depth or to a greater density of spreading roots near the surface may be indicative of the ability of a plant to maintain high water potential by greater extraction of soil water (Turner, 1979; Hurd, 1974; Mac Key, 1968/69). Mac Key (1978) indicated that in wheat the above-ground plant growth somewhat mirrors root development. Tall plants may mean deep roots or long roots. The main culm may be more dependent on the seminal roots while tillers may be more dependent on crown roots, often shallower but spreading, for their moisture supply. Thus, depending on the location of soil moisture supplies, high plant-water potential may be maintained by growing the genotype which has an appropriate rooting pattern. Further, if the number or diameter of xylem vessels is increased, reduced resistance to flow may result and high water potential may be maintained through some moisture stress periods (Turner, 1979). On the other hand, a reduction in the diameter of the main xylem vessel and an increase in the number of smaller metaxylem vessels can have a beneficial effect toward maintaining high water potential by distributing the available soil moisture over a longer period of time (Richards and Passioura, 1981a, b). Further, Fischer and Turner (1978) suggests that excessive rooting should be avoided because soil moisture reserves would be used prematurely.

Drought tolerance with low tissue-water potentials

Solute accumulation (Begg and Turner, 1976) resulting in osmotic potential adjustment (Turner, 1979) may allow plants to function and maintain turgor at lower water potentials. Most of the research has been with crops other than wheat. Turner (1979) reported that significant increases in soluble sugars in both sunflower and sorghum were associated with osmotic adjustment. Small cells, also, may be more effective in maintaining turgor (Turner, 1979).

SELECTION FOR DROUGHT RESISTANCE IN WHEAT

Most of the drought-resistance breeding in wheat has involved the escape mechanism or maintenance of high tissue-water potential. Most of such breeding activity has been indirect (Lewis and Christiansen, 1981). Cultivars that performed well over a range of environments including moisture stress were retained and poor performers discarded. Stability in this case was an important criterion. Finlay (1968) believed that stability over environments and yield potential are more or less independent of each other. Blum (1979) suggested that one means of breeding for increased performance under moisture-stress conditions might be to breed for superior yield under optimum conditions on the assumption that the best lines would perform well also under suboptimum conditions. Worrall et al. (1980) in reviewing wheat breeding progress at CIMMYT showed that there is some practical validity to this assumption. However, increases in yield in wheat are becoming increasingly more difficult to achieve. Sojka et al. (1981) point out that a high yield baseline which allows a cultivar to do well over a range of environments does not imply drought resistance. They say, "Drought resistance is better defined as the ability to minimize yield loss in the absence of optimal soil water availability". The objective should be to combine minimal yield loss due to drought stress with a high-yield baseline in a cultivar, and to provide the greatest possible water-use efficiency.

Lewis and Christiansen (1981) suggest that stress environments be selected at a level which will differentiate between stress-susceptible and stress-resistant genotypes. Christiansen (in Lewis and Christiansen, 1981) proposes that breeding for stress should involve; (a) identification of the critical point where stress occurs in a plant's development cycle and determination of how severe stress has to be before causing losses; (b) the stress responses should be quantified as much as possible; and (c) proper conditions should be created to identify possible genetic variants for response. Productive lines intermediate in their response may be the best choice. Blum (1979) believes that if cultivars are to yield well under stress situations they must be selected under those conditions. However, Gotoh and Chang (with O'Toole, 1979) point out that stress environments are highly variable and difficult to quantify. Therefore, selection for improved response may be very slow. Keim and Kronstad (1981) pointed out that similar yielding abilities under drought stress may be due to entirely different phenomena. The 'Wanser' and 'McDermind' cultivars were classified as drought tolerant because they were able to achieve productivity by maintaining a large number of productive tillers under high internal water stress. In contrast, the drought-resistant 'Yamhill' cultivar retained its productivity due to maintenance of a high water status under the same set of conditions. This result underscores the conclusions by Nass and Sterling (1981) that no one test is sufficient to establish drought resistance or susceptibility. Testing should include measurements of both avoidance and tolerance. Also timing of the tests as it relates to stage of plant development is important (Blum et al., 1981).

In 1974, Moss et al. stated that the information regarding water-use efficiency for use in selecting for specific physiological traits was not available or of practical value for the plant breeder. Blum (1979) was a bit more encouraging by stating that there was adequate information available but that the methodology required at that time was not applicable to the large populations in plant breeders' selection programs. Since then, a number of researchers have suggested that selection for certain physiological traits might be practical.

One physiological trait often mentioned relates to stomatal resistance. Kirkham (1978) and Adjei and Kirkham (1980) suggested that the diffusion porometer was useful in screening for stomatal resistance. Stomatal resistance was considerably higher in drought-resistant wheats then in drought-susceptible wheats. On the other hand, Jones (1979) reported rather small differences among varieties in their stomatal response to leaf water-potential and indicated that the various procedures used may not give straight forward results. Turner (1979) suggests that increases in stomatal resistance may reduce productive processes. Boyer and McPherson (1975) agree that drought-avoidance characters including stomatal resistance may reduce photosynthesis and thus reduce productivity. Centurk, the wheat cultivar classed as drought susceptible by Adjei and Kirkham (1980), may be stress prone under certain situations as shown by Schmidt et al. (1972) in Fig. 1. However, Centurk is a high-yielding cultivar and was the leading wheat cultivar in the United States in 1979 (Briggle et al., 1982). Thus, had Centurk been discarded because of low stomatal resistance early in the breeding program a highly productive and widely grown cultivar would have been lost. Kirkham and Smith (1978), also, suggest that in nonstress conditions short height cultivars were more productive than tall cultivars because of the lower tension with which water was held especially in the top leaves of the short cultivars. 'Pitic 62' (Dedio, 1975), a highly productive cultivar, has a high leaf water content under a range of soil moisture conditions. Pitic 62 is an excellent water retainer (as measured by cut leaf weight) and transmits water retention as a dominant heritable character in crosses. This increases the possibilities in a selection program. Blum (1980) suggests that drought-avoidant genotypes could be identified by measuring leaf-canopy temperature by infra-red techniques since water-stressed but avoidant wheat genotypes had a lower leaf temperature (correlated with higher leaf-water-potential) than nonavoidant genotypes. Overall, however, it does not appear that selection for high stomatal resistance as such will be utilized by many wheat breeding programs although stomatal traits have high heritability (Jones, 1979).

Other traits with high heritability are leaf waxiness and pubescence. Ex-

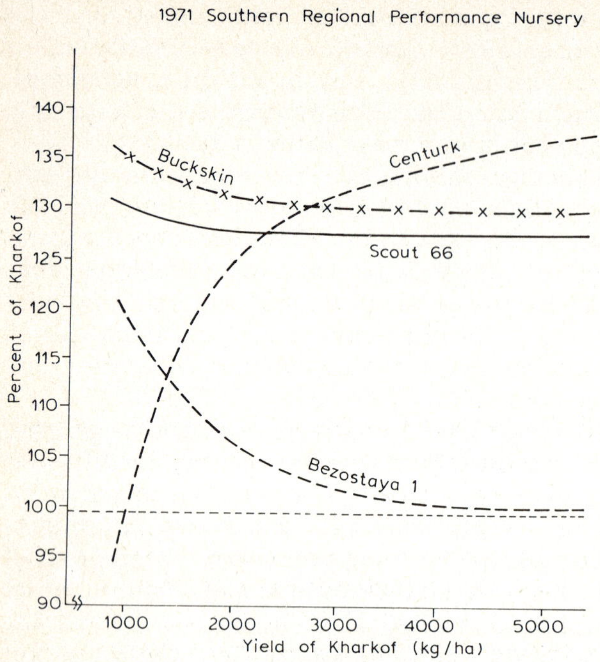

Fig. 1. Yield relationships of four winter wheat cultivars to the Kharkof check cultivar predicted from linear regression of the cultivar yields on nursery mean yields of the Southern Regional Hard Red Winter Wheat Performance Nursery in 1971. Scout 66 and Buckskin cultivars exhibit good stability at high yield levels. The Centurk cultivar is less stable but has a higher yield potential. The poor performance of Centurk under stress conditions often is related to exhaustion of soil moisture supplies by overtilling. From Schmidt et al. (1972).

cessive amounts of waxiness could change the spectral properties of leaves and interfere with photosynthesis. Our experience at the University of Nebraska has indicated to us that a goodly amount of cuticular wax is important. In crosses segregating for this trait, we have never had nonwaxy lines that equalled the performance of their waxy siblings. We have not worked with leaf and spike pubescence. Its presence would aid in heat dissipation, but if too extensive such pubescence could interfere with radiation absorption.

Leaf rolling and leaf senescence are traits readily observable in field plantings. The Nebraska cultivar 'HiPlains' and its progeny respond to drought situations with leaf rolling. With the exception of HiPlains, no other experimental lines with this characteristic have reached cultivaral status in our program. Similarly, early leaf-senescent lines have been dropped from testing early in the program usually for lack of yield.

Two other traits with value in stress-resistance programs and amenable to selection are maturity adjustment and presence of awns. The value of appropriate maturity adjustment was referred to earlier and is a key element in any breeding program. The value of awns under semiarid conditions is well documented by the work of Suneson et al. (1948), Atkins and Norris (1955) and the reviews by Grundbacher (1963) and Arnon (1972). In our program very few awnless lines ever reach the final yield-testing stage. Because this trait is so easily identifiable, it automatically has become a selection criterion for our area where moisture stress is common.

The problem of breeding for the most advantageous root system is compounded by the particular region's seasonal weather pattern and the place ment levels of moisture in the soil. In the Great Plains where we depend on both stored soil moisture (fallow area) and spring rainfall, the situation is quite different from that in other regions. Physiologists and breeders in western Europe and in Australia place much emphasis on the development of a good seminal root system. Mac Key (1966, 1973, 1978) indicates that the type of shoot development is mirrored to a large extent by the root system. Until recently, hard red winter wheat cultivars grown in the Great Plains tended to be tall, tillering types, probably dependent on good crown root development. Tomasovic (1980) examined the seminal root systems of a number of winter wheat cultivars and of these Centurk, a heavily tillering winter wheat, had the most poorly developed seminal root system. It obviously depends heavily on an extensive coronal root system. Fischer and Wood (1979) believe that it may be difficult to combine the drought resistance of the old tall dryland cultivars with the high yielding ability of modern wheats with reduced-height genes. Tandon and Saimi (1978) suggest that semitall lines may be the best compromise. Welsh et al. (1975) failed to establish any relationship between plant height and rooting patterns in studies with tall and semidwarf winter wheat cultivars and lines adapted to the Great Plains. Thus, selection for tall plant height carried no guarantee of greater drought resistance. Richards and Passioura (1981a) believed that under Australian conditions, where spring wheat production depends largely on stored soil moisture, selection for anatomical differences (small xylem vessel diameter) could prove beneficial in developing drought-resistant cultivars. They noted considerable variation in vessel number and diameter in seminal roots even in so-called homozygous lines. Eslick and Hockett (1974) pointed out that many useful genes remain unidentified because of the genotype in which they are located. Identification of such traits as suggested above by Richards and Passioura could be of significant value. Because selection for root characteristics is laborious, we should recall the suggestions by Hurd (1974) that, because of the difficulty involved, breeders should spend more time studying parental germplasm so that this can be combined to the best advantage. He lists these things that breeders should look for in choosing germplasm for breeding programs in semiarid regions: "1) extensive root system especially at lower depths, 2) resist a decay of the photosynthetic operation under stress, 3) have the ability to pursue productivity under water stress, and 4) grow fast at early stages and then build up a reserve of assimilates that can be used at later stages when stress occurs". I would agree but we must remind ourselves as the Scottish poet Robert Burns put it in *To A Mouse*:

"The best laid schemes o'mice and men
Gang aft agley."

After 30 years of wheat breeding, I would have to agree with the poet. Most of my planned 'can't miss' crosses were nonproductive in providing the ex pected superior lines while unique segregates from afterthought crosses made it interesting.

MULTILOCATIONAL BREEDING

Multilocational or regional testing was initiated by the USDA in cooperation with the state experiment stations in the early 1930's. This program has been of unestimable value in the Great Plains of the United States. It has served to identify cultivars that performed well almost every year over a range of environments including moisture-stress situations. This program has been and continues to be the most effective medium of identifying those cultivars described by Finlay (1964) "as having maximum genetic potential in the highest yielding environment and maximum phenotypic stability". W.J.R. Boyd (personal communication, 1980) pointed out that in Western Australia, under a very unpredictable moisture supply, "the result is that often the best variety, over seasons, is seldom the highest yielder in any one trial — there is adaptive advantage in mediocrity". Shebeski and Evans (1973) stated that it is paradoxical in that we realize the value of the regional and interregional testing in order to identify widely adapted cultivars but have failed to realize that the same need to identify adaptability applies to early generations as well. Andrews (1976) of the ICRISAT pearl millet breeding program agrees that "multilocational breeding should precede multilocational yield testing". Such breeding is not easily accomplished at many stations. Eberhart and Russell (1966) stated that, when multilocational testing was not practical, "extra locations" be created at one site by use of planting dates, and the various rates of fertilizer and moisture applications. Hanks et al. (1976) described line sprinkler irrigation systems that can be used to create a continuous water gradient for studying genotype by moisture-level interaction at one location. Or, water can be withheld to varying degrees by placing plastic sheets or rain gutters between rows or by using overhead plastic or fiberglass coverings.

W.J.R. Boyd (personal communication, 1980) in Western Australia selects first for optimum maturity in space-planted F_2 populations. These selected F_2 plants are bulked and yield tested as mass-selected F_3 and F_4 bulks in order to build up to 10 locations-by-season interactions. Lupton (1970) suggested a similar approach by increasing F_2-derived bulks for use in yield testing. Further selection is then restricted to the best performing F_2-derived lines. While at Colorado State University, J.R. Welsh (personal communication, 1975) instituted a system of planting for yield testing early generation bulks at a number of locations with lower elevation and then could make his selections in the base population grown at the higher elevation at Fort Collins after assessing outstate yield data. This neat arrangement is, unfortunately, not available to many breeders. In fact, at the University of Nebraska, plots must be harvested at the main station before any outstate station plots can be harvested because our elevation situation is the exact reverse of that in Colorado.

Because of the limitation already referred to above, and the winter survival problems with space seeding of F_2 populations, we at the University of Nebraska have adopted a limited bulk breeding program with major emphasis on spike selection in the F_3 bulks. Consequently, most cultivars from the Nebraska/USDA cooperative program are F_3-derived lines. This procedure allows us to evaluate from 45 000 to 60 000 F_4 selections and 2 000 to 3 000 of their F_5 progeny at one location. Only by the F_6 generation are we able to obtain up to five location interactions. However, the outstate locations have good area distribution, two of which are in the lower precipitation (400 mm) part of the state. For us, selecting in the F_3 generation has two basic values:

(1) it is financially advantageous for us to select only once in a population; and

(2) if we select in just one generation, we believe the F_3 is the optimum one — a compromise that provides us with (a) a derived population with the largest number of favorable genes (homozygous or heterozygous — Shebeski and Evans, 1973) and with reasonable uniformity; and (b) populational buffering (Allard and Bradshaw, 1964) in our variable Great Plains climate.

SUMMARY

A tremendous amount of research has been reported dealing with drought resistance in crop plants, especially wheat. The results are leading to a much better identification of the many facets in this complex we call drought resistance. There is a growing appreciation by the plant breeders and the plant physiologists for their necessary interaction in moving ahead in this field. Techniques adaptable for large-scale selection methods are still not at a very practical level. Plant breeders must seek to relate the physiological research to their breeding problems and the plant physiologists must expand their plant genotype concept a thousandfold. Plant breeding unfortunately is still largely a numbers game.

REFERENCES

Adjei, G.B. and Kirkham, M.B., 1980. Evaluation of winter wheat cultivars for drought resistance. Euphytica, 29: 155—160.

Allard, R.W. and Bradshaw, A.D., 1964. Implications of genotype-environment interactions in applied plant breeding. Crop Sci., 4: 503—507.

Andrews, D.J., 1976. Multilocational breeding. Paper presented at the All-India Coordinated Millet Improvement Project (AICMIP) Annual Workshop, Gwalior, India (By permission), 4 pp.

Arnon, I., 1972. Crop production in dry regions. In: Systematic Treatment of the Principal Crops, Vol. II. Barnes & Noble, New York, NY, pp. 1—72.

Asana, R.D. and Williams, R.F., 1965. The effect of temperature stress on grain development in wheat. Aust. J. Agric. Res., 16: 1--13.

Atkins, I.M. and Norris, M.J., 1955. The influence of awns on yield and certain morphological characters of wheat. Agron. J., 47: 218—220.

Atsmon, D., 1980. Attempts to breed small grains for increased drought resistance. In: Proc. 3rd International Wheat Conference, 22 May—23 June, Madrid. Nebr. Agric. Exp. Stn Misc. Publ. 41, p. 655.

Begg, J.E. and Turner, N.C., 1976. Crop water deficits. Adv. Agron., 28: 161—217.

Blum, A., 1979. Genetic improvement of drought resistance in crop plants: a case for sorghum. In: H. Mussell and R.C. Staples (Editors), Stress Physiology in Crop Plants. Wiley Interscience, New York, NY, pp. 429—445.

Blum, A., 1980. Drought avoidance in wheat and its rapid estimation by remote infra-red thermal leaf canopy measurements. In: Proc. 3rd International Wheat Conference, 22 May—3 June, Madrid, Nebr. Agric. Exp. Stn. Misc. Publ. 41, p. 643.

Blum, A., Gozlan, G. and Mayer, J., 1981. The manifestation of dehydration avoidance in wheat breeding germplasm. Crop Sci., 21: 495—499.

Boyer, J.S. and McPherson, H.G., 1975. Physiology of water deficits in cereal crops. Adv. Agron., 27: 1—23.

Briggle, L.W., Straus, S.L., Hamilton, D.E. and Howse, G.H., 1982. Distribution of the varieties and classes of wheat in the United States in 1979. USDA Stat. Bull. 676, 107 pp.

Dedio, W., 1975. Water relations in wheat leaves as screening tests for drought resistance. Can. J. Plant Sci., 55: 369—378.

Derera, N.F., Marshall, D.R. and Balaam, L.N., 1969. Genetic variability in root development in relation to drought tolerance in spring wheats. Exp. Agric., 5: 327—337.

Eberhart, S.A. and Russell, W.A., 1966. Stability parameters for comparing varieties. Crop Sci., 6: 36—40.

Eslick, R.F. and Hockett, E.A., 1974. Genetic engineering as a key to water-use efficiency. Agric. Meteorol., 14: 13—23.

Finlay, K.W., 1964. Adaptation — its measurement and significance in barley breeding. In: Barley Genetics I. First International Barley Genetics Symposium, 1963, Wageningen, The Netherlands, pp. 351—359.

Finlay, K.W., 1968. The significance of adaptation in wheat breeding. In: Proc. 3rd International Wheat Genetics Symposium, 5—9 August 1968, Australian Academy of Sciences, Canberra, A.C.T., pp. 403—409.

Fischer, R.A. and Turner, N.C., 1978. Plant productivity in the arid and semiarid regions. Annu. Rev. Plant Physiol., 29: 277—317.

Fischer, R.A. and Wood, J.T., 1979. Drought resistance in spring wheat cultivars. III. Yield associations with morpho-physiological traits. Aust. J. Agric. Res., 30: 1001—1020.

Gotoh, K. and Chang, T.T. (In collaboration with J.C. O'Toole), 1979. Adaptation to drought-prone conditions. In: J. Sneep and A.J.T. Hendriksen (Editors), Plant Breeding Perspective (Crop Adaptation Section). Wageningen, The Netherlands, pp. 243—261.

Grafius, J.E., 1971. Stress: a necessary ingredient of genotype by environment interaction. In: Barley Genetics II. Second International Barley Genetics Symposium. Washington State University Press, Pullman, WA, pp. 346—355.

Grundbacher, E.J., 1963. The physiological function of the cereal awn. Bot. Rev., 29: 366—381.

Hanks, R.J., Keller, J., Rasmussen, V.P. and Wilson, G.D., 1976. Line source sprinkler for continuous irrigation – crop production studies. Soil Sci. Soc. Am. J., 40: 426—429.

Hurd, E.A., 1974. Phenotype and drought tolerance in wheat. Agric. Meteorol., 14: 39—55.

Hurd, E.A., 1976. Plant breeding for drought resistance. In: T.T. Kozlowski, Water Deficits and Plant Growth. Vol. 4, Soil and Water Measurements, Plant Response, and Breeding for Drought Resistance. Academic Press, New York, NY, pp. 317—383.

Jones, H.G., 1977. Aspects of the water relations of spring wheat *(Triticum aestivum* L.) in response to induced drought. J. Agric. Sci., Camb., 88: 267—282.

Jones, H.G., 1979. Stomatal behavior and breeding for drought resistance. In: H. Mussell and R.C. Staples (Editors), Stress Physiology in Crop Plants. Wiley Interscience, New York, NY.

Keim, D.L. and Kronstad, W.E., 1981. Drought response of winter wheat cultivars grown under field stress conditions. Crop Sci., 21: 11—15.

Kirkham, M.B., 1978. Water potential and turgor pressure as a selection basis for wind-grown winter wheat. Agric. Water Manage., 1: 343—349.

Kirkham, M.B. and Smith, E.L., 1978. Water relations of tall and short cultivars of winter wheat. Crop Sci., 18: 227—230.

Lewis, C.F. and Christiansen, M.N., 1981. Breeding plants for stress environments. In: K.J. Frey (Editor), Plant Breeding II. Proc. Plant Breeding Symposium, 12- -16 March 1979, Ames, IA. Iowa State University Press, Ames, IA, pp. 151—177.

Lupton, F.G.H., 1970. Testing and selection procedures in early generations of wheat crosses. In: Proc. 3rd FAO/Rockefeller Foundation Wheat Seminar, 29 April—10 May, Ankara. FAO, Rome, pp. 178—182.

Mac Key, J., 1966. The wheat plant as a model in adaptation to high productivity under different environments. In: Fifth Yugoslav Symposium on Research in Wheat, Novi Sad, pp. 37—48.

Mac Key, H., 1968/69. Ecological adaptation of the yield structure in cereals. Melhoramento, 21: 343—363.

Mac Key, J., 1970. An ecological model for yield in small grains. In: Seminar Series, Department of Agronomy, Iowa State University, Ames, IA, pp. 128—149.

Mac Key, J., 1973. The wheat root. In: Proc. 4th International Wheat Genetics Symposium, Columbia, MO, pp. 827—842.

Mac Key, J., 1979. Wheat domestication as a shoot: root interrelation process. In: Proc. 5th International Wheat Genetics Symposium, 23—28 February 1978, New Delhi, pp. 875—890.

Moss, D.N., Woolley, J.T. and Stone, J.F., 1974. Plant modification for more efficient water use: the challenge. Agric. Meteorol., 14: 311—320.

Nass, H.G. and Sterling, J.D.E., 1981. Comparison of tests characterizing varieties of barley and wheat for moisture stress resistance. Can. J. Plant Sci., 61: 283—292.

Ray, L.L., Wendt, C.W., Roark, B. and Quisenberry, J.E., 1974. Genetic modification of cotton plants for more efficient water use. Agric. Meteorol., 14: 31—38.

Reitz, L.P., 1974. Breeding for more efficient water use — is it real or a mirage? Agric. Meteorol., 14: 3—11.

Richards, R.A. and Passioura, J.B., 1981a. Seminal root morphology and water use of wheat. I. Environmental effects. Crop Sci., 21: 249—252.

Richards, R.A. and Passioura, J.B., 1981b. Seminal root morphology and water use of wheat. II. Genetic variation. Crop Sci., 21: 253—255.

Schmidt, J.W., Johnson, V.A. and Stroike, J.E., 1972. The case for general adaptation. In: Proc. 1st International Winter Wheat Conference, 5—9 June 1972, Ankara. Nebr. Agric. Exp. Stn. Misc. Publ. 28, pp. 47—58.

Shebeski, L.H. and Evans, L.E., 1973. Early-generation selection for wide-range adaptability in the breeding program. In: Proc. 4th International Wheat Genetics Symposium, 6—11 August 1973, Agricultural Experiment Station, University of Missouri, Columbia, MO, pp. 587—593.

Shimshi, D. and Ephrat, J., 1975. Stomatal behavior of wheat cultivars in relation to their transpiration, photosynthesis, and yield. Agron. J., 67: 326—331.

Sojka, R.E., Stolzy, L.H. and Fischer, R.A., 1981. Seasonal drought response of selected wheat cultivars. Agron. J., 73: 838—845.

Specht, J.E. and Williams, J.H., 1978. Testing soybeans for heat and drought tolerance. In: Proc. 8th Soybean Seed Research Conference, Chicago, IL, pp. 17—32.

Suneson, C.A., Bayles, B.B. and Fifield, C.C., 1948. Effects of awns on yield and market qualities of wheat. USDA Circ. 783.

Tandon, J.P. and Saimi, J.P., 1979. Rooting pattern in wheat genotypes. In: Proc. 5th International Wheat Genetics Symposium, 23—28 February 1978, New Delhi, pp. 845—851.

Teare, I.D., Peterson, C.J. and Law, A.G., 1971. Size and frequency of leaf stomata in cultivars of *Triticum aestivum* and other *Triticum* species. Crop Sci., 11: 496—498.

Tomasovic, S., 1980. Number of primary rootlets in wheat, a potentially useful trait for breeding improvement. In: Proc. 3rd International Wheat Conference, 22 May—23 June, Madrid. Nebr. Agric. Exp. Stn. Misc. Publ. 41, pp. 644—654.

Townley-Smith, T.F. and Hurd, E.A., 1979. Testing and selecting for drought resistance in wheat. In: H. Mussell and R.C. Staples (Editors), Stress Physiology in Crop Plants. Wiley Interscience, New York, NY, pp. 447—464.

Turner, N.C., 1979. Drought resistance and adaptation to water deficits in crop plants. In: H. Mussell and R.C. Staples (Editors), Stress Physiology in Crop Plants. Wiley Interscience, New York, NY, pp. 343—372.

Welsh, J.R., Cholick, F., Holbrook, F. and Pepe, J., 1975. Root studies of semidwarf and tall winter wheats. In: Proc. 2nd International Winter Wheat Conference, 9—19 June 1975, Zagreb. Nebr. Agric. Exp. Stn. Misc. Publ. 32, pp. 206—220.

Worrall, W.D., Scott, N.H., Klatt, A.R. and Rajaram, S., 1980. Performance of CIMMYT wheat germplasm in optimum and suboptimum production environments. In: Proc. 3rd International Wheat Conference, 22 May—23 June, Madrid. Nebr. Agric. Exp. Stn. Misc. Publ. 41, pp. 5—29.

Agricultural Water Management, 7 (1983) 195—205
Elsevier Science Publishers B.V., Amsterdam – Printed in The Netherlands

GENETIC AND PHYSIOLOGICAL RELATIONSHIPS IN PLANT BREEDING FOR DROUGHT RESISTANCE

A. BLUM

Institute of Field and Garden Crops, Agricultural Research Organization, The Volcani Center, POB 6 Bet Dagan (Israel)

(Accepted 23 November 1982)

ABSTRACT

Blum, A., 1983. Genetic and physiological relationships in plant breeding for drought resistance. *Agric. Water Manage.*, 7: 195—205.

The genetic improvement of plant drought resistance by the selection for yield under stress is a possible but a prolonged and problematic procedure. Recent developments in the understanding of the physiological responses of plants to water stress and their associations with plant productivity allow to embark upon experimental selection programs that employ physiological selection criteria for drought resistance. This is supported by the recent developments of rapid selection techniques for several physiological components of drought resistance.

It is, however, emphasized that such selection programs should be approached as the best available approximation, rather than established proven routines, for each crop and environment. The deployment of such experimental selection systems is important not only as a probable base for progress in selection but also as a feedback generator of important genetic and physiological information, to be used for further improvement of the selection systems.

Several rapid physiological selection techniques are briefly reviewed, and examples of their logistic incorporation into a routine breeding program are given.

INTRODUCTION

The physiological and genetical aspects of the improvement of drought resistance in crop plants were subjected recently to several comprehensive reviews (e.g., Mussell and Staples, 1979; Kramer and Raper, 1983; Turner and Kramer, 1980; Blum, 1981; Christiansen and Lewis, 1981; Paleg and Aspinall, 1981). The present state of the art allows several conclusions to be drawn.

Yield performance of a genotype under stress is a reflection of both its response to stress and its potential yield level. This conclusion led many plant breeders to believe that the genetic improvement of yield potential will also result in improved yields under stress. While this concept has resulted in apparent empirical improvement of yields also under sub-potential environ-

0378-3774/83/$03.00

ments, its relevance diminishes when breeders face a genetic yield plateau in many of our crops. A direct treatment of sub-potential yield levels by breeding programs is therefore essential, and the question immediately raised is whether drought resistance exists and can be measured. The positive answer is deliberated upon in several of the cited reviews.

DROUGHT RESISTANCE

Drought resistance can be defined in physiological or agronomic terms. The agronomic definition is of major importance to the breeder. One definition of drought resistance in agronomic terms, as derived from wheat research is exemplified here.

Yield data on several wheat varieties were obtained from 31 location × year tests in Israel. Water-use efficiency was regressed against the amount of rainfall (and irrigation) received in each test. Results for three typical varieties are presented in Fig. 1 (see also, Blum, 1981). 'Cajeme' represents varieties with good yield potential, being responsive to the improvement in the water regime, but susceptible to water stress (deterioration of yield as water supply diminishes). 'Miriam' represents varieties that are comparatively drought resistant but deficient in their yield potential. '676' represents varieties with both relatively good drought resistance and yield potential.

When the same yield data are used in a 'stability analysis' (Finlay and Wilkinson, 1963) (Fig. 2), the same relative varietal responses are obtained, with respect to a changing environmental index (mean yields of tests) (see also Blum, 1981). It follows that the major variable affecting the environmental index in these tests was the water regime, as already documented for wheat in Israel (Lomas and Shashoua, 1973). Thus, under our conditions, the common 'stability analysis' can be used to define drought resistance in

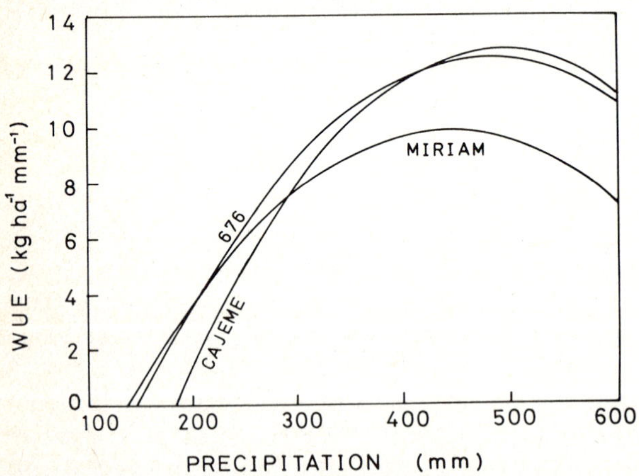

Fig. 1. Water-use efficiency (WUE) as related to amount of precipitation (and irrigation) received over 31 field tests in three common wheat varieties.

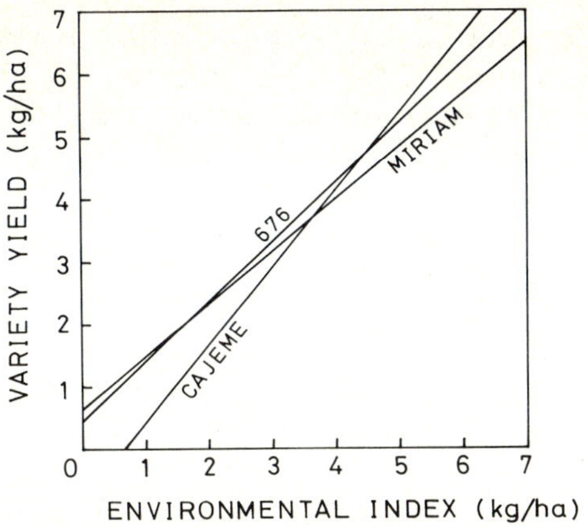

Fig. 2. Stability analysis (linear regression of variety yields on mean yields of 31 tests), in three common wheat varieties.

agronomic terms, independently of the yield potential, of any given wheat variety. The *a* and *b* coefficients of the linear regression are used to indicate relative drought resistance and yield response to improved environments (mainly the water regime), respectively.

Drought resistance, therefore, exists and is independent of the yield potential. The improvement of drought resistance by selection for yield under stress is a possible but prolonged and expensive task, requiring massive plant populations (Townley-Smith and Hurd, 1979). The difficulties are in the low heritabilities for yield and yield components when population yield is reduced by stress. In addition, varieties resulting from yield selection only under stress will most likely yield rather poorly under favorable conditions (Rosielle and Hamblin, 1981).

An alternative is selection based on physiological drought adaptive attributes in conjunction with a selection for high yield potential under a non-stress (potential) environment. The target is a variety with high yield potential, carrying physiological buffering factors against the effects of water deficits.

Physiological components of drought resistance

Recent research (see reviews) indicates that ample genetic variation exists in several crop plants for physiological components of drought resistance, even among well-performing agronomic breeding lines (Blum et al., 1981) or cultivars (Boyer et al., 1980; Sojka et al., 1981). Our instinctive resort to exotic non-adapted germplasm in this respect (e.g., Quisenberry et al., 1981) is not obligatory, unless the existent genetic resources are positively exhausted.

Genetic variation exists in the following drought-adaptive attributes: 'dehydration avoidance' (maintenance of relatively higher leaf water potential under conditions of soil moisture stress), osmotic adjustment, tolerance in plant or organ growth rate, plant recovery upon rehydration, tolerance in the photosynthetic system or its components, tolerance is enzyme activities, tolerance in translocation, stability of the cellular membranes, proline accumulation, root growth attributes and plant developmental or morphological attributes, such as leaf size, leaf area per plant, leaf orientation, tiller survival, epicuticular wax content, and organ pubescence.

However, this volume of information, that currently increases with time, is of limited value to actual breeding work, for two main reasons:

First, while the existance of genetic variation for a given drought-resistance attribute can be indicated by using the standard accurate but slow physiological methodology, this methodology is impossible for routine screening work en masse. The physiological methodology was developed with a major concern for accuracy. Routine screening methodology is heavily concerned with ease and speed of performance even at the expense of accuracy, if necessary. While physiological work requires accurate and absolute terms, selection work involves the probabilities and frequencies of population breakdown, often in relative terms. The breeder's main interest is in efficiently reducing population size, even with a physiologically unacceptable margin of error. This would be partially compensated by the option for repeated testing in subsequent generations. The margin of acceptable error in physiological screening methods can be determined experimentally. Such margin would depend on the method used as well as on the genetic structure of the tested populations.

Second, there is a very serious lack of information on the relationships between given physiological drought adaptive traits and economic yield within defined stress and non-stress environments. It is extremely difficult today to formulate a plant 'ideotype' in terms of the physiological responses required for yield buffering in defined or non-defined drought environments. This is where plant modelling work should attain its glory. In the meanwhile, decision on the physiological selection criteria to be used, in this context, is difficult. In addition, recent evidence (Blum, 1979; Sammons et al., 1980; Nass and Sterling, 1981) suggest that no singular drought-adaptive trait is predictive of plant response to stress and that multiple physiological selection criteria are required.

With this state of understanding, our attempt to incorporate physiological selection criteria for drought resistance in the wheat breeding program in Israel was initiated on an experimental basis as the best possible approximation. The program is based on the assumption that the improvement of drought resistance can be enhanced beyond the empirical approach by applying multiple physiological selection criteria for selection among high yielding families. This program, by all means, consists of a transitional program that changes as results are being developed. This approach has two main advantages:

(1) If selection were based on erroneous assumptions or poor techniques,

no damage is incurred upon the mainstream of the program, because the routine of improving yield by the selection for yield is still performed. This would respond to the serious criticism offered by Reitz (1974) on the dangers in pursuing a mirage due to misconceptions in the selection program.

(2) The results that emerge from each generation of selection constitute a valuable feedback that may be used for further improvements in the program.

The selection criteria chosen for such a selection program depend on the availability of a rapid screening technique and the relationships between attribute evaluated by a particular technique and plant productivity under stress. For example, seed germination in osmoticum is a rapid technique for a response to water stress that does not bear any relationship to plant production (Nass and Sterling, 1981) or even juvenile plant growth under stress (Blum et al., 1980).

Selection for drought resistance

Following is a brief review of some possible selection criteria for drought resistance and associated rapid methodologies, all of which are, or are in the process of being, documented.

DEHYDRATION AVOIDANCE

Superior 'dehydration avoidance' was positively associated with yield performance under stress, across genotypes of wheat (Keim and Kronstad, 1981; Sojka et al., 1981), soybeans (Boyer et al., 1980) and rice (O'Toole and Chang, 1979).

This attribute is commonly measured by midday leaf water potential, under conditions of soil water stress, using the pressure chamber as a field method. However, the pressure chamber is too slow a method for selection work. An indirect method for 'dehydration avoidance' was developed for rice (O'Toole and Chang, 1979), using a visual scoring of the rate of leaf rolling under conditions of soil water stress.

Aerial infrared photography of drought-stressed nurseries was developed as an indirect method for sorghum (Blum et al., 1978). Canopies of relatively stressed ('non-avoidant') genotypes appear in the infrared photograph at relatively lower rate of color saturation, as compared with less stressed genotypes. This color difference is correlated with leaf water potentials and can be readily quantified either visually or instrumentally.

Infrared sensing of midday canopy temperatures of water-stressed wheat plots, using the infrared thermometer, was developed as another indirect method for small grains (Blum et al., 1982). By using an appropriate planting design and by mounting the thermometer and a data recording instrument on a vehicle, hundreds of entries can be evaluated within one hour. The method is used on a routine basis in our wheat breeding program.

The rate of leaf death ('leaf firing') under stress is occasionally employed as a screening test (Castleberry, 1980). This plant response may involve both 'dehydration avoidance' as well as heat tolerance of the dehydrated leaf tissues.

TOLERANCE TO POST-ANTHESIS STRESS

In several of the cereals, grain filling depends partly on transient photosynthesis and partly on stored carbohydrates translocated from the vegetative plant parts. When the photosynthetic source is limited by stress, the relative and absolute proportion of translocated stored stem reserves as a source for grain filling becomes larger with a significant genetic variation in this respect (e.g., Boyer, 1976; Austin et al., 1980).

To estimate the genotypic capacity of wheat for grain filling by translocation of stem reserves in the absence of transient photosynthesis, plots of tested wheats were sprayed with a chemical desiccant. The treatment (magnesium chlorate at 14 days after anthesis) killed all photosynthetic tissues including leaves, leaf-sheaths, glumes and awns. The final kernel weight of treated plots was compared with kernel weight in non-treated (control) plots, and the rate of injury to grain filling by the treatment was estimated. The extent of injury to grain filling by chemical desiccation was correlated across various wheats with the extent of injury to grain filling by post-anthesis drought stress, over 2 years of testing. The method seems therefore predictive of wheat tolerance to post-anthesis drought stress, in terms of grain filling, as based on the mobilization of stem reserves into the growing grains. This method is used on a routine basis in our wheat breeding program.

STABILITY OF THE CELLULAR MEMBRANES

The cellular membranes are a central physiological site mediating the effect of various environmental stresses on the plant cell. A relatively simple and rapid in vitro method is available for assessing tolerance to dehydration (or heat) stress, in terms of stability of cellular membrane (Sullivan and Eastin, 1974; Blum and Ebercon, 1981). The method is based on the measurement of the rate of ion efflux from leaf discs into deionized water, after having been water-stressed by incubation in polyethylene glycol solution. Information is lacking on the selective value of the method, in terms of plant production. Significant genetic variation and high-heritabilities were observed for heat tolerance in soybeans tested by this method (Martineau et al., 1979). Selection for heat tolerance by using this method resulted in improved sorghum yield performance under the hot dryland conditions of Nebraska (Sullivan and Ross, 1979). The method is used in our program for the initial evaluation of dehydration tolerance of potential germplasm resources. Fig. 3 demonstrates the variation observed, in this respect, among populations of a wild wheat progenitor (*T. dicoccoides*), progenies of its crosses onto cultivated wheats, and some Israeli standard wheat cultivars.

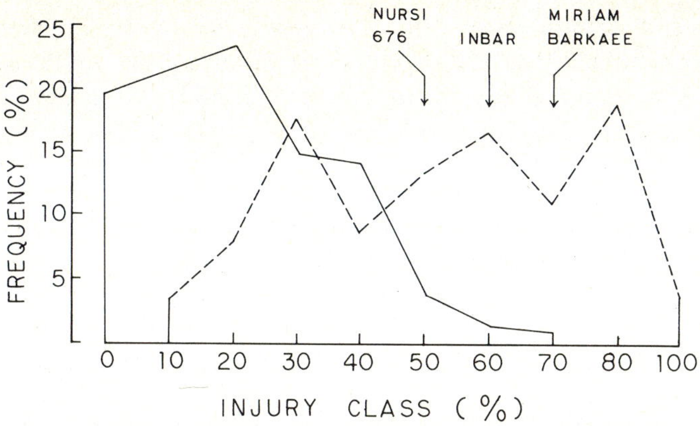

Fig. 3. Frequency distribution of the mean rate of injury to cellular membranes by dehydration of 206 populations of wild emmer (*T. dicoccoides*) (solid line), 91 derivatives of wild emmer × wheat cross (dashed line), four standard Israeli wheat varieties ('Miriam', 'Barkaee', 'Inbar', '676') and one landrace of *T. durum* ('Nursi').

PLANT GROWTH AND RECOVERY

Whatever may be the important physiological components of drought resistance, their occurrence at any given rate and threshold or frequency in a given genotype should be integrated in the plant's growth response to water stress. Simple tests can be developed for assessing plant tolerance in terms of growth rate under stress, by measuring the rate of total dry matter production (Blum et al., 1980) or by linear measurements of plant organs. The test used to some extent in our program, consists of juvenile plant growth response in nutrient solutions containing polyethylene glycol (6000) (stress), as compared with pure nutrient solutions (non-stress).

While tolerance in juvenile plant growth is a relevant test in relation to post-emergence drought stress, tolerance in terms of total vegetative plant production under stress was unrelated in wheat to grain production under stress (Blum et al., 1983). Thus, the use of 'harvest index' alone as a selection index under stress is questionable, because a high 'harvest index' can be obtained with low yielding genotypes that are very susceptible to drought stress during vegetative growth.

Plant growth recovery upon rehydration after stress may constitute an important adaptive trait in certain drought environments. Recovery of growth may proceed from the apical meristems of the desiccated plant or from dormant buds that give rise to new plant branches or shoots, such as basal tillers. Dormant buds and meristems are relatively more drought tolerant than differentiated plant organs (Barlow et al., 1980). The options for recovery from dormant buds and meristems, may or may not exist in given genotypes. The most conspicuous difference in this respect exists between determinate and indeterminate plant types, as well as between early- and late-maturing cereals. The relationships between plant recovery after stress and final crop yield are not simple ones. These associations depend on the subsequent seasonal environment. For example, it is apparent that with a very short frost-free season, plant recovery after a prolonged "dormancy" under stress will not result in any grain yield because flowering or maturation may take place under frost conditions.

Plant recovery after stress, especially in juvenile plants, can be easily tested under controlled conditions. Ample information is documented on the various techniques employed (Sullivan and Eastin, 1974; O'Toole and Chang, 1979; Blum et al., 1980).

A SCHEMATIC SELECTION PROGRAM

Several selection systems can be developed using any of the above physiological selection criteria for drought resistance. Based on the principles of the previous discussion, the proposed schematic outline (Fig. 4) retains the established routine of selection for yield under optimal conditions, and employs multiple physiological selection criteria for stress response.

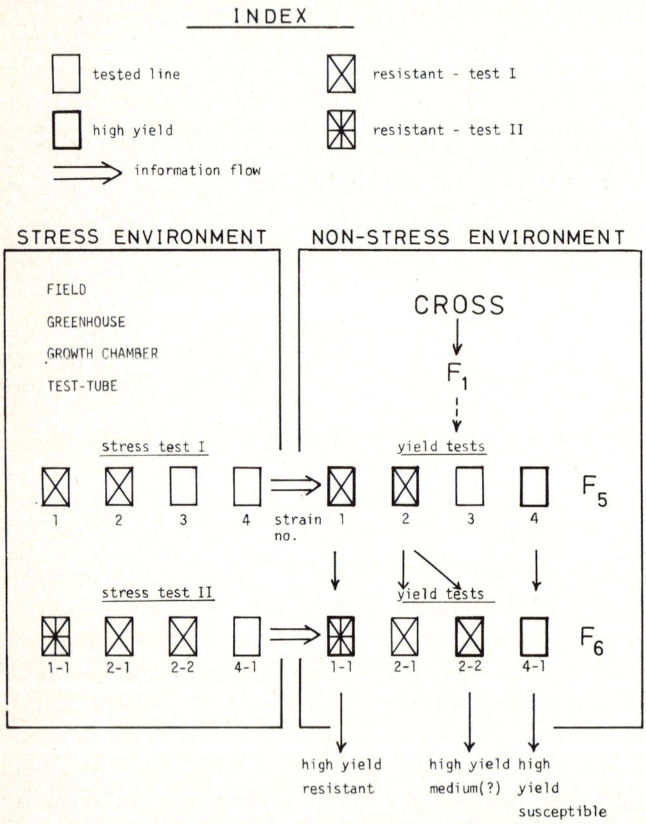

Fig. 4. A schematic outline of a pedigree selection program that combines selection for yield under non-stress conditions (right) and multiple physiological selection criteria for drought resistance (left).

Selection for agronomic types, yield components, and high yield potential is performed under non-stress conditions (Fig. 4, righthand side). Duplicate of F_5 (or F_4) families are then submitted to one or more tests for drought response in the field, greenhouse, growth chamber, or test tube (Fig. 4, lefthand side). Only one test is performed in this example. No selection is performed under stress. The test results are used for deciding on selection within or among families grown under the non-stress conditions. The F_6 selections are grown under non-stress conditions, and the routine is repeated with additional testing for drought resistance. Towards the F_7 (or any other advanced generation), lines are identified as having high yield potential with or without given physiological drought resistance attributes. Only the most resistant lines are then tested in regional tests subjected to drought conditions, while other (non-resistant) lines are retained and included in irrigated tests. Some of the lines may be evaluated in both tests, depending on their numbers and the work load.

In our current wheat breeding program the procedure described above has been adopted, using two selection criteria; infrared thermal sensing of wheat canopies in a stress nursery (F_5) and post-anthesis chemical desiccation of plants (F_6). The limitation to expanding the drought testing program beyond these two physiological screening tests is not lack of knowledge or methodology but rather lack of land and manpower. Initial results indicate that many of the selected lines show a consistent response for each of the two physiological tests. Their value as productive varieties under stress remains to be tested in the next few years.

CONCLUSIONS

Drought resistance can be identified and measured in both agronomic and physiologic terms. While an appreciable number of plant physiological attributes were identified as components of drought resistance, their interrelationships with plant productivity are unclear and should be investigated.

It is believed that the current information on the physiology of drought resistance allows breeders to embark upon selection programs for drought resistance. Such programs would consist of the best available approximation. If the logistics of such programs are well integrated with the mainstream effort to improve or sustain high yield potential, then progress may be expected.

A complete understanding of the physiology and genetics of drought resistance is not foreseen in the near future. Still, breeders should embark upon selection programs for drought resistence, which may be defined as experimental. Even if the expected progress is not attained by such selection experiments, the resulting information should be extremely valuable as guidance for future research and development.

REFERENCES

Austin, R.B., Morgan, R.B., Ford, M.A. and Blackwell, R.D., 1980. Contributions to grain yield from pre-anthesis assimilation in tall and dwarf barley genotypes in two contrasting seasons. Ann. Bot., 45: 309—319.

Barlow, E.W.R., Munns, R.E. and Brody, C.J., 1980. Drought response of apical meristems. In: N.C. Turner and P.J. Kramer (Editors), Adaptation of Plants to Water and High Temperature Stresses. Wiley Interscience, New York, NY, pp. 191—205.

Blum, A., 1979. Genetic improvement of drought resistance in crop plants: a case for sorghum. In: H. Mussell and R.C. Staples (Editors), Stress Physiology in Crop Plants. Wiley Interscience, New York, NY, pp. 429—445.

Blum, A., 1981. Evidence for genetic variability in drought resistance and its implication in plant breeding. Symp. Principles and Methods in Crop Improvement for Drought Resistance: With Emphasis on Rice, May 4—8, International Rice Research Institute, Los Baños, Philippines, 30 pp.

Blum, A. and Ebercon, A., 1981. Cell membrane stability as a measure of drought and heat tolerance in wheat. Crop Sci., 21: 43—47.

Blum, A., Schertz, K.F., Toler, R.W., Welch, R.I., Rosenow, D.T., Johnson, J.W. and Clark, L.E., 1978. Selection for drought avoidance in sorghum using aerial infrared photography. Agron. J., 70: 472—477.

Blum, A., Sinmena, B. and Ziv, O., 1980. An evaluation of seed and seedling drought tolerance screening tests in wheat. Euphytica, 19: 727—736.

Blum, A., Gozlan, G. and Mayer, J., 1981. The manifestation of dehydration avoidance in wheat breeding germplasm. Crop Sci., 21: 495—499.

Blum, A., Mayer, J. and Gozlan, G., 1982. Infrared thermal sensing of plant canopies as a screening technique for dehydration avoidance in wheat. Field Crops Res., 5: 137—146.

Blum, A., Mayer, J. and Gozlan, G., 1983. Associations between plant production and some physiological components of drought resistance in wheat. Plant Cell Environ., 6: 219—225.

Boyer, J.S., 1976. Photosynthesis at low water potentials. Phil. Trans. R. Soc. London B., 273: 501—512.

Boyer, J.S., Johnson, R.R. and Soupe, S.G., 1980. Afternoon water deficits and grain yields in old and new soybean cultivars. Agron. J., 72: 981—986.

Castleberry, R.M., 1980. Breeding programs for stress tolerance in corn. In: Workshop on Crop Reactions to Water and Temperature Stresses in Humid Temperate Climates. October 13—15, Duke University, Durham, NC.

Christiansen, M.N. and Lewis, C.F. (Editors), 1982. Breeding Plants for Less Favourable Environments. Wiley Interscience, New York, NY, 459 pp.

Finlay, K.W. and Wilkinson, G.N., 1963. The analysis of adaptation in a plant breeding programme. Aust. J. Agric. Res., 14: 742—754.

Keim, D.L. and Kronstad, W.E., 1981. Drought response of winter wheat cultivars grown under field stress conditions. Crop Sci., 21: 11—15.

Kramer, P.J. and Raper, C.D., Jr. (Editors), 1983. Crop reactions to water and temperature stresses in humid temperate climate. In: Proc. Conf. Duke University, Durham, NC. West View Press, Boulder, CO (In press).

Lomas, J. and Shashoua, Y., 1973. The effect of rainfall on wheat yields in an arid region. In: R.O. Slatyer (Editor), Plant Responses to Climatic Factors. UNESCO, Paris, pp. 531—537.

Martineau, J.R., Williams, J.H. and Specht, J.E., 1979. Temperature tolerance in soybeans. II. Evaluation of segregating populations for membrane thermostability. Crop Sci., 19: 79—81.

Mussell, H. and Staples, R.C. (Editors), 1979. Stress Physiology in Crop Plants. Wiley Interscience, New York, NY, 510 pp.

Nass, H.G. and Sterling, J.D.E., 1981. Comparison of tests characterizing varieties of barley and wheat for moisture stress resistance. Can. J. Plant Sci., 61: 283—292.

O'Toole, J.C. and Chang, T.T., 1979. Drought resistance in cereals — rice: a case study. In: H. Mussell and R.C. Staples (Editors), Stress Physiology in Crop Plants. Wiley Interscience, New York, NY, pp. 373—405.

Paleg, L.G. and Aspinall, D.D. (Editors), 1981. The Physiology and Biochemistry of Drought Resistance in Plants. Academic Press, New York, NY, 492 pp.

Quisenberry, J.E., Jordan, W.R., Roark, B.A. and Fryrear, D.W., 1981. Exotic cottons as genetic sources for drought resistance. Crop Sci., 21: 889—895.

Reitz, L.P., 1974. Breeding for more efficient water use — is it real or a mirage? Agric. Meteorol., 14: 3—11.

Rosielle, A.A. and Hamblin, J., 1981. Theoretical aspects of selection for yield in stress and non-stress environments. Crop Sci., 21: 943—946.

Sammons, D.J., Peters, D.B. and Hymowitz, T., 1980. Screening soybeans for tolerance to moisture stress: a field procedure. Field Crops Res., 3: 321—335.

Sojka, R.E., Stolzy, L.H. and Fischer, R.A., 1981. Seasonal drought response of selected wheat cultivars. Agron. J., 73: 838—844.

Sullivan, C.Y. and Eastin, J.D., 1974. Plant physiological responses to water stress. Agric. Meteorol., 14: 113—127.

Sullivan, C.Y. and Ross, W.M., 1979. Selecting for drought and heat resistance in grain sorghum. In: H. Mussell and R.C. Staples (Editors), Stress Physiology in Crop Plants. Wiley Interscience, New York, NY, pp. 263—281.

Townley-Smith, T.F. and Hurd, E.A., 1979. Testing and selecting for drought resistance in wheat. In: H. Mussell and R.C. Staples (Editors), Stress Physiology in Crop Plants. Wiley Interscience, New York, NY, pp. 447—464.

Turner, N.C. and Kramer, P.J. (Editors), 1980. Adaptation of Plants to Water and High Temperature Stresses. Wiley Interscience, New York, NY, 482 pp.

Agricultural Water Management, 7 (1983) 207—222
Elsevier Science Publishers B.V., Amsterdam — Printed in The Netherlands

DROUGHT TOLERANT SORGHUM AND COTTON GERMPLASM

D.T. ROSENOW[1], J.E. QUISENBERRY[2], C.W. WENDT[1] and L.E. CLARK[3]

[1] *Texas Agricultural Experiment Station, Lubbock, TX 79401 (U.S.A.)*
[2] *Southern Plains Cotton Research Laboratory, Agricultural Research Service, U.S. Department of Agriculture, Lubbock, TX 79401 (U.S.A.)*
[3] *Texas Agricultural Experiment Station, Vernon, TX 76384 (U.S.A.*

(Accepted 15 March 1983)

ABSTRACT

Rosenow, D.T., Quisenberry, J.E., Wendt, C.W. and Clark, L.E., 1983. Drought tolerant sorghum and cotton germplasm. *Agric. Water Manage.*, 7: 207—222.

Sorghum (*Sorghum bicolor* (L.) Moench) and cotton (*Gossypium hirsutum* L.) are the summer crops traditionally grown in the rainfed or supplementary irrigated areas of the semi-arid southern Great Plains. Both crops have the ability to withstand periods of water deficit and to yield an economic return to the farmer.

Numerous traits contribute to drought tolerance in both sorghum and cotton. Only limited breeding effort has been expended on identifying or combining such traits. Drought tolerance has seldom been a primary breeding objective, usually addressed indirectly under major objectives such as improved yield, pest resistance, and adaptation.

In sorghum, recent emphasis on field screening under severe drought stress has led to significant developments in the understanding of drought tolerance. Two distinct types of stress responses have been identified. One type is expressed when plants are stressed prior to flowering, while the other is expressed when stress occurs during grain fill. Excellent sources of tolerance to each type of stress have been identified, but high levels of both types of tolerance have not been found in the same genotype. Crosses among contrasting types of tolerance are being evaluated.

In cotton, emphasis has been placed on screening primitive photoperiodic-sensitive strains from the world collection. Primary effort has been expended on identifying parental strains that have unique responses to water stress and determining why these responses occur. Physiological evaluations are used extensively to define potentially useful traits. These traits are then incorporated into agronomically desirable germplasm. Germplasm lines chosen through this approach yielded more than commercial varieties under water deficient conditions in preliminary evaluations.

Field screening at more than one location with different water regimes can be used to insure stress at different stages of growth. Rainout shelters and irrigation gradient systems can be used to supplement field screening.

INTRODUCTION

Both sorghum (*Sorghum bicolor* (L.) Moench) and cotton (*Gossypium hirsutum* L.) are recognized as being biologically drought-tolerant. Therein

0378-3774/83/$03.00

lies their adaptation to the southern Great Plains of the U.S.A. and to other semi-arid regions of the world. Most sorghum in the U.S.A. is grown under non-irrigated conditions in marginal rainfall areas, such as the central and southern Great Plains, where water is the major factor limiting yields. A significant proportion of US cotton is grown under similar conditions in the southern Great Plains. Therefore, improved drought tolerance through breeding would be a major contribution to increasing or stabilizing production. The declining water table in underground aquifers in irrigated areas, greater demand for irrigation of various agricultural crops to increase production, increased water use in urban areas, and increased pumping costs all focus attention on the overall water problem. Development of crops with improved drought tolerance is receiving increased emphasis from various agricultural research agencies, but procedures to accomplish this are complex and progress is slow. In this contribution, we will discuss some of the breeding and screening techniques being used to identify and select for drought tolerance in sorghum and cotton.

It is important to know the major physiological and morphological traits associated with drought tolerance in order to develop effective screening techniques and to know which traits may be utilized in a breeding program. Several excellent reviews have been published on the various physiological and morphological traits and their association with drought tolerance (Downes, 1972; Jordan and Monk, 1980; Jordan and Sullivan, 1981; Kramer, 1980; Sullivan, 1972; Sullivan and Ross, 1979; Turner, 1979). They will not be discussed here in detail. Some of the traits studied and discussed relative to breeding potential include leaf conductance and osmotic adjustment, leaf rolling, leaf-water potential, proline accumulation, epicuticular wax, root growth and soil water extraction, and heat and desiccation tolerance. Technology exists for evaluation of these traits. Some methods are simple, rapid, and readily adaptable to screening large numbers of plants, whereas others are time consuming and expensive. However, plant breeders have made little use of these techniques to select sorghum or cotton wih improved drought tolerance.

Drought escape (the ability of a plant to grow and complete its life cycle before serious water deficits occur) is an important mechanism in most crop plants (Blum, 1970; Downes, 1972). Use of early maturing genotypes, as well as adjustment of planting dates to avoid periods of low rainfall probabilities is a common practice. Genetic manipulation of maturity is relatively simple in both sorghum and cotton, but the problem with early maturing genotypes is that a negative relationship generally exists between yield and earliness.

In many areas, in spite of peaks of high rainfall probability at specific times of the year, significant rainfall can occur throughout the growing season, with some years being nearly ideal for production. To take advantage of favorable years, cultivars with sufficiently high yield potential are needed. Yield data over a period of years often show that later maturing sorghum hybrids or cotton varieties have the highest average yields. However, yield stability under less than optimum conditions is usually more important to producers. Medium-early or early-maturing sorghum or cotton is generally considered to be most stable over years when grown under low rainfall, dryland conditions.

SORGHUM

Sorghum is native to parts of Africa and Asia where it has been grown for more than 2000 years. Sorghum is used in these areas primarily as a food crop. During the middle of the last century, sorghum was first brought into the United States and grown along the Atlantic Coast. From there, sorghum moved westward to drier regions and was well established in the southern Great Plains and California before 1900. Grain sorghum is generally grown in areas too hot or too dry for corn (*Zea mays* L.).

Since its introduction to the United States, sorghum has greatly changed. These changes resulted from naturally occurring mutations combined with the work of plant breeders. Original sorghum introductions were typically tall, late maturing, and unadapted to the U.S.A. With the development of early-maturing varieties, sorghum production extended northward into Nebraska and South Dakota. Expensive hand-harvesting of sorghum was eliminated by the development of dwarf, erect plant types which could be handled by the combine harvester. In the early 1950's cytoplasmic male sterility was found and used to produce sorghum hybrids that yielded 20 to 40% more than conventional varieties. During the 1960's and 1970's, emphasis was on the use of exotic sorghums converted to temperate adaptation (Stephens et al., 1967) to improve disease and insect resistance and increase yield potential. These improvements in sorghum result from a few intensive and closely coordinated breeding programs carried out in major sorghum-producing states.

Recent sorghum research in Texas has placed an increased emphasis on understanding drought response. We have identified two distinct stress responses thru the use of large field screening nurseries at several locations with differing stress environments. The 'pre-flowering' or 'early season' response occurs when plants are under significant moisture stress prior to flowering. Specifically, this growth period is from panicle differentiation, or very shortly thereafter, until flowering. The other distinct response, called 'post-flowering' or 'late-season', occurs when plants are under severe water stress during the grain filling stage.

Plant symptoms indicating either a desirable or undesirable reaction to these two types of stress can be visually rated in the field. Symptoms of pre-flowering drought stress susceptibility include: leaf rolling; uncharacteristic leaf erectness, leaf bleaching; leaf tip and margin burn; delayed flowering; 'saddle effect' — only end plants next to alleyways produce a panicle; poor panicle exsertion; panicle blasting and floret abortion; and reduced panicle size. Tolerance to pre-flowering drought stress is indicated by the alternative condition in each instance.

Symptoms of post-flowering drought stress susceptibility include premature plant (leaf and stem) death or plant senescence, stalk lodging, stalk rot (charcoal rot, *Macrophomina phaseolina*), and sometimes a significant reduction in seed size, particularly at the base of the panicle. Tolerance is indicated when plants remain green and fill grain normally. Such green stalks are resistant to charcoal rot and stalk lodging. The cultivars are sometimes referred to as having good 'stay-green' or as 'non-senescing'. The post-flowering response is most obvious and distinct in plants which have been grown under favorable soil moisture conditions until flowering, with a severe water deficit developing during grain fill. When water stress develops gradually and occurs over the entire season, distinct stress responses may not be as obvious. Often there is a blending of the two types of stress responses.

The relationship of post-flowering drought stress to sink size, stalk rot, and lodging has been discussed by several workers (Dodd, 1977, 1980a,b; Frederiksen and Rosenow, 1980; Rosenow, 1980; Rosenow et al., 1981; Rosenow and Frederiksen, 1981). High yielding genotypes with a large grain sink size relative to the vegetative portion of the plant are more susceptible to post-flowering drought stress than are low grain yielding genotypes. Susceptibility to charcoal rot is predisposed by severe water stress during the latter stages of grain fill. Because this strong relationship exists, charcoal rot is treated as a post-flowering drought stress problem. We believe that a major part of charcoal rot resistance can be explained by tolerance to post-flowering drought stress. Rather than inoculating stalks with charcoal rot infested toothpicks, genotypes are rated for presence or absence of premature leaf and plant death when they are under water stress at or near maturity.

Excellent sources of tolerance to each type of stress have been identified. High levels of both types of tolerance have not been identified in the same genotype, although some genotypes possess moderate tolerance of both types. Many genotypes with excellent pre-flowering drought tolerance are susceptible to post-flowering drought stress. Likewise, most lines with high levels of tolerance to post-flowering drought are susceptible to pre-flowering drought stress.

Breeding approaches

Significant variability exists in the drought response of different sorghum genotypes (Blum, 1974a,b; Lewis et al., 1974; Foster and Weng, 1979; Sullivan and Ross, 1979; Woodfin et al., 1979; Rosenow and Clark, 1981). A number of breeding and screening techniques have been tried or proposed for sorghum (Maunder, 1975; Blum, 1979; Clark et al., 1979). Several of these involved screening for certain physiological or morphological traits using laboratory or field screening methods, in combination with selection of other desirable plant and grain traits. Only limited breeding effort has been placed on drought tolerance as a primary breeding objective. In most cases, selection is first practiced for yield, adaptation, and disease and insect resistance. This is often done under favorable moisture conditions, with only the surviving genotypes evaluated for drought tolerance or adaptation under severe drought stress.

Many sorghum breeders believe that the most rapid breeding progress can be made by practicing early generation selection under optimum or near optimum conditions where reliable expression of yield can be obtained. This concept is supported by studies in other crops on heritability of yield and rate of genetic advance. Extreme variability in timing and amount of rainfall, along with variability in soil type, depth, and water content contribute to a strong genotype × year or genotype × location interaction for drought response, especially under low rainfall conditions. Present knowledge of the distinctly different reactions to drought stress at different growth stages explains much of the lack of progress in the past in selecting sorghum under dryland, low rainfall conditions.

Quisenberry (1982) discusses several approaches to breeding for drought resistance in crop plants. Of particular interest is the evaluation of total dry matter production rather than grain yield under drought. This technique is currently under evaluation in sorghum using photoperiod sensitive, nonflowering exotic lines and their respective converted nonphotoperiod-sensitive, short counterparts.

Our approach to breeding for drought tolerance in sorghum is to practice early generation selection under naturally occurring dryland, low-rainfall conditions. Duplicate nurseries are evaluated under irrigation where full yield potential may be expressed. Large amounts of diverse germplasm are evaluated in field screening nurseries at several locations having different stress environments, different planting dates, and different water regimes. This approach helps to insure stress at different stages of growth.

Selected progeny from the field nurseries are further evaluated under rainout shelters and a line-source irrigation gradient system. Separate shelters are used for pre-flowering stress and post-flowering stress treatments. Regulation of moisture stress at different growth stages is essential for significant progress in breeding for drought tolerance.

Major field drought screening locations in Texas are:

(1) Halfway (clay soil) — limited irrigation
(2) Lubbock (mixed soil) — limited irrigation
(3) Lubbock (mixed soil) — dryland
(4) Big Spring (sandy soil) — dryland
(5) Chillicothe (mixed soil) — dryland.

The Halfway and Lubbock limited irrigation nurseries are used to screen for post-flowering drought tolerance. The Lubbock dryland nursery commonly shows water stress during much of the season. At Lubbock, portions of the nursery are often duplicated with one planted early (May) and one late (mid- or late June) to increase the probability of stress occurring at different growth stages. At Big Spring, the combination of sandy soil and high summer temperatures contributes to good levels pre-flowering drought stress. The Chillicothe location, with extreme heat and low rainfall during midsummer, consistently produces severe pre-flowering stress, especially in plantings made in mid- or late June. Early plantings at Chillicothe in late March or early April often result in severe post-flowering drought and heat stress.

In the limited irrigation nursery at Halfway and Lubbock where post-flowering stress is studied, abundant irrigation is applied during the early growth stages to produce good growth and yield expression. Irrigation is terminated prior to anthesis which allows moisture stress to develop beginning near the flowering stage. We have obtained good post-flowering drought stress in at least one location per year in 8 out of the past 10 years.

In nurseries where post-flowering drought tolerance is evaluated, we subjectively rate each entry or plot for the amount of premature leaf and stem death. The nursery is often left standing for an extended period following maturity to allow stalk lodging to occur. This facilitates the identification of entries which have stalks weakened by water stress. Entires can also be rated for yield and other traits, and selections can be made. Maturity is critical because sorghum has a period just prior to physiological maturity when it is most susceptible to post-flowering stress. Plants a few days earlier or later in maturity may show little senescence. Therefore, flowering notes are taken on all plots and comparisons are made only among similar maturities.

In nurseries where severe water deficits occur prior to flowering, subjective ratings are recorded whenever distinct differences appear. Prior to heading, ratings are made on leaf stress symptoms of rolling, excessive erectness, bleaching, and firing. Ratings can be made separately, but they are often combined into a single overall drought susceptibility rating. Leaf rolling is normally the first visible symptom of drought stress. Excessive leaf erectness usually follows. Some cultivars have erect leaves in the absence of stress, so care must be used when evaluating this trait. The leaf angle of the lower leaves generally indicates if a cultivar has genetically controlled normally erect leaves. Leaf bleaching refers to a loss in green color during the hottest portion of the day, causing a bleached effect. Care must also be used when scoring this trait because there is a range from dark to light green leaf color even in the absence of stress. Leaf margin and tip burn is usually the last vegetative drought response to appear. Some cultivars are very susceptible and show considerable early leaf firing, but this early reaction does not always correlate well with final yield. Drought-induced leaf necrosis is characterized by the absence of anthocyanin pigment and is thus distinctly different from coloration due to other causes, such as disease or insect injury.

Later appearing symptoms caused by water stress prior to flowering include delay in flowering, panicle and floret blasting, poor panicle exsertion, reduced panicle size, and the 'saddle' effect. These symptoms can be rated individually or in combination. Delay in flowering is evaluated by comparison with nonstressed plantings. These late appearing symptoms are probably the best evaluation of pre-flowering drought tolerance even though ratings may be made at or after maturity. Evaluation of pre-flowering drought tolerance in early maturing genotypes is difficult because they often escape water stress.

In field screening nurseries, we use standard checks every five or ten plots. Alternating every fifth plot with a pre-flowering tolerant (post-flowering susceptible) line such as TX 7078 or TX 7000 and a post-flowering tolerant (pre-flowering susceptible) line like SC35-6 provides a reference for comparison. Then by comparing ratings with those of the adjacent checks, adjustment for variability within the field can be made.

In addition to field screening, we use a sprinkler irrigation gradient system at Chillicothe to manipulate timing and quantity of water applied. The advantage is two-fold: (a) the evaluation of plant response to a wide range of stress under otherwise identical conditions; and (b) manipulation of onset, cessation, and degree of stress.

In these evaluations, it is important to recognize the two drought stress responses before interpreting results from the gradient system. At Chillicothe, cultivars with pre-flowering drought tolerance give a positive response in the dry segment of the plot row, while cultivars with a high level of post-flowering stress tolerance have poor panicle development and very low yield in the dry segment of plot row. If one is not familiar with the two types of reactions, cultivars with high levels of post-flowering stress tolerance will be discarded.

Disadvantages of the system are the influence of wind on water distribution, and the inability to control precipitation. The amount and frequency of irrigation may be less than ideal. However, reaction under the system correlates well with field evaluation.

Rainout shelters are also used at Lubbock to supplement evaluations made in field nurseries. Often, untimely rains prevent evaluation or restrict these evaluations to only short periods during the growing season. From our field studies, sorghum genotypes have been identified which are more tolerant to water stress. However, 7—10 years have been required for these evaluations. Rainout shelters can be used to greatly accelerate and improve the efficiency of selection by controlling both timing and amount of water applied, while otherwise maintaining the environment.

Pre-flowering stress ratings under shelters correspond to known field reactions. Simple-row plots of 400 breeding selections can be evaluated for pre-flowering drought stress in one 40 ft × 60 ft (about 12 m × 18 m) shelter.

Through the use of these approaches, excellent sources of pre- and post-flowering drought tolerance have been identified (Table I). Converted lines from the sorghum conversion program have been evaluated for new and improved sources of drought tolerance. Many more lines have been identified

TABLE I

Pre- and post-flowering drought-tolerant sorghums

Pre-flowering tolerant	Post-flowering tolerant
T×7078	SC33-14 (IS12553C)
T×7000 (Caprock)	SC35-6 (IS12555 deriv.)
BT×623	SC56-6 (IS12568 deriv.)
T×430	SC56-14 (IS12568C)
T×2536	SC599-6 (R9188) Rio deriv.
BT×3197 (C.K.-60)	1790E (SC56 × SC33)
Early Hegari	1778 (SC56 × SC33)
RS610	R1584 (SC56 × SC170)
Many commercial hybrids	NSA440
	BKS9

which possess good pre-flowering tolerance than possess good post-flowering tolerance. Most of the post-flowering tolerant source materials are partially converted or converted exotic lines.

Crosses have been made among many of the pre- and post-flowering tolerant lines and elite yielding germplasm. Early generation progeny (F_4 and F_5) are being evaluated simultaneously at the five nursery sites. Using this technique, we are attempting to combine both types of drought tolerance into the same genotype.

COTTON

Cultivated cottons originated from wild perennial plants in semi-arid, semi-tropical environments, where long days and seasonal summer rains encouraged vegetative growth. As the season progressed and day-lengths decreased, flowering was induced, and plants matured their seeds and fiber during the short days of the dry season. Seeds had hard, water-impermeable seedcoats preventing premature germination, and extending the period of germination to enhance the probability of seedling survival. These wild plants were adapted to survive in their environments, which included periodic droughts, temperature extremes, predacious insects, and diseases (Kohel, 1974).

From these wild perennial ancestors, cotton has been domesticated over eons of geologic time because of the value of the plant's fiber to make clothing. Domestication consisted of selecting seeds that rapidly germinated, and plants that fruited quickly to produce abundant fruit regardless of day length.

Today, cotton is grown as an annual and is structurally modified to facilitate mechanical harvesting and processing equipment and to produce large amounts of a specific-type fiber. In recent years, rigid selection criteria, especially for fiber quality, have been applied by plant breeders. The use of these criteria on cotton in a relatively short period of evolutionary time have produced dramatic changes in commercial cotton varieties and has reduced the amount of genetic variability existing within commercial germplasm. Cotton is now grown primarily in temperate climates and under environmental conditions far removed from those in which cotton originally evolved (Stephens et al., 1972).

The first cottons established in the United States were grown on the Eastern Seaboard and were apparently domesticated types of *Gossypium hirsutum* L. (Lewis and Richmond, 1968). How the seed of these first cottons was obtained is not clear. Cotton is not indigenous to the Atlantic Seaboard or the southeastern United States, thus the seed must have been introduced. Cotton was likely brought in from either Mexico, Central America or, most probably, the Caribbean islands. Studies on primitive or 'dooryard' stocks of upland cotton collected in these regions have shown that three-fourths, or more, of these stocks are photoperiod-sensitive and will not flower in the long summer days of temperate climates. Because a temperate climate forces cotton to be grown as an annual, the introduced daylength-sensitive stocks were lost to the present cultivated upland cotton germplasm pool. Thus, from the beginning of cotton culture in the United States, there has been a decisive natural selection pressure against a large amount of cotton germplasm.

Cotton culture began in the southeastern United States and moved progressively West. Today, three major centers of cotton production exist in the United States: the Mississippi Delta, the Southern Great Plains, and the irrigated valleys of Arizona and California. The climatic conditions of these three regions differ greatly. The Mississippi Delta is a high rainfall area, Arizona-California is arid and all water is applied by irrigation, and the Southern High Plains is semi-arid and irrigation (where available) is used to supplement rainfall. Seasonal temperatures, length of growing season, and rainfall have envoked the development of cotton cultivars uniquely adapted to each region.

The southern Great Plains region consists of the High Plains of Texas and New Mexico, the Rolling Plains of Texas and Oklahoma, and the Blackland areas of Texas. This region grows approximately 3 million ha of cotton, most of it without irrigation. On the High Plains, supplemental irrigation water is derived primarily from the Ogallala aquifer. Because this aquifer is not readily rechargeable, the amount of water available for irrigation is declining and the cost of pumping the remaining water is increasing. As water availability has declined, the land area planted to cotton has increased. This increase is due primarily to the high value of cotton lint compared to the alternative grain or oil crops available to the farmer.

In the southern Great Plains, we are evaluating cotton germplasm for drought tolerance. The purposes of this research are: (a) to acquire new knowledge about plant adaptation to semi-arid climates; (b) to identify basic germplasm; and (c) to devise breeding methodology for the development of cultivars better adapted to semi-arid climates.

Initial research suggests that, apart from fruiting habit, little variability exists within the commercial cotton germplasm pool for traits which may be exploited to enhance drought resistance or improve water-use efficiency (Quisenberry et al., 1981). Even in case of fruiting habit, which is difficult to quantify (Roark and Quisenberry, 1977), crop performance is closely tied to the seasonal pattern of soil-water availability. For example, Quisenberry and Roark (1976) found that early maturing "determinate" cultivars used water less efficiently than "indeterminate" cultivars when water deficits developed early in the growing season. However, early maturing genotypes were more efficient under more optimal water conditions or when the growing season was abbreviated by low temperatures. This response may be partially explained by the inability of the more determinate types to compensate for early, water deficit-induced fruit loss, if rainfall occurred later in the growing season. Murray and Verhalen (1970) and Jordan (1979) have pointed out the importance of early season fruit set to lint yield and, therefore, water-use efficiency of dryland cotton in situations where insect pressures render late-season fruit unproductive.

The observation that little variability for drought resistance or water-use efficiency traits exist in commercial cotton germplasm caused us to look at the world collection of primative *G. hirsutum* stocks. Over 1000 accessions of this exotic germplasm from southern Mexico and Central America are maintained by USDA-ARS scientists at College Station, TX. The vast majority of these stocks are photoperiod-sensitive and will not flower in the long summer days of the southern Great Plains. The use of this type of germplasm has both advantages and disadvantages. The advantage is the easier evaluation of the material for traits that increase either water-use efficiency or resist the detrimental effects of prolonged drought. Because the germplasm does not produce fruit, the tremendous drain on soil moisture associated with fruit does not occur on the plant. Cotton is indeterminate in fruiting habit and the source-sink relations of the plant are in constant flux. This changing relationship tends to override the expression of other potentially useful traits and responses. Growth, in response to water deficits, can more easily be evaluated in this photoperiodic sensitive type of material. Quisenberry et al. (1981) demonstrated significant differences among 15 exotic stocks for dry weights, water-use efficiency, heat tolerance, and root growth. Our program has spent most of its effort on evaluating this exotic cotton germplasm because of the diversity of germplasm available, the potential utility of this germplasm in other research endeavors, and the potential for identifying truly unique germplasm.

The disadvantage of using photoperiodic material is the necessity to introduce day-neutral genes into the stocks before they will flower and set fruit in the temperate zones and the long period of time necessary before agronomically desirable types are available for release to breeding programs.

Breeding approaches

Our research program consists of a germplasm development program coupled with a physiological evaluation program (Fig. 1). For the program to function efficiently, it must have a strong interface with basic physiology and biochemistry programs. The initial portion of the program consists of observation and preliminary evaluation of the world collection of *Gossypium* germplasm grown in irrigated and dryland field plots at several locations and in rainout shelters. We use rainout shelters because of the unpredictable timing and amount of rainfall. Under field conditions one cannot depend on having water stress at appropriate stages of growth to evaluate crops for drought tolerance. Unique observations and responses are noted and potential germplasm is identified for further evaluation. The primary approach is devoted to identifying parental germplasm that contains known drought tolerance traits or responses. Effort is expended on primary germplasm evaluation and not on massive screening and evaluation of a large number of progeny from numerous crosses.

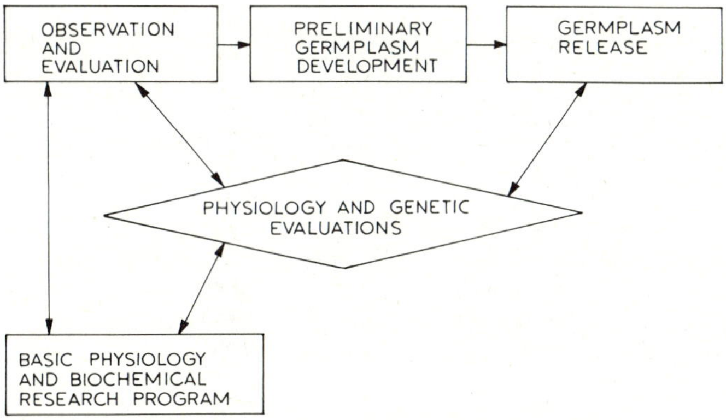

Fig. 1. Diagramic representation of the USDA-ARS research program to develop drought-tolerant cotton germplasm.

After initial observations and preliminary evaluations of gross water-use efficiency (g of total plant dry weight per unit water used), potentially useful exotic stocks are catalogued and advanced into a preliminary germplasm development program and into a physiological evaluation program. In the preliminary germplasm development program and into a physiological evaluation program. In the preliminary germplasm development program, the stocks are crossed with a day-neutral donor strain and a backcrossing program is begun to recover as many of the genes from the exotic stock as possible. In the physiological evaluation program, standard measurements of water status, photosynthesis, stomatal behavior, and growth rates are measured in rainout shelters in an attempt to explain why a plant response occurred. To develop an understanding of why a particular plant response

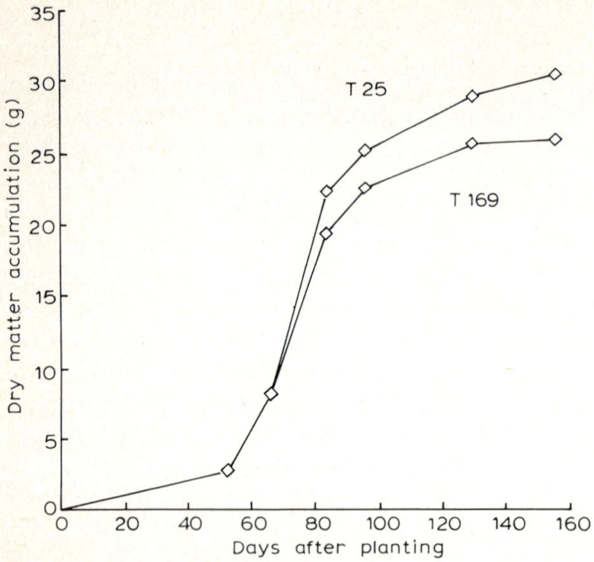

Fig. 2. Dry matter accumulation of T25 and T169 exotic cotton strains grown in a rainout shelter during 1981.

occurred, attempts are made to relate the response to productivity and to determine the mode of inheritance.

An example of this approach is the research conducted on two exotic stocks (T25 and T169) with extreme phenotypic expression of leaf wilting. The T25 stocks appeared, in field observation plots, to resist the usual leaf wilting normally observed in water stressed cotton plants. The T169 stock appeared to wilt sooner than most cotton cultivars. These observations coupled with growth data (Fig. 2) suggested that these two stocks were responding differently to plant water stress. Peterschmidt and Quisenberry (1982) evaluated the water status and stomatal behavior of these two stocks grown under water deficit field conditions. Their results suggested that these two stocks were adapted to different environments. Traits such as stomatal control of transpiration and extensive root growth were present in T25 but not in T169. They suggested that the highly negative water potential measured in T169 probably allowed extraction of the more tightly bound water held in the limited root zone. This ability of T169 enabled maintenance of some growth throughout the season, even though daily wilting occurred. These same two stocks were used by Kelly et al. (1983) to evaluate their response with and without fruit (plants were shaded to provide the proper day length) when grown under water deficit field conditions. They showed that T25, both with and without fruit, was able to maintain a more favorable water status throughout the season than was T169. More dry matter was produced by flowering types than by non-flowering types and T25 produced more dry matter than did T169. The flowering plants used

TABLE II

Morphological, physiological, and anatomical parameters measured on non-flowering T25 and T169 exotic cotton strains under irrigated field conditions

Parameters	T25	T169
Leaf area (dm^2)	53.4[a]	60.6
Leaf number	104	98
Mainstem nodes, No. of	20	17
Leaf size (cm^2)	51	62
Relative water content (%)	83	78
Water potential (MPa)	1.7	1.9
Osmotic potential (MPa)	1.8	1.8
Turgor potential (MPa)	0.1	−0.1
Gross photosynthesis (mg CO_2 dm^{-2} h^{-1})	26	26
Conductance (cm s^{-1})	0.8	1.4
Transpiration (mg H_2O dm^{-2} h^{-1})	4896	6768
Stomata (mm^{-2})		
Top	170	151
Bottom	342	354
Epidermal cells (mm^{-2})		
Top	928	896
Bottom	1554	1552

[a]Data taken between 09.00 and 18.00 h. Means shown are averages over these times.

less soil water than did the non-flowering plants and, therefore, had a much higher water-use efficiency. Differences between the two stocks in seasonal water use were not measured, but due to the higher dry matter production of the T25 stock, it apparently had a higher water-use efficiency. Studies on the relationship between stomatal conductance and photosynthesis show that T25 transpires less water than does T169, but their photosynthetic rates are equal (Table II). The improved photosynthesis/conductance ration association with T25 may partially explain the higher dry matter per unit soil water used since no additional water (the profile was filled prior to planting) was added during the growing season.

As a better understanding of the drought resistance and water-use efficiency traits of a stock is gained, a decision is made whether to continue the development of that stock into advanced germplasm strains (Fig. 1). We decided to continue the development program on T25. Yield tests were conducted comparing the new germplasm to conventional commercial cultivars. When T25 derived strains were compared with Paymaster 303, a widely-used commercial cultivar, over a range of environmental conditions in paired rows, the T25 germplasm produced more than the commercial cultivar in the more stressful environments while the commercial cultivar was superior in the better environments.

SUMMARY

Several traits contribute to drought tolerance in both sorghum and cotton. Only limited breeding effort has been placed on combining such traits.

In sorghum, recent emphasis on field screening under severe drought stress has led to significant developments in understanding drought tolerance. Two distinct types of stress responses have been identified. One type is expressed when plants are stressed prior to flowering, while the other is expressed when the stress occurs during grain fill. Excellent sources of each type of tolerance have been identified. Plant symptoms indicating tolerance or susceptibility to these two types of stress can be visually rated in the field. High levels of both types of tolerance have not been found in the same genotype. Progeny from crosses among lines in contrasting types of tolerance are being evaluated.

In cotton, primary attention is on selection of parental strains which possess traits that enhance the performance of cotton in water deficient field conditions. Physiological evaluations are used extensively to determine why an observed response occurred. Preliminary evaluations of germplasm chosen through this approach have performed better than commercial germplasm under water stress conditions.

Field screening at more than one location with different stress environments, different planting dates, and different moisture regimes is used to insure stress at different stages of growth. Rainout shelters and irrigation gradient systems are used to supplement field screening.

REFERENCES

Blum, A., 1970. Effect of plant density and growth duration on grain sorghum yield under limited water supply. Agron. J., 62: 333—336.

Blum, A., 1974a. Genotypic responses in sorghum to drought stress. I. Response to soil moisture stress. Crop Sci., 14: 361—369.

Blum, A., 1974b. Genotypic responses in sorghum to drought stress. II. Leaf tissue water relations. Crop Sci., 14: 691—692.

Blum, A., 1979. Genetic improvement of drought resistance in crop plants: a case for sorghum. In: H. Mussell and R.C. Staples (Editors), Stress Physiology in Crop Plants. Wiley Interscience, New York, NY, pp. 429—445.

Clark, L.E., Rosenow, D.T. and Hamburger, A.J., 1979. Screening for stress tolerance in sorghum. In: Proc. 11th Biennial Grain Sorghum Res. and Util. Conf., Grain Sorghum Prod. Assoc., Wichita, KS, p. 52.

Dodd, J.L., 1977. A photosynthetic stress-translocation balance concept of corn stalk rot. Proc. Annu. Corn Sorghum Res. Conf., 32: 112—130.

Dodd, J.L., 1980a. The role of plant stresses in the development of corn stalk rot. Plant Dis., 64: 533—537.

Dodd, J.L., 1980b. Grain sink size and predisposition of *Zea mays* to stalk rot. Phytopathology, 70: 534—535.

Downes, R.W., 1972. Discussion: Physiological aspects of sorghum adaptation. In: N.G.P. Rao and L.R. House (Editors), Sorghum in the Seventies. Oxford and IBH Publ. Co., New Delhi, pp. 265—274.

Foster, K.W. and Weng, V.K., 1979. Response of sorghum genotypes to drought induced by differential irrigation. Sorghum. Newsl., 22: 24—25.

Frederiksen, R.A. and Rosenow, D.T., 1980. Breeding for disease resistance in sorghum. In: M.K. Harris (Editor), Biology and Breeding for Resistance to Arthropods and Pathogens in Agricultural Plants. Proc. Int. Short Course in Host Plant Resistance, 22 July—4 August, 1979, Texas A&M University, College Station, TX. Tex. Agric. Exp. Stn. MP-1451, pp. 137—167.

Jordan, W.R., 1979. Role of plant water deficit. In: J. Mc D. Steward (Editor), Cotton Physiology — A Treatise. Section I, Flowering, Fruiting, and Cut Out. Proc. Beltwide Cotton Prod. Res. Conf., National Cotton Council, Memphis, TN, pp. 84—89.

Jordan, W.R. and Monk, R.L., 1980. Enhancement of drought resistance in grain sorghum and methods of measurement. Proc. Annu. Corn Sorghum Res. Conf., 35: 185—204.

Jordan, W.R. and Sullivan, C.Y., 1981. Reaction and resistance of grain sorghum to heat and drought. In: Sorghum in the Eighties, Proc. Int. Sorghum Workshop, November 1981, Hyderabad. Oxford and IBH Publ. Co., New Delhi (In press).

Kelly, C.A., Wendt, C.W., Isbell, V.R., Mabry, J.L. and Phillips, W.O., 1983. Growth and water relations of exotic cotton strains with and without fruit. Crop Sci., 23 (In press).

Kohel, R.J., 1974. Influence of certain morphological characters on yield. Cotton Grow. Rev., 51: 281—292.

Kramer, P.J., 1980. Drought, stress, and the origin of adaptations. In: N.C. Turner and P.J. Kramer (Editors), Adaptation of Plants to Water and High Temperature Stresses. Wiley Interscience, New York, NY, pp. 7—20.

Lewis, C.F. and Richmond, T.R., 1968. Cotton as a crop. In: F.C. Elliot, M. Hoover and W.K. Porter (Editors), Advances in Production and Utilization of Quality Cotton: Principles and Practices, Iowa State University Press, Ames, IA, pp. 1—73.

Lewis, R.G., Hiler, E.A. and Jordan, W.R., 1974. Susceptibility of grain sorghum to water deficit at three growth stages. Agron. J., 66: 589—591.

Maunder, A.B., 1975. Breeding for drought and heat resistance, Mimeo, Paper presented at Seminar, Texas Tech University, 26 pp. (Unpublished).

Murray, J.C. and Verhalen, L.M., 1970. Genotype by environment interaction of study of cotton in Oklahoma. Crop Sci., 10: 197—199.

Peterschmidt, N.A. and Quisenberry, J.E., 1982. Plant water status among cotton genotypes. In: Proc. Beltwide Cotton Prod. Res. Conf., National Cotton Council, Memphis, TN, pp. 108—111.

Quisenberry, J.E., 1982. Breeding for drought resistance and plant water use efficiency. In: M.N. Christiansen and C.F. Lewis (Editors), Breeding Plants for Less Favorable Environments. Wiley, New York, NY, pp. 289—327.

Quisenberry, J.E. and Roark, B., 1976. Influence of indeterminate growth habit on yield and irrigation water use efficiency in upland cotton. Crop Sci., 16: 762—765.

Quisenberry, J.E., Jordan, W.R., Roark, B.A. and Fryrear, D.W., 1981. Exotic cottons as genetic sources for drought resistance. Crop Sci., 21: 889—895.

Roark, B. and Quisenberry, J.E., 1977. Evaluation of cotton germplasm for drought resistance. In: Proc. Beltwide Cotton Prod. Res. Conf., National Cotton Council, Memphis, TN, pp. 49—50.

Rosenow, D.T., 1980. Stalk rot resistance breeding in Texas. In: Sorghum Diseases — A World Review. Proc. International Workshop on Sorghum Diseases, December 1978, Hyderabad, India, ICRISAT and Texas A&M University, pp. 306—314.

Rosenow, D.T. and Clark, L.E., 1981. Drought tolerance in sorghum. Proc. Annu. Corn Sorghum Res. Conf., 36: 18—30.

Rosenow, D.T. and Frederiksen, R.A., 1981. Breeding for disease resistance in sorghum. In: Sorghum in the Eighties. Proc. Int. Sorghum Workshop, November 1981, Hyderabad, India. Oxford and IBH Publ. Co., New Delhi (In press).

Rosenow, D.T., Clark, L.E. and Woodfin, C.A., 1981. Breeding for drought resistance in sorghum. In: Proc. 12th Biennial Grain Sorghum Res. and Util. Conf., Lubbock, TX, p. 111.

Stephens, J.C., Miller, F.R. and Rosenow, D.T., 1967. Conversion of alien sorghums to early combine genotypes. Crop Sci., 7: 396.

Stephens, S.G., Lewis, C.F., Miller, P.A. and Richmond, T.R., 1972. Cotton. In: Genetic Vulnerability of Major Crops. National Academy of Science, Washington, DC, pp. 269—281.

Sullivan, C.Y., 1972. Mechanisms of heat and drought resistance in grain sorghum and methods of measurement. In: N.G.P. Rao and L.R. House (Editors), Sorghum in the Seventies. Oxford and IBH Publ. Co., New Delhi, pp. 247—263.

Sullivan, C.Y. and Ross, W.M., 1979. Selecting for drought and heat resistance in grain sorghum. In: H. Mussell and R.C. Staples (Editors), Stress Physiology in Crop Plants. Wiley Interscience, New York, NY, pp. 263—281.

Turner, N.C., 1979. Drought resistance and adaptation to water deficits in crop plants. In: H. Mussell and R.C. Staples (Editors), Stress Physiology in Crop Plants. Wiley Interscience, New York, NY, pp. 343—372.

Woodfin, C.A., Rosenow, D.T., Clark, L.E. and Johnson, J.W., 1979. Differential response of sorghum cultivars to drought stress. Am. Soc. Agron., Agron. Abstr., 82.

Agricultural Water Management, 7 (1983) 223—236
Elsevier Science Publishers B.V., Amsterdam — Printed in The Netherlands

DROUGHT TOLERANCE IN US MAIZE

STANLEY D. JENSEN[1] and ANTHONY J. CAVALIERI[2]

Department of Corn Breeding, Pioneer Hi-Bred International, Inc.,
[1] *Box 505, York, NE 68467,* and [2] *Johnston, IA 50131 (U.S.A.)*

(Accepted 23 November 1983)

ABSTRACT

Jensen, S.D. and Cavalieri, A.J., 1983. Drought tolerance in US maize. *Agric. Water Manage.*, 7: 223—236.

Soil water deficits, especially when accompanied by excessively high temperatures, are probably the most common yield limiting factors in maize production. Hybrid yield test programs have emphasized mean yield as the main criterion for advancement. An important component of that mean is the stability of yield across a range of environments. Regression analysis allows us to quantify yield stability, and was used to evaluate 339 commercial maize hybrids grown in the U.S.A. under a variety of conditions. Considerable variation existed among hybrids studied. Hybrids studied had b values equal to, significantly above, and significantly below 1.0. In addition, for each b value, hybrids existed with above and below average yields. The number of replicates necessary to obtain precision in estimates of b values was necessarily large ($\pm$0.05 units required 372 replicates). Variation in hybrid response between geographical regions of the country was considerable, as was that between different years of testing. Many plant breeders feel that stress resistance and high yield are negatively correlated; however, this negative relation is not always the case as correlation of mean yield and b value was not high (r = 0.37 overall). Breeding programs which include use of physiological and morphological traits, control of water stress, side-by-side stress and nonstress selection programs and extensive testing should result in increased efficiency in developing genotypes which perform well under water deficit. Incorporation of the regression analysis technique in evaluating field data will allow identification of hybrids with the performance desired under droughts of various types.

INTRODUCTION

Soil water deficits, especially when accompanied by excessively high temperatures, are probably the most common yield limiting factors in maize production. Significant reductions in US maize yields due to drought stress have occurred one year out of four or five over the last 50 years (Newman, 1978). In the last 49 years, the Palmer Drought Index has registered a negative value in either July or August, a critical period for the maize crop, 43 times in central Iowa, 49 times in central Illinois, and 36 times over the same period in central Indiana (National Climatic Center, 1982).

0378-3774/83/$03.00

Maize is not one of the more drought tolerant crops and will not compete with sorghum or wheat in the drier areas of the West. Areas in which the threat of drought is predictable and significant are increasingly being brought under irrigation (USDA, 1980). Maize under irrigation in the U.S.A. has increased steadily since the 1940's and dramatically since 1970. Extensive pumping from the Ogallala aquifer, which provides irrigation water to a significant portion of the High Plains of the U.S., has reduced water levels significantly. The compounding effects of lower water tables and higher energy costs are causing serious economic problems for maize farmers in that area (Buller, 1982). Pressures of limited water supply and increased costs have resulted in efforts to use water more efficiently, and farmers are working to schedule irrigations at critical times in the crop life cycle. Questions of varietal response to this kind of irrigated maize culture are largely unanswered and need attention.

There is a region of dryland maize production in eastern South Dakota, Nebraska, and Kansas where drought tolerance is very important. Drought stress ranges from moderate to severe every year because soil profiles are seldom saturated and short periods without rain can cause serious stress. In the rest of the maize growing region in the U.S., water deficit problems vary considerably from year to year. Problems can range from severe, as in 1980, to an intermittent drought of several weeks in duration, to the midday water potential deficits that occur even in well watered plants. Soil type influences the extent of water deficit and regions with sandy soils, such as the southeastern Coastal Plain, may be particularly susceptible to short periods without rain. High temperatures, low relative humidities, and wind all add to problems of water stress. Regions that experience one or more of these weather factors are more vulnerable, but too little is known about their individual or compounded effects on drought stress.

Several factors have reduced the emphasis of maize geneticists and breeders on maize production under dry conditions. These problems include: (a) the lack of control over the stress condition; (b) yields that are too low or too high to be useful in selection when trials are planted in uncontrolled drought-prone environments. (c) high error variances that are normal in testing under drought stress environments; (d) high genotype by environment interactions resulting from different rainfall and temperature patterns from location to location or year to year; (e) the apparent negative correlation between drought resistance and yield potential of a hybrid under more optimum conditions. Despite limited direct effort in breeding for drought resistance in maize, evidence exists to suggest that modern maize hybrids are genetically superior to hybrids of past decades in their ability to withstand drought stress.

Russell (1974) compared double cross hybrids used in the 1930's with modern commercial hybrids, and found that the modern hybrids performed better at all yield levels, but their relative advantage was considerably greater at the lower yielding locations where drought stress was a strong limiting factor. Russell (1974) concluded that evaluations over the years have added important genes for drought resistance or characteristics needed in a stress environment. Duvick (1977) compared 48 hybrids grouped according to year of release. Average yield per location ranged from 4.5 to 8.0 t/ha. The newer the release date for a group of hybrids, the higher was the yield whether at low, or high yielding locations relative to other materials in the test. In contrast to Russell's data, the subsets containing recently released hybrids had their greatest advantage at the highest yielding locations. Considerable gain has been made even in the most severe drought environments, and one can only speculate on the contribution that improved drought resistance makes toward the higher yields at the higher yielding locations. Logic suggests that periodic stress was a factor in limiting yields at those locations as well.

A key then to hybrid stress tolerance is stability in maintaining yield levels across a range of environmental conditions. Regression analysis is a useful tool for identifying those genotypes that are stable in their yield in response to drought stress relative to the mean of materials being tested (Finlay and Wilkinson, 1963, Finlay, 1964, 1971; Eberhart and Russell, 1966). Also, this statistical technique is useful for evaluating yield stability in general; where yields may be limited by disease, temperature, and light, etc. We have used regression analysis to evaluate a large number of commercial maize hybrids grown in the U.S. under a variety of conditions.

MATERIALS AND METHODS

Regression analysis as described by Finlay and Wilkinson (1963) was used to calculate phenotypic stability (b value) for a number of widely tested hybrids. Phenotypic stability (b value) for a hybrid is the slope of a linear regression of the yield of that hybrid at a given location against the mean yield of all hybrids grown at the location. The mean yield of a hybrid was expressed as a percent of mean yield of the location to characterize its relative yield level. The average b value for a group of hybrids is 1.0 while a hybrid with $b > 1.0$ has less than average stability and $b < 1.0$ greater than average. If yield at low yielding sites is reduced by drought then b values are indications of relative performance of hybrids under drought conditions. Values of r^2 for the regression may also indicate stability.

Hybrids were yield tested in replicated, small plot (approximately 8.12 m^2) experiments at test locations scattered over the entire corn growing area of the U.S. and Canada in 1979, 1980, and 1981. Cultural conditions included those used by progressive farmers in each area. Some hybrids were tested at as few as five locations and 15 replications while others were tested at 700 locations and over 2000 replications over the 3-year period. At least 25 hybrids and as many as 75 were grown at any one location in a given year. A total of 339 hybrids was evaluated, including Pioneer commercial and experimental hybrids, commercial hybrids of other private companies, and crosses of important public inbred lines. The set of hybrids used included commercial hybrids released in the last 10 years as well as a number of experimental hybrids that may or may not have potential for commercial sales. Obviously, this is not a random sample of genetic material but, rather, a selected sample of elite crosses and were not expected to show very different behavior. A given hybrid was predominately tested in the area that was appropriate for its maturity.

RESULTS AND DISCUSSION

Genetic variation among modern maize hybrids

Considerable variation exists in yield stability among the 339 hybrids studied (Table I). Twenty-two percent of the 339 hybrids had b significantly <1.0 and 12% had b values significantly >1.0. If only those hybrids that were tested in over 500 replications are considered, 40% had b values <1.0 and 29% had b values >1.0. In Fig. 1, four hybrid responses are presented. Data are from all regions during 3 years. Hybrids were found that exhibit low yield and low b value (3388) and high yield and high b values (3323). In addition to these types of expected performance, some hybrids had average b values and above average yields (3377, 3358), suggesting that performance in low yielding environments and yield potential in favorable environments are not mutually exclusive.

Regardless of the number of replicates, more hybrids with b values significantly less than 1.0 were present than with b values significantly greater than 1.0 (Table I). This distribution indicates recent selection has been for stable or stress-resistant hybrids rather than hybrids with very high yield potential.

The testing program required to obtain precision in estimates of b values is discouragingly large (Table I). In the group of hybrids with 100 or fewer replicates, only 18% of the hybrids had b values significantly different from 1.0. In the 101 to 250 replicate group, 47% were significantly different; in

TABLE I

Percent of hybrids tested at different replicates with b values significantly ($P>0.95$) different from 1.0; data are for all hybrids tested over a 3-year period in the U.S.

No. Replicates	b			No. Hybrids
	<1.0	≈ 1.0	>1.0	
<100	11	82	7	186
101–250	30	53	17	83
251–500	37	48	15	35
>500	40	31	29	35
Total	22	66	12	339

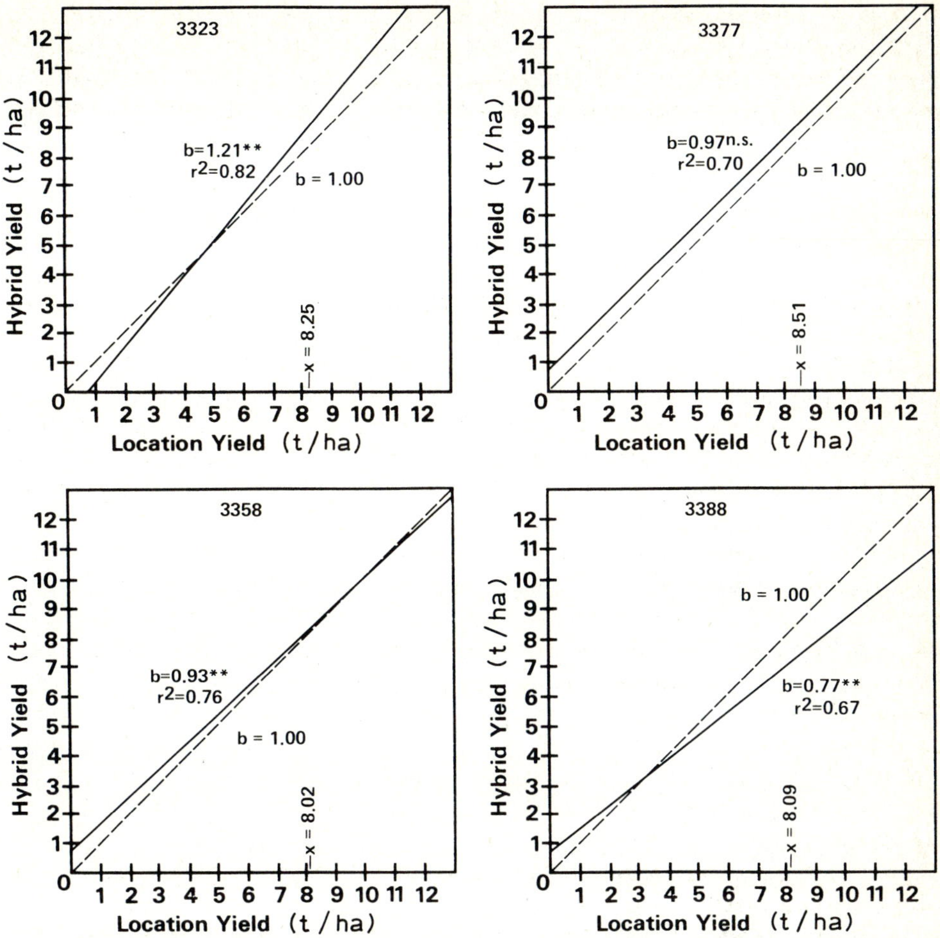

Fig. 1. Stability regressions of four hybrids grown at locations throughout the U.S. over 3 years (*$P>0.95$, **$P>0.99$).

the 251 to 500 replicate group, 52% were significant; and in the over 500 replicate group, 69% were significantly different from 1.0. These data suggest that a large number of replicates are required to classify hybrids according to their true b value rather than concluding that hybrids in the highest replicate group have a different response. However, selection of the desired hybrids, those in the top group, may require fewer replications.

Determining optimal sample size is supported by a formal technique described by Snedecor and Cochran (1967). The method requires an estimate of the population standard deviation for the parameter in question — in this case the population b value — and a decision on how large an error can be tolerated in the estimates of the parameter. This can be calculated using the formula: Number of replicates = $4S^2/L^2$, where S equals the estimate of the population standard deviation, in this case obtained from the standard errors of a number of paired stability regressions, and L equals the acceptable error. Required replications necessary to be 95% confident of being beyond a precision level of ±0.05 units are 372 while precision at the ±0.10 level requires 93 replications and the ±0.01 level requires 9306. With a range in b values of 0.80 to 1.20 a confidence interval of ±0.05 would seem like a reasonable level of precision to strive for in classifying hybrids.

These results suggest that the necessary testing to obtain meaningful regression values when compared to requirements for measuring most agronomic traits will be considerably more extensive. However, as was pointed out earlier this sample of maize hybrids was made up of elite Corn Belt genetic material. In other material, if the genetic variation were greater, the breeder might be able to arrive at a similar level of precision with fewer replicates, or may be willing to accept a somewhat lower precision level in the selection process and make acceptable progress. Targeting the development and testing program at a smaller geographical region might also reduce the number of replicates required.

Phenotypic yield stability interactions with regions and years

Twelve hybrids were selected for more in-depth study to get a better understanding of the influence of geographical testing area and year of testing on b value. Hybrids selected had all been tested in over 600 replications at many test sites over the entire U.S. corn growing region in the three years (Table II). The hybrids represent a range of b values and are grouped as $b<0.95$, $0.95 \leqslant b \leqslant 1.04$, and $b>1.04$ with four hybrids in each group. Relative maturity ratings (Minnesota system) range from 99 to 131 with a range of maturities represented in each group. R^2 values are high and indicate a good fit of the regression line in all cases.

A hybrid's stability as predicted by regression response varies depending on geographical regions of the country. Table III illustrates the variations in b values calculated for the twelve hybrids tested in the eastern (Ontario, New York, Pennsylvania, Ohio, Michigan, Indiana, Delaware), central (Illinois, Missouri, Iowa, Wisconsin, Minnesota) or western (South Dakota, North Dakota, Nebraska, Colorado, Texas) regions. Some hybrids (3323, 3541, 3509, 3901) have b values that are reasonably consistent regardless of the region in which they were tested. A high interaction between b value and region was found with the majority of the hybrids. Pioneer brand 3377, which had a b value of 0.97 when tested over all areas, was significantly greater than 1.0 (central) and less than 1.0 (west) when calculated from regional tests.

Variation in hybrid b values for different regions may arise from a number of sources. Drought was felt to be a major determining factor of b values in all regions, but there are confounding elements including humidity, wind intensity, temperature, drought duration and timing, diseases, insects, and

TABLE II

Relative maturity (RM), b values and r^2 values of twelve widely tested commercial hybrids; hybrids were tested over 3 years throughout the U.S. (**$P>0.99$ that b is different than 1.0)

Hybrid	RM	Replicates	b	r^2
Hybrids with $b > 1.04$				
3323	127	1072	1.21**	0.82
B73/Mo17	124	887	1.13**	0.80
3541	115	1936	1.07**	0.73
3183	131	1326	1.06**	0.75
Hybrids with $0.95 \leqslant b \leqslant 1.04$				
3389	121	651	1.04	0.74
3780	107	2047	0.98	0.70
3509	119	816	0.97	0.77
3377	122	725	0.97	0.70
Hybrids with $b < 0.95$				
3358	124	939	0.93**	0.76
3901	99	1448	0.93**	0.67
3720	109	1334	0.90**	0.75
3382	124	2017	0.93**	0.72

TABLE III

Phenotypic stabilities (b values) of twelve widely tested hybrids (>100 replicates per region) when tested in different geographical regions of the U.S. (*$P>0.95$, **$P>0.99$)

Hybrid	East	Central	West	Overall
Hybrids with $b > 10$				
3323	1.21**	1.19**	1.28**	1.21**
B73/Mo17	0.98	1.08*	1.26**	1.13**
3541	1.08	1.05*	1.06*	1.07**
3183	1.05	1.05	0.94	1.06**
Hybrids with $b \approx 1.0$				
3389	1.13	1.08*	1.04	1.04
3780	0.97	0.72**	1.03	0.98
3509	1.03	0.98	0.97	0.97
3377	1.17	1.09*	0.79**	0.97
Hybrids with $b < 1.0$				
3358	1.08	0.90**	0.82**	0.93**
3901	0.70**	0.71**	0.94**	0.93**
3720	1.04	0.77**	0.90**	0.90**
3382	0.96	0.92**	0.98	0.93**

weeds. Environmental effects would vary considerably from region to region in the U.S., and hybrid sensitivity to them would also differ.

Calculation of r^2 values for different regions suggests that b values of hybrids tested in the Western region show less deviation from the regression line, which indicates that the environmental and biological factors causing variation may be fewer or more uniform in time of occurrence (Table IV). However, variation in location means was greater in the West which could provide more favorable r^2's.

Hybrid b values for tests run in each of the 3 years of the study are shown in Table V. The b value × year interaction appears to be similar to that which occurs between regions. Selection within regions and across several years would be most useful since regional interactions are high and yearly variation must be tolerated by the crop. The data do not rule out selection for crosses that have wide interregional adaptation.

TABLE IV

Values for r^2 calculated for hybrids tested in different regions and different years

Region	Year		
	1979	1980	1981
Eastern	0.35	0.49	0.55
Central	0.59	0.55	0.47
Western	0.67	0.74	0.75

TABLE V

Phenotypic stabilities (b values) of twelve widely tested hybrids when tested in different years (*P>0.95, **P>0.99)

Hybrid	1979	1980	1981	Overall
Hybrids with $b > 1.0$				
3323	1.14**	1.20**	1.28**	1.21**
B73/Mo17	—	1.12**	1.12**	1.13**
3541	1.03	1.14	0.95	1.07**
3183	1.09*	1.03	1.10**	1.06**
Hybrids with $b \approx 1.0$				
3389	—	0.99	0.99	1.04
3780	0.96	1.01	0.96	0.98
3509	1.03	1.02	0.92	0.97
3377	—	1.04	0.93*	0.97
Hybrids with $b < 1.0$				
3358	—	0.94*	0.97	0.93**
3901	0.94	0.91**	0.93**	0.93**
3382	0.89**	0.98	1.01	0.93**
3720	0.96	0.90**	0.92**	0.90**

Paired comparisons

The most critical comparison between two hybrids is obtained when both hybrids are always grown together at the same locations and at the same time. Such paired comparison allow direct comparisons of two hybrids over a range of environmental conditions. Pioneer brand 3377 and 3358 were compared using this technique (Fig. 2). The two hybrids both have desirable yield potential and perform well under stress. Performance of the two hybrids over all geographical regions is similar; however, significant differences in performance are apparent when the data are compared by regions. The paired comparisons are very useful in making critical compari-

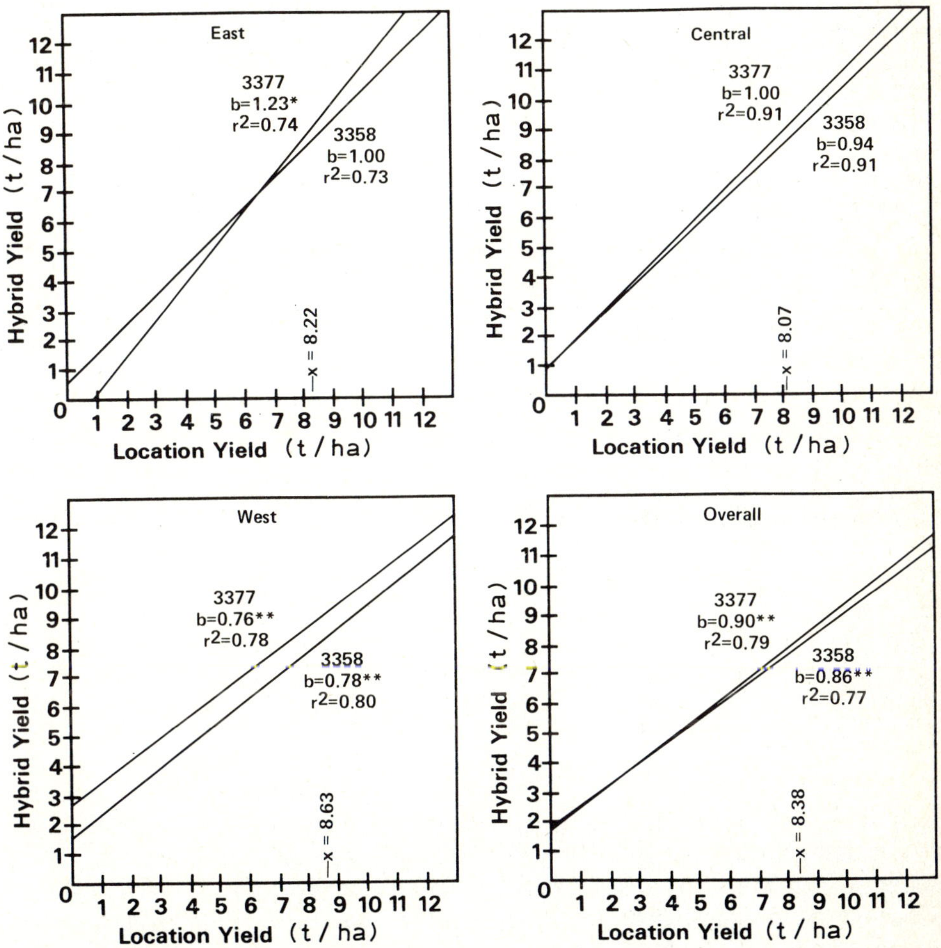

Fig. 2. Paired stability regressions for 3377 and 3358 by region. Stability regressions plotted when 3377 and 3358 were grown together in the same location and year. Data are divided by geographical region ($^{*}P>0.95$, $^{**}P>0.99$).

sons between hybrids which appear to have similar average yield potentials. The technique also provides a valid picture of the relative response to stress of each hybrid and how those responses are affected by regional interactions.

Mean yield and b values

Many plant breeders feel that stress resistance is negatively correlated with high yield. Pioneer 3323 and 3388 fit this pattern, but performance of hybrids such as 3377 and 3358 (Fig. 1) suggest that this is not always the case. How much of a compromise exists between these two traits? Correlations between mean hybrid yield as a percent of location mean and *b* values of the 339 hybrids studied provides some insight into this question (Table VI). Hybrids were grouped by geographical region. Correlations range from $r = 0.15$ in the West to $r = 0.45$ in the South (Virginia, Kentucky, Tennessee, North Carolina, South Carolina, Georgia). The overall correlation was $r = 0.37$. Although a positive relationship between yield and *b* value was confirmed, it is encouraging that the correlation is relatively low. Thus selection for high yield and drought resistance, although somewhat difficult, should be achievable.

TABLE VI

Correlations between mean hybrid yield and *b* value of 339 hybrids tested in geographical regions over 3 years (**$P > 0.99$)

Region	*n*	*r*
East	167	0.26**
North Central	93	0.19
South Central	120	0.28**
West	124	0.15
South	59	0.45**
Overall	339	0.37**

Physiological and morphological traits in breeding

A number of investigators have outlined programs that would incorporate selection for physiological traits into breeding programs for superior crop performance during drought (Hurd, 1971; Boyer and McPherson, 1975; Fischer et al., 1981; Simpson, 1981; Turner and Begg, 1981). Preliminary evaluation of physiological traits for use in maize breeding programs for drought tolerance include proline accumulation (Pintèr et al., 1979), chromosome length (Chokairi and Gorenflot, 1979), high temperature in artificial environments (Hunter et al., 1936; Williams et al., 1967), germination and growth in solutions of polyethylene glycol (Parmar and Moore,

1966), seedling wilting (Williams et al., 1967), and chlorophyll stability (Kilen and Andrew, 1969). Despite these sometime promising approaches, only one group (Fischer et al., 1981) has incorporated the use of physiological traits into a breeding program for performance of maize under drought. Fischer et al. (1981) have used recurrent selection to improve a maize population (Tuxpeño) using an index of yield under drought and well watered conditions, reproductive synchrony, leaf area loss during grain fill, leaf elongation under dry conditions, and leaf temperature as selection criteria.

Because sensitivity of yield to dry conditions varies with developmental stage, it is important that timing and intensity of drought imposed in breeding programs are appropriate for the location and the type of water deficit experienced. Plant characteristics which may be desirable in a line or hybrid adapted to different types of water deficit are different and need individual attention. Maize is most susceptible to reduction in yield if drought occurs at flowering (Salter and Goode, 1967; Claassen and Shaw, 1970a,b) and, if the timing of water deficits is not predictable, as is the case for the Corn Belt, attention should be focused on the most sensitive portion of the life cycle.

Regression analysis provides in-depth information on the performance of a genotype. Hybrid stability estimates may be useful in providing a set of varieties or hybrids which have different responses to stress. Those materials might be further characterized to identify physiological or morphological traits associated with the type of performance of interest.

As is apparent from the stability regressions, it is possible to develop hybrids that yield well in all soil water conditions. It is necessary to improve performance of maize under drought without reducing yield potential under good conditions. More information is needed on physiological mechanisms that give drought tolerance or allow avoidance, but are not detrimental to yield potential. Physiological traits used in breeding for specific traits in maize must always be considered in light of yield reduction under optimal conditions. Material with exceptional drought tolerance may not have direct commercial value, but would certainly have a place in a breeding program.

Strategies for improving the genetic tolerance to drought

A significant factor in our evaluations for drought resistance has been the lack of a good test system for measuring the characteristic. Genotype × environment interactions caused by uncontrolled water stress have made it difficult to generate repeatable response slopes in most field studies. Selection for certain genes related to drought resistance in one year may not be helpful in the next. Similarly, selection for those same genes in one area of the country may not be helpful in another.

Special facilities that can aid in controlling soil water include rainout shelters in areas that receive adequate rainfall for a normal crop (Teare et al., 1973). In drier climates, line source irrigation systems have been used to test large numbers of genotypes at variable water levels (Hanks et al., 1976). Other types of controlled irrigation can be used in low rainfall areas for selection at the level of stress that the breeder desires.

The successful breeding effort of Fischer et al. (1981) utilizing several physiological traits in a recurrent selection program to improve drought resistance is encouraging. The emphasis of this work was on field screening in an area of Mexico where there is no appreciable rainfall during the growing season. The complete dependency on applied water was, no doubt, important to the genetic gains that were achieved. An important question regarding this improvement in drought resistance is whether it will be expressed in other areas, especially at different elevations and temperature regimes and higher yield levels.

The side-by-side stress and nonstress selection programs as used by Fischer et al. (1981) should be very helpful in removing the confounding effects of other factors that tend to mask genetic differences in drought resistance. Material selected under these conditions would require follow-up, into multiple location and year evaluations over the target area to establish overall performance records and estimates of b values. A large number of replicates are necessary to estimate those b values with the necessary precision to make effective decisions.

The relative efficiencies of various breeding systems for improving drought resistance in maize will be difficult to determine without a better understanding of the inheritance of the characteristic. We will not deal with the question of inheritance in a specific way but will list some of the factors to consider in making decisions on breeding systems:

(1) Drought resistance is a complex factor involving a number of physiological and morphological characteristics with unknown inheritance. If gene action for drought resistance is additive as are most other complex traits in maize (Hallauer and Miranda Fo., 1981), then good progress for that character can be made using S1 or S2 per se testing.

(2) If selection is successful, the relative portion of the nonadditive gene effects should increase and a system designed to utilize nonadditive gene effects may be warranted in advanced cycles or in more elite material. This approach would involve the use of a tester to take advantage of both additive and nonadditive effects.

(3) The subtle nature of drought resistance, the problem with high genotype × environment interactions, and the higher than normal error variances in drought stress experiments will require a considerably higher than normal commitment to testing. The breeder may very well be forced into multiple year testing and resultant longer cycle time in recurrent selection programs for drought resistance.

(4) Some form of pedigree selection where large numbers of replicates can be accumulated in characterizing a genotype may be a more effective technique than traditional recurrent selection programs.

The potential for the genetic improvement of drought resistance in maize clearly exists (Jensen, 1971; Christiansen et al., 1981; Castleberry and LeRette, 1979). Conventional breeding programs have been successful in making progress (Russel, 1974; Duvick, 1977). A continuation of those procedures, since drought is a significant factor in the selection process, will virtually assure that improvement will continue. Even with a large and expensive effort aimed specifically at selecting for drought tolerance, progress beyond that expected from a conventional breeding program would not be assured. However, because small differences in yield under water stress will translate into large increases in production when the total maize acreage is considered, the goal of increased drought tolerance would be well worth pursuing.

ACKNOWLEDGEMENT

The authors wish to thank Nancy Albertsen, Corn Data Manager, and John Wagner, Research Statistician, Pioneer Hi-Bred Int., Inc., for their assistance with this paper.

REFERENCES

Boyer, J.S. and McPherson, H.G., 1975. Physiology of water deficits in cereal crops. Adv. Agron., 27: 1—23.

Buller, O., 1982. High plains agriculture in transition. Kansas Agric. Exp. Stn. Conf., 5—7 January 1982, 36 pp. (Unpublished).

Castleberry, R.M. and LeRette, R.J., 1979. Latente, a new type of drought tolerance? Proc. Annu. Corn Sorghum Res. Conf., 34: 46—56.

Chokairi, M. and Gorenflot, R., 1979. Caryological study among maize populations in Morocco in relation with drought and salt resistance. Rev. Cytol. Biol. Veg. Bot., 2: 365—375.

Christiansen, M.N., Jensen, S.D. and Shannon, M.C., 1981. Discussion: Stress Environments. In: K.J. Frey (Editor), Plant Breeding II. Proc. Plant Breeding Symposium, 12—16 March 1979, Ames, IA. Iowa State University Press, Ames, IA, pp. 164—177.

Claassen, M.M. and Shaw, R.H., 1970a. Water deficit effects on corn. I. Vegetative components. Agron. J., 62: 649—652.

Claassen, M.M. and Shaw, R.H., 1970b. Water deficit effects on corn. II. Grain components. Agron J., 62: 652—655.

Duvick, D.N., 1977. Genetic rates of gain in hybrid maize yields during the past 40 years. Maydica, 22: 187—196.

Eberhart, S.A. and Russell, W.A., 1966. Stability parameters for comparing varieties. Crop Sci., 6: 36—40.

Finlay, K.W., 1964. Adaptation — its measurements and significance in barley breeding. In: Barley Genetics I. First International Barley Genetics Symposium, 1963, Wageningen, The Netherlands, pp. 351-359.

Finlay, K.W., 1971. Breeding for yield in barley. Barley Genetics II. Second International Barley Genetics Symposium, 1969, Washington State University Press, Pullman, WA, pp. 338—345.

Finlay, K.W. and Wilkinson, G.N., 1963. The analysis of adaptation in a plant breeding programme. Aust. J. Agric. Res., 14: 742—754.

Fischer, K.S., Johnson, E.C. and Edmeades, G.O., 1981. Breeding and selection for drought resistance in tropical maize. In: Symp. Principles and Methods of Crop Improvement for Drought Resistance: With Emphasis on Rice, May 4—8, International Rice Research Institute, Los Baños, Philippines, 30 pp.
Hallauer, A.R. and Miranda Fo., J.B., 1981. Quantitative Genetics in Maize Breeding. Iowa State University Press, Ames, IA, 468 pp.
Hanks, R.J., Keller, J., Rasmussen, V.P. and Wilson, G.D., 1976. Line source sprinkler for continuous variable irrigation-crop production studies. Soil Sci. Soc. Am. J., 10: 426—429.
Hunter, J.W., Laude, H.H. and Brunson, A.M., 1936. A method for studying resistance to drought injury in inbred lines of maize. J. Am. Soc. Agron., 28: 694—698.
Hurd, E.A., 1971. Can we breed for drought resistance? In: K.L. Larson and J.D. Eastin (Editors), Drought Injury and Resistance in Crops. Crop Sci. Soc. Am. Spec. Publ. 2, pp. 77—88.
Jensen, S.D., 1971. Breeding for drought and heat tolerance in corn. Proc. Annu. Corn Sorghum Res. Conf., 26: pp. 198—208.
Kilen, T.C. and Andrew, R.H., 1969. Measurement of drought resistance in corn. Agron. J., 61: 669—672.
National Climatic Center, 1982, Palmer Index Environmental Data Service, NOAA, Ashville, NC, 28801.
Newman, J.E., 1978. Drought impacts on American agricultural productivity. In: N.J. Rosenberg (Editor), North American Droughts. AAAS Selected Symposia Series 15, Westview Press, Boulder, CO, pp. 43—62.
Parmar, M.T. and Moore, R.P., 1966. Effects of simulated drought by polyethylene glycol solutions on corn (*Zea mays* L.) germination and seedling development. Agron. J., 58: 391—392.
Pintèr, L., Kàlmàn, L. and Pàlfi, G., 1979. Determination of drought resistance in maize by proline test. Maydica, 24: 155—159.
Russell, W.A., 1974. Comparative performance for maize hybrids representing different eras of maize breeding. Proc. Annu. Corn Sorghum Res. Conf., 29: 81—101.
Salter, P.J. and Goode, J.E., 1967. Crop responses to water at different stages of growth. Res. Rev. 2, Commonwealth Bureau pf Horticultural and Plantation Crops, Commonwealth Bureau of Horticultural and Plantation Crops, Commonwealth Agricultural Bureau, Farnham Royal, Bucks., 246 pp.
Simpson, G.M., 1981. The value of physiological knowledge of water stress in plants. In: G.M. Simpson (Editor), Water Stress on Plants. Praeger, New York, NY, pp. 235—265.
Snedecor, G.W. and Cochran, W.G., 1967. Statistical Methods, 6th Edition. Iowa State University Press, Ames, IA, 593 pp.
Teare, I.D., Schimmelpfennig, H. and Waldren, R.P., 1973. Rain out shelter and drainage lysimeters to quantitatively measure drought stress. Agron. J., 65: 544—549.
Turner, N.C. and Begg, J.E., 1981. Plant-water relations and adaptation to stress. Plant Soil, 58: 97—131.
USDA, 1980. America's Soil and Water: Conditions and trends. U.S. Department of Agriculture, Washington, DC, 32 pp.
Williams, T.V., Snell, R.S. and Ellis, J.F., 1967. Methods of measuring drought tolerance in corn. Crop Sci., 7: 179—182.

Introduction to Part 4

PHYSIOLOGY OF STRESS

Plants growing in water-limited environments adapt and/or acclimate in a variety of ways via mechanisms that originate at the subcellular level. Boyer discusses dehydration-induced accumulation of solutes in cells and organelles. He speculates on the role of osmoregulation in maintenance of turgor, chloroplast activity and organelle integrity. Krieg extends portions of Boyer's discussion to the whole-plant level and contrasts the ability of determinate and indeterminate plants to maintain photosynthetic activity during and following periods of water stress.

The two remaining papers deal with aspects of the application of quantitative measures of physiological and morphological characters to crop improvement programs. Passioura describes recent efforts to enhance the drought resistance of dryland wheat by increasing the hydraulic resistance of seminal roots. Higher axial root resistance restricts vegetative growth and results in the availability of a greater fraction of the stored soil water during the grain filling period. Jordan et al. illustrate the use of crop simulation models for grain sorghum and wheat to predict potential increases in yield arising from specific genetic modifications. Using daily weather data for a 30-year period from three Great Plains locations, they examine yield benefits from early maturity date, greater rooting depth, and enhanced osmoregulation. Higher grain yields were associated with additional water supplies provided by greater rooting depth. This 'water spending' strategy differs greatly from the 'water conserving' strategy proposed by Passioura, but examination of the basic underlying principles of each strategy suggests both may be appropriate for specific regions.

WAYNE R. JORDAN
Texas Agricultural Experiment Station
Temple, TX, U.S.A.

Agricultural Water Management, 7 (1983) 239—248
Elsevier Science Publishers B.V., Amsterdam — Printed in The Netherlands

SUBCELLULAR MECHANISMS OF PLANT RESPONSE TO LOW WATER POTENTIAL

JOHN S. BOYER

Agricultural Research Service, U.S. Department of Agriculture, Department of Botany and Department of Agronomy, University of Illinois, 289 Morrill Hall, 505 S. Goodwin Avenue, Urbana, IL 61801 (U.S.A.)

(Accepted 23 November 1983)

ABSTRACT

Boyer, J.S., 1983. Subcellular mechanisms of plant response to low water potential. *Agric. Water Manage.*, 7: 239—248.

Understanding the metabolic basis of the resistance of plants to limited water may hasten the development of superior techniques for increasing crop yields. Three promising areas where understanding has recently advanced are osmotic adjustment, photosynthesis, and tolerance to severe desiccation.

Osmotic adjustment, i.e., large change in cell solute content, occurs in response to soil water availability and permits plants to maintain growth on limited soil water when growth otherwise would not occur. Grain yields are enhanced in wheat genotypes having superior osmotic adjustment as soils become dry. Osmotic adjustment is controlled by nonsimultaneous changes in the rate of solute uptake and the rate of solute use by cells in response to limited water availability.

Photosynthesis is generally inhibited by losses in activity of leaves as well as by losses in viable leaf surface as the availability of soil water decreases. These losses are caused by stomatal closure, decreased chloroplast activity, and factors controlling the persistence of viable leaves. Each of these factors appear to be under metabolic control: stomatal aperture is determined by solute retention by the guard cells; chloroplast activity is altered by concentrations of regulatory ions in the cells; and viable leaf area is controlled by metabolic factors accelerating the normal senescence of the leaves. By contrast, the translocation of photosynthetic products is less sensitive than photosynthesis to limited water supply, which has the effect of maintaining yield by mobilizing stored reserves. The loss of viable leaves contributes only a small amount to this mobilization, and selection against accelerated senescence is probably desireable.

Tolerance to severe desiccation can occur in nonsenescing leaves of some native species but not in many crops. The tolerance appears to be associated with the persistence of the nucleus in the leaf cells, because considerable degradation and loss of integrity occurs in all other cell structures. Mechanisms may exist that preserve nuclear integrity when cells are severely desiccated.

INTRODUCTION

The response of plants to a limited water supply involves a broad range of developmental and cellular events that, for native species, are advanta-

mature tissue by the stem was inhibited. The production of mature tissue is a measure of the rate of biosynthesis for the stem, and therefore the accumulation of dry matter by the stem exceeded the utilization of dry matter for biosynthesis between 4 and 12 h after transplanting. As a consequence, dry matter unused for biosynthesis 'piled up' in the elongating region and this region became heavier. This result was in contrast to the controls, which used dry matter for production of mature tissue at a rate about equal to the accumulation of dry matter by the stem.

The behavior of the seedlings in dry vermiculite suggests that the control of osmotic adjustment occurs at two sites: (a) there is cellular control of the rate of biosynthesis, which in turn determines solute use (we may disregard respiratory effects in this analysis because respiratory activity accounts for equivalent components of solute uptake and use, and the net effect is zero); (b) there is control of the rate of uptake of solute by the cell, perhaps at the plasmalemma. If changes in these two control sites are not simultaneous, solute accumulates or is depleted from the solute pools in the cells.

In soybean stems, osmotic adjustment involved decreased biosynthesis rather than increased uptake of solute. However, this behavior may not apply to other tissues. In maize roots, for example, biosynthesis may continue as osmotic adjustment occurs, because dry matter accumulates (Sharp and Davies, 1979). Thus, the regulation of solute uptake and use may differ in different organs. Regardless of the balance between accumulation and utilization, however, control at both sites appears to be central to the process of osmotic adjustment.

PHOTOSYNTHESIS

Photosynthesis usually decreases as leaf water potential decreases (Boyer, 1976). Although the decreases in photosynthetic activity were attributed to stomatal closure, recent evidence shows that chloroplast activity is also inhibited (Boyer, 1976). In some plants, these losses in chloroplast activity may be more limiting than stomatal closure at low water potentials (Boyer, 1971; Ackerson et al., 1977).

Stomatal closure occurs because the guard cells no longer retain high solute concentrations at low water potentials (Ehret and Boyer, 1979). Even when leaf dehydration occurs over times as short as a few minutes, solute release can be observed in the guard cells, and closure follows soon thereafter. Often there is a lag before stomata reopen after water has been resupplied to the plants. This lag is associated with a lag in the accumulation of solute by the guard cells. Therefore, guard cell closure and reopening in response to leaf water potentials are unlikely to be a hydraulic consequence of the change in leaf turgor but rather a result of changes in the osmotic potential necessary for guard cell opening. Thus, stomatal behavior at low water potential is likely to be under some form of metabolic control.

The losses in chloroplast activity that accompany losses in photosynthesis at low water potentials involve decreases in electron transport and photophosphorylation (Keck and Boyer, 1974) and are associated with changes in conformation of the thylakoid membranes rather than chloroplast degradation or the loss of structural integrity (Fellows and Boyer, 1976). These changes in conformation are in turn associated with conformational changes in subunits of the thylakoid membranes. A recent study (Younis et al., 1979) of the losses in photophosphorylation that take place in chloroplasts exposed to low leaf water potentials showed that coupling factor (ATP synthetase in thylakoid membranes) changed conformation in such a way that the binding of ADP to the enzyme was restricted.

Attempts to simulate the effects of low leaf water potentials on chloroplast photophosphorylation were made using Mg^{2+} concentrations likely to occur in chloroplasts of cells dehydrated to varying degrees. Table II shows that Mg^{2+} concentrations usually present in chloroplasts (1 to 3 mmol l^{-1} — Portis and Heldt, 1976; Portis, 1981) were likely to have increased during leaf dehydration and were associated with an inhibition of photophosphorylation. When Mg^{2+} was supplied to isolated chloroplasts at concentrations expected during dehydration, a similar inhibition of photophosphorylation occurred (Table III). Furthermore, this result extended to chloroplast coupling factor assayed as Ca^{2+}-ATPase (Table III).

TABLE II

Phosphorylation activity of isolated spinach chloroplasts, approximate water content of spinach leaves, and calculated stromal Mg^{2+} concentrations at various leaf water potentials

Water potential (MPa)	Photophosphorylation activity (μmol h^{-1} (mg chloroplast)$^{-1}$)	Approximate water content (% of turgid)	Calculated stromal concentration of Mg^{2+} (mmol l^{-1})
—0.2	1 060	100	3
—1.5	740	55	6
—2.5	475	35	9

Chloroplasts were isolated from spinach leaves that had been dehydrated to levels shown. Assays were conducted as in Younis et al. (1979). Stromal Mg^{2+} concentrations were calculated from water content of tissue assuming no changes in compartmentation of the Mg^{2+} or H_2O, and 3 mmol Mg^{2+} l^{-1} in chloroplast stroma of turgid leaves.

Since Mg^{2+} was supplied during preincubation of the isolated chloroplasts, the experiment resembled the preincubation conditions that occurred in dehydrating leaves. The concentrations of Mg^{2+} necessary to bring about inhibition were similar to the increases in concentration that would be expected simply from water loss by the dehydrated leaf tissue (Table II). It is therefore possible that the concentration of cellular constituents to

TABLE III

Magnesium concentration during preincubation, and phosphorylation activity of isolated spinach chloroplasts and Ca^{2+}-ATPase activity of chloroplast coupling factor during subsequent assay

Mg^{2+} concentration during preincubation ($mmol\ l^{-1}$)	Photophosphorylating activity ($\mu mol\ h^{-1}$ (mg chloroplast)$^{-1}$)	Ca^{2+}-ATPase activity ($\mu mol\ h^{-1}$ (mg chloroplast)$^{-1}$)	
		Before activation	After activation
0	1500	21	21
5	1050	23	7
10	750	14	5

Preincubation was carried out in the presence of buffer, and a small aliquot of the preparation was transferred to the assay medium after 30 min at room temperature. The assay medium diluted the Mg^{2+} concentration to 1% of the preincubation concentration. Heat activation of coupling factor is necessary to demonstrate high ATPase activities. The Mg^{2+} preincubation was therefore done prior to heat activation or after heat activation, followed by assay in the usual way. Details of methods are given in Younis et al. (1979).

which specific photosynthetic reactions are sensitive could have an effect on chloroplast activity as leaf water potentials decline.

Translocation of photosynthate continues at low water potentials despite the loss of photosynthetic activity in leaves. Experiments conducted both in laboratory (McPherson and Boyer, 1977) and field (Jurgens et al., 1978) environments showed that leaf water potentials low enough to result in the cessation of dry matter accumulation by the whole plant nevertheless allowed the accumulation of dry matter in developing grain in corn plants. The translocation consisted of stored reserves mobilized primarily from stems. The result was grain production in excess of that contributed by drought-inhibited photosynthesis during the grain filling period. Thus, translocation is less sensitive than photosynthesis to low soil water (McPherson and Boyer, 1977; Jurgens et al., 1978; Sung and Krieg, 1979) and acts to preserve grain production under dry soil conditions.

An additional contributor to the losses in photosynthetic activity that occur at low water potentials is accelerated leaf senescence. It is commonly observed that leaf senescence occurs more rapidly at low than high water potentials and this has the result of removing photosynthetic surface from the crop canopy. In determinate plants, like corn, there often is no regrowth of leaves if water is supplied after flowering takes place and, thus, accelerated leaf senescence can permanently inhibit the potential photosynthetic activity of the crop (Legg et al., 1979).

We recently explored the significance of accelerated leaf senescence in corn. Plants were dehydrated during the grain filling period and remained at low soil water potentials until grain maturity. During the treatment, all the leaves senesced. In the controls, however, senescence did not occur until the end of the season. Table IV shows that the controls lost water from the leaves at a rate of 42 mg m^{-2} s^{-1} and, after senescence, at 2.8 mg m^{-2} s^{-1}. The plants growing in dry soil lost water at a rate of 6.9 mg m^{-2} s^{-1} and after the leaves senesced, at a rate of 0.7 mg m^{-2} s^{-1}. Because leaf water potentials had decreased to —1.8 to —2.0 MPa in the plants in dry soil, photosynthesis was negligible (McPherson and Boyer, 1977) and, consequently, grain fill was dependent on stored reserves. At maturity, the grain had accumulated approximately 40 g per plant of dry material from reserves in the vegetative organs (about 25% of control yields). Of this amount, about 8 g per plant was supplied from the leaves and the remainder was supplied by the stems. Thus, the leaves contributed relatively small amounts of dry matter to the grain.

TABLE IV

Transpiration in maize plants having viable and senescent leaves after water had been withheld during the grain filling period

Leaf water potential (MPa)	Transpiration (mg m^{-2} s^{-1})	
	Viable leaves	Senescent leaves
—0.5	42	2.8
—1.8 to —2.0	6.9	0.69

Transpiration was measured in the whole plant before and after leaf senescence and is expressed on a unit leaf area basis. At —0.5 MPa, leaf senescence occurred at the end of the growing season. At —1.8 to —2.0 MPa, leaf senescence was complete 3 weeks after water had been withheld from the soil.

In view of the small amount of reserves made available to the grain and the small amount of transpiration that was prevented by accelerated senescence of the leaf tissue, it appears that accelerated leaf senescence under dry conditions is an undesirable feature for agriculture. It therefore seems that genetic improvement of crop performance under dry conditions could be based at least in part on genotypes that retain leaf tissue during drought but undergo the usual senescence at maturity.

SEVERE DESICCATION

Leaf tissue that does not undergo accelerated senescence must be capable of withstanding significant dehydration. Most higher plants are capable of withstanding desiccation to the air-dry state when they are embryos in seed. After germination, however, metabolic changes cause the seedling to become susceptible to severe dehydration, which usually is lethal. Recently, native species have been found in Africa that can withstand desiccation of leaves to the air-dry state. Hallam and Gaff (1978a, b) and Wellburn and

Wellburn (1976) have described the characteristics of these plants at the ultrastructural level. Table V shows that there is loss of integrity or damage in virtually all organelles except the nucleus of leaf cells of these plants during severe desiccation. In contrast, the nucleus is damaged and, indeed, disappears during severe leaf desiccation in sunflower, which is not desiccation tolerant. These results suggest that the maintenance of many types of cellular organelles is not necessary for severe desiccation, but that maintenance of the nucleus is essential. This implies that survival of severe desiccation requires the cell to repair and/or reconstruct organelles outside the nucleus. Thus, the metabolic basis of cell tolerance to severe desiccation may reside in mechanisms that protect nuclear integrity.

TABLE V

Altered ultrastructure in cells of leaves of higher plants in the air-dry state; *Xerophyta*, *Talbotia*, and *Myrothamnus* are desiccation tolerant, *Helianthus* is desiccation sensitive

Structure	*Xerophyta villosa*[a]	*Talbotia elegans*[b]	*Myrothamnus flabellifolia*[c]	*Helianthus annuus*[d]
Nucleus	0	--	0	++
Chloroplast envelope	+	++	++	++
Chloroplast thylakoids	++	+	+	+
Mitochondria	++	++	0	++
Golgi bodies	--	++	+	++
Tonoplast	--	--	++	++
Plasmalemma	--	++	--	++

0, shape change; +, moderate change in integrity; ++, major change in integrity; —, no observation.
[a]Hallam and Gaff (1978b); [b]Hallam and Gaff (1978a); [c]Wellburn and Wellburn (1976); [d]Fellows and Boyer (1976).

CONCLUSIONS

Several opportunities for improving the drought tolerance of plants are suggested by these results. First, the accelerated senescence of leaf tissue so frequently observed under dry conditions in crop plants could probably be selected against without sacrificing much yield. If accelerated senescence did not occur, photosynthesis could resume when water was resupplied to the crop. Second, the survival of leaf tissue in the air dry state in certain species indicates that drought tolerance can be attained by plants during vegetative growth. In fact, the embryos of most seed plants can withstand dehydration to the air dry state, but this ability is only rarely expressed at later stages of growth. The significance of nuclear stability in desiccation tolerant cells remains to be assessed, but the stability implies that there are cellular properties that can protect organelles during severe desiccation. Third, with evidence accumulating that osmotic adjustment permits growth under dry conditions, and that genetic variability exists in this trait, the improvement of osmotic adjustment in crop species may enhance production under dry conditions. The beneficial effects of osmotic adjustment are likely to be modest, because rapid growth is often sacrificed to allow solute to accumulate in the growing tissue. However, even a modest gain might translate into significant economic benefits.

REFERENCES

Acevedo, E., Fereres, E., Hsiao, T.C. and Henderson, D.W., 1979. Diurnal growth trends, water potential, and osmotic adjustment of maize and sorghum leaves in the field. Plant Physiol., 64: 476—480.

Ackerson, R.C., Krieg, D.R., Haring, C.L. and Chang, N., 1977. Effects of plant water status on stomatal activity, photosynthesis, and nitrate reductase activity of field grown cotton. Crop Sci., 17: 81—84.

Bernstein, L., 1961. Osmotic adjustment of plants to saline media. I. Steady state. Am. J. Bot., 48: 909—918.

Black, R.F., 1960. Effects of NaCl on the ion uptake and growth of *Atriplex cesicaria* Heward. Aust. J. Biol., 13: 249—266.

Boyer, J.S., 1971. Nonstomatal inhibition of photosynthesis in sunflower at low leaf water potentials and high light intensities. Plant Physiol., 48: 532—536.

Boyer, J.S., 1976. Water deficits and photosynthesis. In: T.T. Kozlowski (Editor), Water Deficits and Plant Growth. Vol. 4, Soil and Water Measurements, Plant Response, and Breeding for Drought Resistance. Academic Press, New York, NY, pp. 153—190.

Cutler, J.M., Rains, D.W. and Loomis, R.S., 1977. Role of changes in solute concentration in maintaining favorable water balance in field-grown cotton. Agron. J., 69: 773—779.

Cutler, J.M., Shahan, K.W. and Steponkus, P.L., 1980. Influence of water deficits and osmotic adjustment on leaf elongation in rice. Crop Sci., 20: 314—318.

Eaton, F.M., 1927. The water requirement and cell-sap concentration of Australian saltbush and wheat as related to the salinity of the soil. Am. J. Bot., 14: 212—226.

Ehret, D.L. and Boyer, J.S., 1979. Potassium loss from stomatal guard cells at low water potentials. J. Exp. Bot., 30: 225—234.

Fellows, R.J. and Boyer, J.S., 1976. Structure and activity of chloroplasts of sunflower leaves having various water potentials. Planta, 132: 229—239.

Fereres, E., Acevedo, E., Henderson, D.W. and Hsiao, T.C., 1978. Seasonal changes in water potential and turgor maintenance in sorghum and maize under water stress. Physiol. Plant., 44: 261—267.

Greacen, E.L. and Oh, J.S., 1972. Physics of growth. Nat. New Biol., 235: 24—25.

Hallam, N.D. and Gaff, D.F., 1978a. Regeneration of chloroplast structure in *Talbotia elegans*: a desiccation-tolerant plant. New Phytol., 81: 657—662.

Hallam, N.D. and Gaff, D.F., 1978b. Re-organization of fine structure during rehydration of desiccated leaves of *Xerophyta villosa*. New Phytol., 81: 349—355.

Jones, M.M. and Rawson, H.M., 1979. Influence of rate of development of leaf water deficits upon photosynthesis, leaf conductance, water use efficiency and osmotic potential in sorghum. Physiol. Plant., 45: 103—111.

Jones, M.M. and Turner, N.C., 1978. Osmotic adjustment in leaves of sorghum in response to water deficits. Plant Physiol., 61: 122—126.

Jones, M.M., Osmond, C.B. and Turner, N.C., 1980. Accumulation of solutes in leaves of sorghum and sunflower in response to water deficits. Aust. J. Plant Physiol., 7: 193—205.

Jurgens, S.K., Johnson, R.R. and Boyer, J.S., 1978. Dry matter production and translocation in maize subjected to drought during grain fill. Agron. J., 70: 678—682.

Keck, R.W. and Boyer, J.S., 1974. Chloroplast response to low leaf water potentials. III. Differing inhibition of electron transport and photophosphorylation. Plant. Physiol., 53: 474—479.

Legg, B.J., Day, W., Lawlor, D.W. and Parkinson, K.J., 1979. The effects of drought on barley growth: models and measurements showing the importance of leaf area and photosynthetic rate. J. Agric. Sci., Camb., 92: 703—716.

McPherson, H.G. and Boyer, J.S., 1977. Regulation of grain yield by photosynthesis in maize subjected to a water deficiency. Agron. J., 69: 714—718.

Meyer, R.F. and Boyer, J.S., 1972. Sensitivity of cell division and cell elongation to low water potentials in soybean hypocotyls. Planta, 108: 77—87.

Meyer, R.F. and Boyer, J.S., 1981. Osmoregulation, solute distribution, and growth in soybean seedlings having low water potentials. Planta, 151: 482—489.

Morgan, J.M., 1977. Differences in osmoregulation between wheat genotypes. Nature, 270: 234—235.

Munns, R., Brady, C.J. and Barlow, E.W., 1979. Solute accumulation in the apex and leaves of wheat during water stress. Aust. J. Plant Physiol., 6: 379—389.

Portis, A.R., Jr., 1981. Evidence of low stromal Mg^{2+} concentration in intact chloroplasts in the dark. Plant Physiol., 67: 985—989.

Portis, A.R., Jr. and Heldt, H.W., 1976. Light-dependent changes of the Mg^{2+} concentration in the stroma in relation to the Mg^{2+} dependency of CO_2 fixation in intact chlotoplasts. Biochim. Biophys. Acta, 449: 434—446.

Sharp, R.E. and Davies, W.J., 1979. Solute regulation and growth by roots and shoots of water-stressed maize plants. Planta, 147: 43—49.

Sung, F.J.M. and Krieg, D.R., 1979. Relative sensitivity of photosynthetic assimilation and translocation of ^{14}Carbon to water stress. Plant Physiol., 64: 852—856.

Turner, N.C., Begg, J.E., Rawson, H.M., English, S.D. and Hearn, A.B., 1978a. Agronomic and physiological responses of soybean and sorghum crops to water deficits. III. Components of water potential, leaf conductance, $^{14}CO_2$ photosynthesis, and adaptation to water deficits. Aust. J. Plant Physiol., 5: 179—194.

Turner, N.C., Begg, J.E. and Tonnet, M.L., 1978b. Osmotic adjustment of sorghum and sunflower crops in response to water deficits and its influence on the water potential at which stomata close. Aust. J. Plant Physiol., 5: 597—608.

Wellburn, F.A.M. and Wellburn, A.R., 1976. Novel chloroplasts and unusual cellular ultrastructure in the "resurrection" plant *Myrothamnus flabellifolia* Welw. (Myrothamnaceae). Bot. J. Linn. Soc., 72: 51—54.

Younis, H.M., Boyer, J.S. and Govindjee, 1979. Conformation and activity of chloroplast coupling factor exposed to low chemical potential of water in cells. Biochim. Biophys. Acta, 548: 328—340.

Agricultural Water Management, 7 (1983) 249—263 249
Elsevier Science Publishers B.V., Amsterdam — Printed in The Netherlands

PHOTOSYNTHETIC ACTIVITY DURING STRESS

DANIEL R. KRIEG

Plant and Soil Science Department, Texas Tech University, and Texas Agricultural Experiment Station, Box 4169, Lubbock, TX 79409 (U.S.A.).

(Accepted 23 November 1982)

ABSTRACT

Krieg, D.R., 1983. Photosynthetic activity during stress. *Agric. Water Manage.*, 7: 249—263.

Developmental and physiological limitations imposed by water stress upon photosynthetic activity of green plants and the effects on growth rate and yield are addressed. Photosynthetic activity is defined as a function of leaf area development, photosynthetic rate, and source-sink ratio responses. Leaf area development response to stress is defined as a function of growth habit differences on leaf number and blade size. Derived relationships suggest that whole plant leaf area is very sensitive to stress with differences between determinate and indeterminate plants in the source of the reductions. Indeterminate plants such as cotton (*Gossypium hirsutum* L.) tend to reduce leaf numbers; whereas, determinate plants reduce blade area. Genotypic variation exists in sorghum (*Sorghum bicolor* L. Moench) as to the extent of the reduction in leaf area development and retention.

The photosynthetic rate is also sensitive to water stress. The cause of the response is explained on a biochemical rather than a diffusion-dependent basis. Genotypic variation in the photosynthesis per unit leaf conductance ratios for sorghum and cotton are suggested as possible means to increase water use efficiency.

The relative sensitivity of the source compared with the sink is explored. Both leaf area and photosynthetic rate are reduced by stress, resulting in large reductions in total dry matter accumulation rates and final yields. In sorghum, stress prior to flowering results in a greater reduction in panicle differentiation and potential seed number than in leaf area. Additionally, stress reduces the efficiency of the existing leaf area through photosynthetic rate reductions. In cotton, leaf area is reduced more than fruit numbers per plant. The efficiency of the existing leaf area actually increases and the harvest index increases.

Large genetic variation exists in almost every photosynthetic activity parameter examined, offering promise for yield improvement for dryland conditions.

INTRODUCTION

Terrestrial seed plants have evolved and survived by balancing carbon assimilation with transpirational water loss during development to allow sufficient reproductive development for insuring viable seed production and perpetuation of the species. Inadequate water supply throughout the life cycle of green plants has constituted a major selection pressure during the

0378-3774/83/$03.00

course of evolution. In many currently used agronomic species, the wild progenitors have higher photosynthetic rates per unit leaf area and less leaf area per plant reflecting the need to conserve available water supplies through vegetative development for use during the reproductive growth stages. Modern crop improvement programs have selected against this trait because of the direct relationship between leaf area and yield. In many species the photosynthetic rate per unit leaf area has actually decreased with crop improvement and water use efficiency has declined. Many adaptive mechanisms exist in germplasm of current crop species which confer a degree of drought resistance (either avoidance or tolerance) many of which are negatively related to productivity and are simply survival related. Many of these adaptations differentially alter photosynthetic activity and water use on a whole plant basis.

The effects of water deficits and plant stress on carbon assimilation and use by higher plants are addressed in this paper. Plant response is defined as a function of leaf area development and retention. The limitations of the component processes on the photosynthetic system are discussed. The relative sensitivity of the source as compared with the sinks of stressed plants as to major yield limitations on a whole plant basis are described as currently understood.

LEAF AREA DEVELOPMENT

Most seed plants have three rather distinct growth and development phases. The first phase is concerned with development of leaf area and vegetative tissues. The second phase involves the attainment of maximum leaf area, maximum rates of dry matter accumulation and the concurrent development of the potential yield container. The third portion is a period associated with filling the yield container, i.e., seed development. The size and numbers of the reproductive structures, and thus yield potential, are directly related to conditions allowing maximum growth rates and leaf area development.

The development of canopy leaf area is a function of the rate of leaf initiation, the rate and extent of expansion of new leaf tissue, and the retention of fully expanded leaf tissue. All three processes are subject to water stress effects with the final result on whole plant leaf area being a function of the type of growth habit (determinate vs indeterminate), the growth stage when the stress occurs, its intensity, and its duration (Schulze and Hall, 1981).

The sensitivity of the leaf initiation process appears to differ between determinate and indeterminate plant types. Determinate plants are relatively insensitive to moderate to severe stress in terms of rate of leaf initiation and total number of leaves produced (Begg and Turner, 1976; Rawson et al., 1980; Takami et al., 1981). Indeterminate plants appear to be much more sensitive than determinate species with respect to leaf production under stress. In cotton (*Gossypium hirsutum* L.), significant reductions in total leaf

TABLE I

Whole plant leaf area component changes in sorghum and cotton due to long term water stress[a]

Plant parameter[b]	Nonstressed	Stressed	Δ (%)[c]
Sorghum			
Plant height (m)	1.13	0.70	−39
Leaf area (max.) (m^2 per plant)	0.20	0.14	−30
Leaf number (remaining)	9.3	8.7	− 7
Mean blade area (m^2)	0.023	0.015	−33
LAI (max.)	3.73	2.03	−46
Cotton			
Plant height (m)	0.70	0.45	−35
Leaf area (max.) (m^2 per plant)	0.23	0.09	−60
Leaf number (remaining)	52	28	−48
Mean blade area (m^2)	0.004 4	0.003 2	−27
LAI (max.)	2.2	0.9	−60

[a]Nonstressed, irrigated conditions. Stressed, dryland conditions.
[b]Mean of four sorghum hybrids and four cotton cultivars planted at three different dates during the 1980 growing season.
[c]Δ, (nonstressed — stressed)/nonstressed, expressed as a percentage.

numbers are observed under soil water deficit conditions as compared to nonlimiting soil water conditions (Table I). The leaf initiation and development processes associated with the axillary branches are most severely affected as compared with main stem leaf development, resulting in whole plant leaf area reductions. In sorghum (*Sorghum bicolor* L. Moench) blade area changes were largely responsible for the whole plant leaf area reductions.

The leaf expansion process has long been recognized as being extremely sensitive to water stress largely because of the dependence upon turgor pressure (Boyer, 1970; Bunce, 1977). However, the sensitivity of the expansion process under field conditions is considerably less than under controlled environmental conditions largely due to the vast differences in the rate of stress development (Acevedo et al., 1979; Krieg, 1981). Takami et al. (1982) have indicated that the rate of leaf expansion of sunflower was reduced by water stress but the duration of the expansion period was lengthened. Leaf water potentials of −0.1 MPa at dawn were required to completely inhibit the expansion process. Leaf expansion requires not only adequate turgor pressure but also sufficient organic supplies to build the new tissue. Therefore, if photosynthesis is reduced on a whole plant basis or if competition from other sinks reduces the assimilate available to expanding leaves, their growth rate may be reduced. Michelena and Boyer (1982) have examined the roles of turgor pressure and assimilate supply in the elongating zone of maize leaves. Extension rates were greatly reduced in stressed leaves even though turgor maintenance and solute accumulation were evident. They concluded that some factor, other than turgor pressure and assimilate availability, was responsible for the growth rate reductions. A key observation of this study was that the water relations of the exposed blade were sufficiently different from that of the elongating zone, that inferences about the water status, especially the turgor pressure, of this growth zone could not be made from measurements of the existing blade. In both sorghum (McCree and Davis, 1974) and cotton (Berlin et al., 1981) blade area reductions resulting from water and high temperature stress are due to a combination of fewer cells and smaller cells per blade. These results suggest that the cell division process as well as the expansion process are being affected by stress. In sunflower, however, leaves that were initiated during stress, but expanded after stress was relieved, were unaffected as compared to the nonstressed controls (Takami et al., 1982). No determination of cell number or size was made to establish the cause of the recovery response.

The diurnal course of the expansive growth pattern is also altered by stress, at least in broad leaf plants such as cotton and sunflower. Much of the daily growth occurs during the early morning and during the dark period in stressed plants (Krief, 1981; Takami et al., 1982). The leaf expansion rate was not directly related to available turgor pressure in these two species similar to the results for maize.

The plasticity of the leaf area development process provides an important mechanism for seed producing plants to conserve water during vegetative development for use during seed development. However, excessive reduction in leaf area in the vegetative phase of development will impose an irreversible limitation on yield due to the close association between leaf area and seed number in most determinate plant species. Indeterminate plants are able to offset the adversities of short term stress effects on leaf area development by continued initiation and expansion of new leaves if they are not restricted by length of the growing season (Day, 1981; Schulze and Hall, 1981).

The third component of whole plant leaf area is the duration of maximum leaf area. The importance of duration to yield is dependent upon leaf area index (LAI), radiation interception, and functionality of the existing leaf area (Monteith, 1981). At LAI's less than that required for maximum interception of solar radiation, both the rate of leaf development and the rate of senescence are critical in determining dry matter production rate. Monteith (1981) has stated that the crop growth rate per unit intercepted radiation does not change much during the growing season for a particular crop and is surprisingly similar across crop types.

Our data indicate that production of above ground biomass was directly related to leaf area for both cotton and sorghum. The rate of dry matter production per unit leaf index (LAI) area is slightly greater for cotton than for

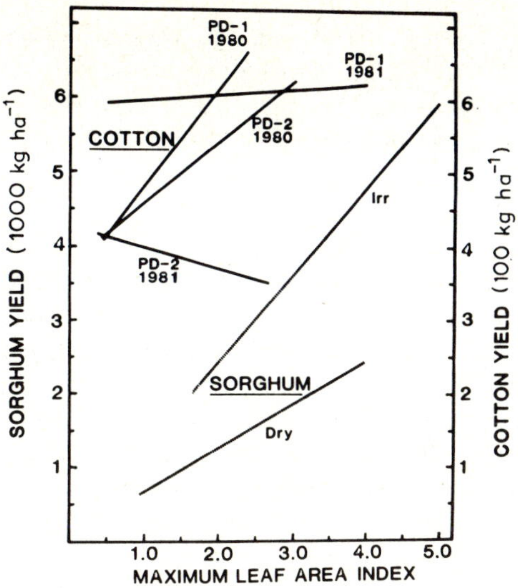

Fig. 1. Cotton and sorghum yields as a function of leaf area index. Planting date and year influences are depicted for cotton yield-LAI relationships. Irrigation effects are depicted for sorghum yield-LAI responses.

sorghum and is slightly greater for dryland cotton than for irrigated cotton.. Harvestable yield, however, was not always highly correlated with maximum LAI in either crop (Fig. 1). Planting date and cultivar had a major effect on the correlation. In cotton, the length of the growing season was a major factor altering the correlation. In sorghum, both genotype and environment were major factors altering the seed number-leaf area relationship which directly affected the yield-LAI relationships. These results emphasize the relative sensitivity of source-sink activity to environment and the resultant partitioning of assimilates. McPherson and Boyer (1977) have emphasized the use of existing assimilate for grain yield of maize, which is quite significant under stress conditions during grain filling. Under field conditions severe lodging usually accompanies this type of response in sorghum.

PHOTOSYNTHETIC RATE CONTROL

The photosynthetic process consists of a series of integrated systems independently subjected to various controls from environmental and genetic factors. Diffusion processes are involved in the concurrent, but opposite direction, movement of CO_2 and H_2O vapor between the leaf mesophyll and the atmosphere. Gas exchange occurs through a series of variable resistances with the stomatal component being of major concern as a control mechanism. The effects of various environmental conditions, such as water stress, on each of the component processes and the overall regulatory control on whole plant carbon assimilation continue to be the subject of extensive research efforts worldwide (Boyer, 1976; Lawlor, 1979; Mansfield and Wilson, 1981; Hanson and Hitz, 1982; Krieg, 1983). This review will build on the existing information using examples of sorghum and cotton subjected to slowly developing stress under field conditions, where applicable.

Stomatal control

Farquhar and Sharkey (1982) have stated that three concepts are prevalent throughout the existing literature concerning photosynthetic rate and stomatal limitations. These are: (a) stomata impose a large limitation on the rate of CO_2 assimilation; (b) the stomatal limitation is more severe when a plant is stressed than when it is not; (c) stomata limit the CO_2 assimilation rate of C_4 species more than that of C_3 species.

The literature is replete with examples of parallel changes in stomatal conductance and photosynthetic rate which has been used to suggest a major control due to gas exchange. However, parallel changes do not define cause and effect relationships.

Farquhar and Sharkey (1982) have suggested that the major limitations to photosynthesis are biochemical and that the stomata respond to an internal stimulus to 'optimize' water use efficiency, or more explicitly, the photosynthesis : transpiration ratio. They have theoretically developed and experimentally demonstrated that the photosynthetic rate is linearly related to the partial pressure of CO_2 in the intercellular spaces, $p(O_2)_i$. In the linear portion of the response curve the activity of RuBP carboxylase is proposed to be rate limiting. At high $p(CO_2)_i$ the ability to regenerate the CO_2 acceptor. RuBP, is proposed to be the rate limiting step (Farquhar and Sharkey, 1982; Von Caemmerer and Farquhar, 1981).

The relationship that we routinely observe between photosynthetic rate and stomatal conductance of CO_2 for sorghum and cotton is depicted in Fig. 2. A linear phase exists where photosynthetic rate and conductance are positively associated. However, as conductance increases beyond 0.01 m s^{-1} photosynthetic rate remains essentially constant suggesting other limitations become dominant. The photosynthesis-conductance relationship is altered in stressed plants and is especially apparent at high leaf conductance (> 0.01 m s^{-1}) in the nonlinear phase of the response. Approximately a 10 to 25% reduction in photosynthesis exists at any given level of leaf conductance in stressed leaves compared to the nonstressed leaves, with sorghum being affected to a greater extent than cotton. Analyses of the relationship between leaf water status, stomatal conductance and photosynthesis indicates that reductions are observed in the photosynthetic rate prior to measurable reductions in leaf conductance.

Using Farquhar and Sharkey's (1982) analyses and rationale we can analyze the stomatal conductance-photosynthesis responses for cause and effect relationships. If stomatal conductance imposed a major limitation on the photosynthetic process, $p(CO_2)_i$ should initially decline and then possibly

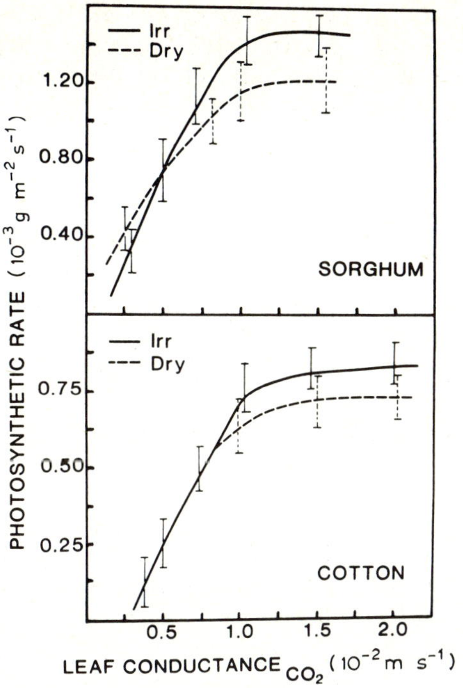

Fig. 2. Photosynthetic rates of sorghum and cotton as a function of leaf conductances for CO_2.

remain constant in the CO_2 limiting range. If stomatal conductance responds to limitations in the photosynthetic process then the $p(CO_2)_i$ should remain constant or slightly increase.

Our data for photosynthetic rate changes due to growth stage and growth condition for both sorghum and cotton indicate that the $p(CO_2)_i$ is only slightly responsive to large differences in photosynthetic rates (Table II). At each growth stage the higher photosynthetic rates were associated with slightly lower $p(CO_2)_i$ within an irrigation treatment. In sorghum, lower $p(CO_2)_i$ levels were observed prior to flowering as compared with the levels during grain filling. Considerable genetic variation was observed in the photosynthetic rates and the substomatal $p(CO_2)$. In cotton, less variation was observed in photosynthesis and $(p(CO_2)_i$ among cultivars than observed for sorghum. In order to determine whether the observed $p(CO_2)_i$ levels were rate-limiting to photosynthesis in each treatment, a photosynthetic rate response to external CO_2 concentration changes was developed. Photosynthetic rate and leaf conductance of single leaves were measured and $p(CO_2)_i$ calculated. The results suggest that photosynthesis of nonstressed plants reached a threshold at $p(CO_2)_i$ of 275 to 280 $\mu l\ l^{-1}$ for sorghum and about 290 to 300 $\mu l\ l^{-1}$ for cotton. Under stressed conditions, the slope of the response curve was altered at $p(CO_2)_i$ less than ambient suggesting carboxylase activity changes. Saturation occurred at slightly lower $p(CO_2)_i$ than in non-

TABLE II

Growth stage and plant stress effects on photosynthetic rates and substomatal CO_2 concentrations of sorghum and cotton

Plant parameter		Growth stage			
		Preflower		Post flower	
		Irrigated	Dry	Irrigated	Dry
Sorghum					
Photosynthetic rate	$\bar{x}$	0.98	0.32	0.83	0.72
(10^{-3} g m^{-2} s^{-1})	range	0.71—1.17	0.17—0.52	0.63—1.02	0.56—0.89
$p(CO_2)_i$	$\bar{x}$	267	258	285	297
(μl l^{-1})	range	248—296	203—277	271—299	269—309
Cotton					
Photosynthetic rate	$\bar{x}$	0.75	0.68	0.47	0.42
(10^{-3} g m^{-2} s^{-1})	range	0.63—0.96	0.55—0.88	0.38—0.59	0.33—0.56
$p(CO_2)_i$	$\bar{x}$	315	308	302	301
(μl l^{-1})	range	307—321	300—314	293—308	294—308

Data generated from 13 sorghum genotypes and four cotton cultivars grown under field conditions.

stressed plants although this difference was not large. A significant reduction in photosynthesis at high $p(CO_2)_i$ was observed in both species when stressed.

If stomatal conductance is not the major limitation to photosynthesis then the possibility of increasing water use efficiency at the single leaf level exists. We have evaluated a wide range of both sorghum and cotton genotypes for this very basic measure of water use efficiency (Ackerson and Krieg, 1977; Ackerson et al., 1980; Hutmacher and Krieg, 1982). Our premise has been that the major limitation to photosynthesis is at the chloroplast level and not at the stomatal level. In both cotton and sorghum, we observe rather significant differences in the photosynthetic rate per unit leaf conductance. In sorghum the major source of the variation appears to be in the photosynthetic rate rather than in leaf conductance (Table III). As much as a 60% difference was observed among the genetic material with which we have been working. In cotton, we have observed a lesser degree of difference among genetic types for the photosynthesis:conductance ratios (Karami et al., 1980; Hutmacher and Krieg, 1982). Additionally, what differences do exist have resulted largely from the differences in leaf conductance rather than photosynthetic rate. The worth of these differences in terms of productivity and water use efficiency are currently being evaluated. Additionally, the heritability of the systems involved in determining this ratio is largely unknown but certainly worthy of exploration.

TABLE III

Midday photosynthetic rates and leaf conductance characteristics of sorghum and cotton as a function of growth stage and soil water status

Plant parameter	Growth stage			
	Preflower		Post flower	
	Irrigated	Dry	Irrigated	Dry
Sorghum[a]				
Photosynthetic rate (10^{-3} g m^{-2} s^{-1})	0.99	0.81	0.69	0.55
Leaf conductance (10^{-2} m s^{-1})	1.29	1.12	0.98	0.69
Cotton[b]				
Photosynthetic rate (10^{-3} g m^{-2} s^{-1})	0.64	0.58	0.66	0.58
Leaf conductance (10^{-2} m s^{-1})	1.19	0.80	1.38	1.15

[a] Data generated from eight lines, 16 hybrids grown under field conditions. Genetic variation was in excess of two fold for photosynthesis and about 50% for leaf conductance.
[b] Data generated from ten cultivars grown under field conditions. Genetic variation was less than 20% for photosynthesis and about 50% for leaf conductance.

Nonstomatal rate limitations

Nonstomatal limitations to photosynthetic rates can be due to photochemical limitations, increased mesophyll resistance and carboxylation activity changes (Boyer, 1976; Lawlor, 1979). In the model proposed by Farquhar and Sharkey (1982) essentially all the photosynthetic rate changes are due to biochemical limitations. At low $p(CO_2)_i$, RuBP carboxylase activity is proposed to be rate limiting. At high $p(CO_2)_i$, RuBP regeneration is considered rate limiting due to photochemical limitations. The RuBP carboxylase limitation was determined as a function of the in vitro response of the enzyme extracted from leaves subjected to a range of atmospheric CO_2 concentrations. The RuBP regeneration limitation was proposed to be due to photochemical limitations as measured by O_2 production and methylviolegen reduction. Although this hypothesis is founded on sound theoretical principles, definitive experimental proof is limited. Only recently have methods been proposed to estimate 'active' carboxylase activity as a function of 'total' carboxylase activity (Jensen and Bahr, 1977). Results should be rapidly forthcoming to define the $p(CO_2)_i$ required for complete enzyme activation as contrasted with the activity of the enzyme with increasing CO_2 concentrations as the substrate for photosynthetic reduction. Concurrently, the

RuBP regeneration limitation should be quantitatively defined and clarified. Sicher and Jensen (1979) have reported that when the RuBP concentration is less than 15 to 20 mmol (g chlorophyll)$^{-1}$, photosynthetic rates are limited by RuBP. In intact spinach leaves, the RuBP concentrations were determined to be 80 to 100 mmol (g chlorophyll)$^{-1}$. Concentrations of RuBP in actively photosynthesizing tissue and that subjected to various environmental conditions are largely unknown at present. RuBP availability can be controlled by limitations in photochemical activity and thus availability of ATP. Water stress has been demonstrated to reduce photosystem II activity in several plant species (Boyer, 1976, Fry, 1972; Sullivan and Ross, 1979). In cotton, Fry (1972) has reported a slight linear reduction in photosystem II activity as leaf water potentials declined from -1.2 to -1.4 MPa. No estimates of actual rates of photophosphorylation and ATP use requirements were made to determine whether this reduction was responsible for photosynthetic limitations.

Another possible problem associated with photophosphorylation potential is the availability of inorganic phosphorus (P_i). Triose-P is exported from the chloroplast to produce sucrose for export from the cell. Co-transport of P_i is required from the cytoplasm to the chloroplast. The rate of this transport could be a possible limitation to the availability of P_i, although diversion of the triose-P to starch in the chloroplast should minimize the possible P_i limitations. Excessive starch accumulation has been suggested as a photosynthetic inhibitor due to increased CO_2 diffusion length. However, starch concentrations greatly in excess of those normally found in leaves are required for measurable reductions in gas exchange (Mauney et al., 1979).

Another biochemical system that reduces net carbon assimilation in C_3 plants is photorespiration. Much has been learned about the various effectors of photorespiration from both chemical and environmental standpoints. Temperature plays a major role in controlling $CO_2 : O_2$ concentration in the stroma due to the differential solubilities of CO_2 and O_2 with increasing temperature. Our results (Perry and Krieg, 1981) with cotton indicate that temperature is the major effector of photorespiration. Water stress, independent of temperature, had little effect as did increasing photon flux density above 1000 μmol m^{-2} s^{-1}.

Some discrepancy exists concerning the regulatory aspects and even the magnitude of the resistance to CO_2 diffusion imposed by the liquid phase in moving CO_2 from the substomatal cavity to the chloroplast. At present, we consider liquid phase conductance to not be a major determinant of photosynthesis.

The proposal being made therefore, is that biochemical limitations are the major regulators of photosynthesis. Under normal atmospheric conditions, the biochemical machinery is operating in the transition zone between RuBP carboxylase limiting and RuBP regeneration limiting conditions according to the Farquhar and Sharkey (1982) model. The stomata then respond to maintain the internal CO_2 concentration at a constant level that will allow maximum rates of CO_2 fixation per unit of transpirational water loss. The cause of the photosynthetic rate changes are biochemical in origin and not related to the availability of CO_2 caused by stomatal conductance limitations.

SOURCE-SINK RELATIONS

Whole plant growth rate is a function of source activity (effective leaf area × photosynthetic rate) and sink activity (number of growing sinks × rate of incorporation of dry matter). These two functions are highly integrated such that assimilation capacity is regulated to a large extent by the utilization rate. It is generally agreed that crop growth rate can be source limited or sink limited depending upon the growth stage and environment. Little data currently exist which examines the integration of source and sink activity and the limitations in each response to various environmental stresses like water or temperature.

The system is even further complicated by the presence of different types of sinks existing at any one point in time. During the very early growth stages, leaf growth and root growth represent the major sinks. As growth progresses, stem development begins to compete for assimilate. Later in the development process, stem growth, leaf growth and even root growth either cease or become rather weak sinks in relation to seed development. Simultaneously, daily maintenance costs exist for every tissue. The maintenance costs are ill-defined at best, but are largely related to tissue mass with cost differences due to the protein concentration of the tissue.

Several measures of source-sink ratio responses on a whole plant basis have been evaluated to begin to establish the relative sensitivity of the individual component processes and the extent of genetic variability in the response. Dryland yields are highly variable across years dependent upon rainfall amount and growth stage of the plant. Sorghum grain yield can be reduced as much as 70% under dryland conditions (Table IV). a large degree of genetic variation exists with dryland yields ranging from over 3 500 kg ha^{-1} to less than 100 kg ha^{-1} in a common nursery. Analyses of leaf area, dry weight at flowering, and leaf area index were not highly associated with the yield reduction. The number of seed per panicle was reduced to a much greater extent than was the leaf area, emphasizing the extreme sensitivity of the panicle differentiation process which is occurring simultanously with leaf area development. No significant correlation was established between leaf area reductions and the extent of seed number reductions. The efficiency of the existing leaf area was greatly reduced by stress as indicated by the grain weight per unit leaf area measurements. The yield response of sorghum was largely a function of reduced effectiveness of the existing leaf area.

Cotton responded quite differently than sorghum. Leaf area development was highly sensitive to soil water deficits (Table V). Fruit number per plant was reduced by stress, but the fruit number and the fruit dry weight per unit leaf area increased. The efficiency of the leaf area of stressed plants actually

TABLE IV

Source-sink relations of sorghum as related to soil water availability[a]

Plant parameter	Irrigated	Dry	Δ (%)[b]
Leaf area (m^2 per plant)	0.21	0.19	−9
Total dry weight at flowering (g per plant)	60.49	45.60	−25
Leaf area index, LAI	3.73	2.03	−46
Yield (kg ha^{-1})	5 046.9	1 447.9	−71
No. seed per head	992.8	490.7	−51
1000 seed weight (g)	27.98	25.26	−10
Grain weight (g) per panicle	27.92	12.29	−56
Grain weight (g (m leaf area)$^{-2}$)	144	70	−51
No. seed per m^2 leaf area	5.15×10^4	2.80×10^4	−46
kg (grain)/LAI	1 436.9	699.7	−51

[a]Data generated from eight lines and 16 hybrids grown under irrigated and dryland conditions during 1980. The range among genotypes for each parameter was 1.5 to 4 times.

[b]Δ, (Irrigated —dry)/irrigated, expressed as a percentage.

TABLE V

Source-sink relationships of cotton as related to growth stage and soil water status

Plant parameter	Plant age (days after emergence)					
	60		85		110	
	Irrigated	Dry	Irrigated	Dry	Irrigated	Dry
Total leaf are (< A) (m^2 per plant)	0.17	0.11	0.28	0.19	0.27	0.19
No. fruit per plant	0.37	0.51	5.13	4.45	9.83	7.02
Fruit weight (g per fruit)	0.08	0.10	2.25	2.02	3.21	3.53
No. fruit per m^2 leaf area	1.0	7.0	36.0	29.0	39.0	42.0
Fruit dry weight (g m^{-2} (leaf area)$^{-2}$)	2.0	9.0	79.0	127.0	126.0	169.0
Whole plant dry weight	156.0	175.0	245	331	299	369

Date generated from 1980 and 1981 growing seasons using three planting dates and four commercial cultivars each year. 1980 was extremely hot and dry throughout the season. 1981 was hot and dry through July then began raining and became cooler through August and September.

increased compared to irrigated plants probably reflecting a high degree of shading of the inner canopy leaves of the nonstressed plants.

The ratio of harvestable dry weight per unit total dry weight is termed harvest index. This ratio has been useful as a selection criterion in breeding programs. In many seed producing plants, especially cereals, improvements in the harvest index have been largely responsible for the yield increases of modern cultivars compared to the wild varieties. Total dry matter production rates have not improved, just partitioning of that dry matter into harvestable yield. Many crop physiologists currently believe that in many of our crop species we are approaching the upper limit of harvest index improvement and further yield increases will have to be made through increasing the photosynthetic rate per unit leaf area.

Environmental stresses and environmental conditions, in general, can alter the harvest index. In sorghum, for example the harvest index increases as planting date is delayed from early May until early June on the Texas High Plains. The rate of development also speeds up, indicating that metabolic efficiency and possibly water use efficiency are increased. Indeterminate plants, such as cotton, have very low harvest indices and considerable improvement is possible. Additionally water stress normally increases the harvest index. This phenomenon has been suggested as a management tool especially in short growing season conditions.

SUMMARY AND CONCLUSIONS

The photosynthetic activity of a green plant is a prime determinant of its growth rate and productivity. Leaf area per plant and the photosynthetic rate per unit leaf area are the two components of photosynthetic activity. The leaf area development component is very sensitive to stress in most plants and constitutes a major limitation to productivity. The photosynthetic rate is also quite sensitive. The cause of the sensitivity has been related to stomatal conductance limitations to gas exchange. Recent evidence and careful analyses of existing data, however, indicate that biochemical reductions are largely responsible for the photosynthetic rate reductions. The stomata then respond to the reduced photosynthetic activity to maintain an optimum exchange of H_2O vapor and CO_2. The primary site of stress on a whole plant basis has not been carefully defined. Some plant species, such as the cereals, may be source limited even under nonstressed conditions, whereas, other plant types such as cotton may be largely sink limited. This area of whole plant physiology is certainly deserving of research attention in order to resolve the major limitations to crop yield and the interactions of genetic and environmental controls.

Several opportunities are available for increasing crop productivity, especially for drought prone environments. The relative sensitivity of leaf expansion and yield container differentiation among genotypes offers opportunity for selection. The genetic differences in stomatal conductance and photosynthetic rates offers a means to increase water use efficiency if this trait is heritable and can be translated to the whole canopy. The large degree of genetic variation in photosynthetic rate per unit leaf area should also be exploited and matched with increased sink : source alterations. In addition, drought resistance mechanisms such as osmotic adjustment and more extensive root systems need exploitation (Turner, 1979). If these two mechanisms are truly adaptations to stress and not just symptoms of growth inhibitions in other sinks, then they may be related to continued productivity.

Much work remains to be done, but considerable promise for genetic and management related improvements exists and needs to be developed.

ACKNOWLEDGEMENTS

Funding for the author's research was provided through Grant No. 5901-0410-8-0095-0 from the Competitive Research Grants Office, USDA, and through Grant No. 58-7830-0-160 from USDA Cooperative Research Service.

REFERENCES

Acevedo, E., Fereres, E., Hsiao, T.C. and Henderson, D.W., 1979. Diurnal growth trends, water potential and osmotic adjustment of maize and sorghum leaves in the field. Plant Physiol., 64: 476—480.

Ackerson, R.C. and Krieg, D.R., 1977. Stomatal and nonstomatal regulation of water use in cotton, corn and sorghum. Plant Physiol., 60: 850—853.

Ackerson, R.C., Krieg, D.R. and Sung, F.J.M., 1980. Leaf conductance and osmoregulation of field grown sorghum genotypes. Crop Sci., 20: 10—14.

Begg, J.E. and Turner, N.C., 1976. Crop water deficits. Adv. Agron., 28: 161—217.

Berlin, J.D., Quisenberry, J.E., McMichael, B.L., Woodworth, M. and Bailey, F., 1981. Morphometric analyses of water stressed cotton leaves. In: J.M. Brown (Editor), Proc. Beltwide Cotton Prod. Res. Conf., 4—8 January 1981, New Orleans, LA. National Cotton Council of America, Memphis, TN, p. 45.

Boyer, J.S., 1970. Leaf enlargement and metabolic rates in corn, soybean and sunflower at various leaf water potentials. Plant Physiol., 46: 233—235.

Boyer, J.S., 1976. Water deficits and photosynthesis. In: T.T. Kozlowski (Editor), Water Deficits and Plant Growth. Vol. 4, Soil and Water Measurements, Plant Response, and Breeding for Drought Resistance. Academic Press, New York, NY, pp. 153—190.

Bunce, J.A., 1977. Effects of water stress on leaf expansion, net photosynthesis, and vegetative growth of soybeans and cotton. Can. J. Bot., 56: 1492—1498.

Day, W., 1981. Water stress and crop growth. In: C.B. Johnson (Editor), Physiological Processes Limiting Plant Productivity. Butterworths, London, pp. 199—216.

Farquhar, G.D. and Sharkey, T.D., 1982. Stomatal conductance and photosynthesis. Annu. Rev. Plant Physiol., 33: 314—315.

Fry, K.E., 1972. Inhibition of ferricyanide reduction in chloroplasts prepared from water-stressed cotton leaves. Crop Sci., 12: 698—701.

Hanson, A.D. and Hitz, W.D., 1982. Metabolic responses of mesophytes to plant water deficits. Annu. Rev. Plant Physiol., 33: 163—203.

Hutmacher, R.B. and Krieg, D.R., 1982. Stomatal and non-stomatal regulation of photosynthesis. In: Proc. Beltwide Cotton Prod. Res. Conf., 3—7 January 1982, Las Vegas, NV. National Cotton Council of America, Memphis, TN, p. 50.

Jensen, R.G. and Bahr, J.T., 1977. Ribulose 1,5-bisphosphate carboxylase-oxygenase. Annu. Rev. Plant Physiol., 28: 379—400.

Karami, E., Krieg, D.R. and Quisenberry, J.E., 1980. Water relations and carbon-14 assimilation of cotton with different leaf morphology. Crop Sci., 20: 421—426.

Krieg, D.R., 1981. Leaf development and function as related to water stress. In: J.M. Brown (Editor), Proc. Beltwide Cotton Prod. Res. Conf., 4—8 January 1981, New Orleans, LA. National Cotton Council of America, Memphis, TN, pp. 41—42.

Krieg, D.R., 1983. Feedback inhibition and stress effects on photosynthesis. In: Mc.D. Stewart and J.R. Mauney (Editors), Cotton Physiology: A treatise. USDA Publ. (in press).

Lawlor, D.W., 1979. Effects of water and heat stress on carbon metabolism of plants with C_3 and C_4 photosynthesis. In: H. Mussell and C.R. Staples (Editors), Stress Physiology in Crop Plants. John Wiley, New York, NY, pp. 303—326.

Mansfield, T.A. and Wilson, J.A., 1981. Regulation of gas exchange in water stressed plants. In: C.B. Johnson (Editor), Physiological Processes Limiting Plant Productivity, Butterworths, London, pp. 237—252.

Mauney, J.R., Guinn, G., Fry, K.E. and Hesketh, J.D., 1979. Correlation of photosynthetic carbon dioxide uptake and carbohydrate accumulation in cotton, soybean, sunflower and sorghum. Photosynthetica, 13: 260—266.

McCree, K.J. and S.D. Davis, 1974. Effect of water stress and temperature on leaf size and number of epidermal cells in grain sorghums. Crop Sci., 14: 754—755.

McPherson, H.G. and Boyer, J.S., 1977. Regulation of grain yield by photosynthesis in maize subjected to water deficiency. Agron. J., 69: 714—718.

Michelena, V.A. and Boyer, J.S., 1982. Complete turgor maintenance at low water potentials in the elongating region of maize leaves. Plant Physiol., 69: 1145—1149.

Monteith, J.L., 1981. Does light limit crop production? In: C.B. Johnson (Editor), Physiological Processes Limiting Plant Productivity. Butterworths, London, pp. 23—38.

Perry, S.W. and Krieg, D.R., 1981. Gross net photosynthesis ratios of cotton as affected by environment and genotype. In: J.M. Brown (Editor), Proc. Beltiwde Prod. Res. Conf., 4—8 January 1981, New Orleans, LA. National Cotton Council of America, Memphis, TN, pp. 51—52.

Rawson, H.M., Constable, G.A. and Howe, G.N.M., 1980. Carbon production of sunflower cultivars in field and controlled environments. II. Leaf growth, Aust. J. Plant Physiol., 7: 575—586.

Schulze, E.D. and Hall, A.E., 1981. Short term and long term effects on steady-state and time-integrated plant processes. In: C.B. Johnson (Editor), Physiological Processes Limiting Plant Productivity. Butterworths, London, pp. 217—236.

Sicher, R.C. and Jensen, R.G., 1979. Photosynthesis and ribulose 1-5 biphosphate levels in intact chloroplasts. Plant Physiol., 64: 880—883.

Sullivan, C.Y. and Ross, W.M., 1979. Selecting for drought and heat resistance in grain sorghum. In: H. Mussell and R.C. Staples (Editors), Stress Physiology in Crop Plants. Wiley Interscience, New York, NY, pp. 263—282.

Takami, S., Rawson, H.M. and Turner, N.C., 1981. Leaf expansion of four sunflower (*Helianthus annuus* L.) cultivars in relation to water deficits. I. Patterns during plant development. Plant Cell Environ., 4: 399—407.

Takami, S., Rawson, H.M. and Turner, N.C., 1982. Leaf expansion of four sunflower (*Helianthus annuus* L.) cultivars in relation to water deficits II. Diurnal patterns during stress and recovery. Plant Cell Environ., 5: 279—286.

Turner, N.C., 1979. Drought resistance and adaptation to water deficits in crop plants. In: H. Mussell and R.C. Staples (Editors), Stress Physiology in Crop Plants. John Wiley, New York, NY, pp. 343—372.

Von Caemmerer, S. and Farquhar, G.D., 1981. Some relationships between the biochemistry of photosynthesis and the gas exchange of leaves. Planta, 153: 376—387.

Agricultural Water Management, 7 (1983) 265—280
Elsevier Science Publishers B.V., Amsterdam — Printed in The Netherlands

ROOTS AND DROUGHT RESISTANCE

J.B. PASSIOURA

CSIRO, Division of Plant Industry, P.O. Box 1600, Canberra City, A.C.T. 2601 (Australia)

(Accepted 22 November 1982)

ABSTRACT

Passioura, J.B., 1983. Roots and drought resistance. *Agric. Water Manage.*, 7: 265—280.

The influence of roots on the yield of water-limited crops is analysed with the help of the identity:

yield = water used × water-use efficiency × harvest index

Despite being severely water-stressed, many droughted crops leave substantial amounts of apparently available water in the subsoil at maturity. The factors influencing this amount are outlined, particularly those concerning the morphology of the root system. Prospects for improving yield by extracting the residual water are discussed.

Because roots are difficult to harvest, water-use efficiency is usually defined as above-ground-biomass/water-used. It follows that the more assimilate a plant transfers to its roots the lower will be its water-use efficiency. There is presumably an optimal root/shoot ratio (in terms of water relations) at which above-ground biomass is maximal for a given water supply. This ratio appears to exceed the optimum in many cases.

For a given biomass, the yield of a grain crop depends in part on the pattern of water use during the season, because harvest index is often related to the proportion of the total water supply that is used after anthesis. For crops relying on a limited supply of stored water, a high axial resistance to flow in the roots may ensure that water in the subsoil is not used so quickly that too little remains at anthesis for the plants to set and fill an adequate number of grains. A breeding program aimed at changing this resistance in wheat roots is described.

Finally, the principles are discussed on which physiological research can be useful in improving drought resistance. The need to dissect water-limited yield into largely independent components is emphasised, for such a dissection greatly improves the focus of the research.

INTRODUCTION

Drought resistance is a nebulous term that appears to become more nebulous the more closely we look at it. Few of us would have trouble listing physiological and morphological traits that we associate with plants growing naturally in arid environments, and that we imagine confer drought resistance on those plants. But when we come to compare plants within a

0378-3774/83/$03.00

species, and particularly within a highly developed crop species, our confidence usually disappears. There are several reasons for this. One is that many of the traits that we associate with the plants of arid regions are concerned with survival, whereas in a crop we are concerned with production. Another is that dealing with crop plants forces us to face squarely the variability of droughts and to recognize that a trait that helps in one type of drought may hinder in another. A third is that differences between plants within a species are much more subtle than those between species, so we are forced to consider small quantitative differences in a character rather than simply its presence or absence.

When we think of roots in relation to drought resistance, we usually do so in the context of the simple but implicit idea that the more water the roots of a plant collect, the greater will be its yield. Few would argue with Kramer (1969) when he says that an essential feature of a drought-resistant plant is a "deep, wide-spreading, much-branched root system such as that of sorghum". Yet the connection between yield and water used is subtle, particularly in a grain crop, whose yield may depend as much on when the water is used as on the amount.

When thinking about the connection between the behavior of roots and the yield of a water-limited crop, it is useful to do so within the framework provided by the identity:

$$\text{yield} = \text{water used} \times \text{water-use efficiency} \times \text{harvest index}$$

This identity is useful largely because the components are to a first approximation independent of each other. This approximate independence is illustrated by a simple glasshouse experiment (Passioura, 1977) in which wheat plants were grown in large pots containing about as much soil as they would have untrammelled access to if they were growing in the field, and to which limiting amounts of water were added, either totally, at sowing, or in small aliquots at frequent intervals throughout the lives of the plants. The results are shown in Figs. 1 and 2, where it can be seen that harvest index (i.e. grain/above-ground-biomass) depended strongly on the pattern of water supply but not on the amount. The same data show, although it is not readily apparent from the figures, that water-use efficiency, i.e. the biomass produced per unit water transpired, was largely independent of both harvest index and water supply. All plants extracted water to the same final water content in the soil irrespective of treatment.

I plan now to discuss each of the components of the yield identity in turn, exploring how the behavior of roots affects the value of each, and how genetic or cultural manipulation of root systems could help increase water-limited yield, while remaining aware of possible interactions between them. Although the components may be largely independent of each other, they are not completely so, and there is the danger that a small increase in one component may lead to a large decrease in another, with a consequential fall in yield. A good hypothetical example of a negative interaction between the

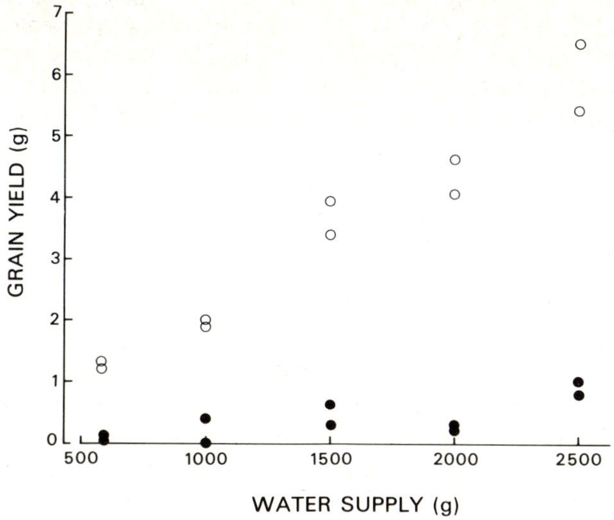

Fig. 1. Grain yield as a function of total water supply for individual wheat plants grown in pots; ●, water supplied totally at sowing; ○, water metered out throughout the life of the plant. Redrawn from Passioura (1977).

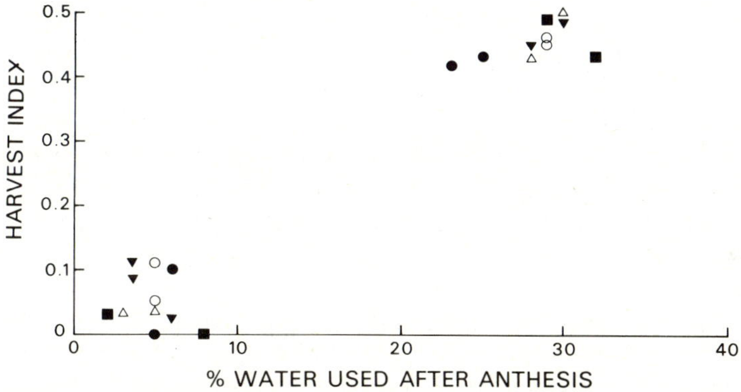

Fig. 2. Harvest index as a function of the percentage of the total water supply which the plants of Fig. 1 used after anthesis. The different symbols designate different total water supplies as follows: ■, 600 g; ●, 1000 g; ○, 1500 g; △, 2000 g; ▼, 2500 g. Redrawn from Passioura (1977).

components is that of a plant that responds to drought by diverting a lot of assimilate to root growth. The greater growth of roots may result in their extracting more water from the soil, but this advantage may be more than offset by a fall in harvest index because there is much less assimilate available for grain growth.

The terms 'water-use' and 'water-use efficiency' are ambiguous in that they can be based on either transpiration or evapotranspiration, and 'water-use efficiency' can be based on either grain yield or above-ground biomass.

In the following discussion 'water-use' means cumulative water transpired, and 'water-use efficiency' means above-ground-biomass-produced/cumulative-water-transpired and is identical with the '*T*-efficiency' of Tanner and Sinclair (1983).

AMOUNT OF WATER COLLECTED BY THE ROOTS

The amount of water available to a crop depends both on the amount of water stored in the soil profile at sowing and on the precipitation during the life of the crop. Some of the precipitation will run off, some of the stored water will drain beyond the reach of the roots, some of the water in the topsoil will evaporate directly. This section discusses the role of roots in maximizing the amount of water used by a droughted crop by maturity, or, alternatively, the role of roots in minimizing both the water evaporated directly from the soil during the growing season, and that left behind in the soil at the end of the season.

For the purposes of this discussion it is useful to focus on three different types of yield-limiting water supply, namely: (a) most of the supply is contained in the soil at sowing and occupies a deep profile that extends well beyond the reach of the roots; (b) as for (a), but the wetted profile is shallow and is well within the reach of the roots; and (c) almost all of the supply comes as rain during the growing season, particularly frequent light falls that are prone to evaporate directly from the soil surface before the roots can extract the water. These three types of supply are in fact points on a continuum, but they do serve to highlight particular principles, some of which are important in some kinds of drought but not in others, and some of which are sometimes even contradictory.

UPTAKE VERSUS EVAPORATION

Despite the importance of the problem, we know little about the ability of roots to extract water from the surface soil before it evaporates directly. Depending on the type of drought the proportion of the plants' total water supply that evaporates directly varies from close to zero, where the supply is exclusively water stored in the soil at the time of sowing, to perhaps 50%, where the supply is many light falls of rain during the growing season and where the leaf area index of the crop is low (Fischer and Turner, 1978). The worth of light falls of rain depends greatly on the texture of the soil: 10 mm of rain may penetrate deeply into a sandy soil and so be protected against rapid evaporation, or may penetrate only 20 mm into a clay soil whose surface was air-dry, unless the channelling of rain by the leaves to the stems, or the presence of influential macropores, results in deep localized infiltration.

Where the rain does moisten surface soil that was previously close to air-dry and in which the roots would be at best dormant, the speed of recovery of the roots will influence the proportion of the rain that is extracted by the roots before it has evaporated directly from the soil. Little is known about the responses of roots to alternate wetting and drying, and possibly it is only new root length that is effective in taking up water. If this is so then the recovery may depend on the abundance of viable root tips in the dry soil, or perhaps on the abundance of the short fleshy roots that Vartanian (1981) has reported forming in response to drought and which seem particularly able to start growing rapidly after rain. The partition between uptake and evaporation, despite its importance, is very poorly understood, and nothing at all seems to be known about associated varietal variation. The problem is very difficult to explore and will remain so until accurate methods are devised for measuring evaporation from the soil under a crop.

RESIDUAL WATER AT MATURITY

Agronomists have found it convenient to assume that there is an amount of water in soil which is 'available' to plants. This amount is the difference between the water content in the soil a couple of days after a soaking rain, the 'field capacity', and the water content below which the plants can extract no water, the 'permanent wilting point'. While this idea is certainly useful in some circumstances it is misleading in others. 'Field capacity' is arbitrary, for a soil continues to drain forever unless equilibrium is established with a water table, in which case the water content depends on the depth of the water table. 'Permanent wilting point' depends in practice not only on the ultimate suction to which the plant can dry the soil but also on the rate at which the roots can extract the water. The problem of the availability of water is a dynamic one.

If the availability of water does depend on how fast the roots are extracting the water, what does this rate depend on? Much has been written on this subject, and Tinker (1976) has provided the most thorough reasonably recent review of it. Water flows through the soil to an absorbing root along gradients of matric potential, and various idealizations of the geometry of the root system and the conditions at its surface have been used to enable differential equations to be devised and solved to give a quantitative view of what is going on. To a good approximation, the rate of uptake of water from a given volume of soil is proportional to the rooting density, L (cm (root) cm (soil)$^{-3}$), the diffusivity of soil water, D (cm^2 s^{-1}), and the difference in soil water content θ (cm^3 cm^{-3}) between the average θ in the soil and θ at the root surface. D varies markedly with θ in any given soil, but, in many soils is about 1 to 2×10^{-5} cm^2 s^{-1} over the range of -5 to -20 bar in soil water potential, ψ, and is about 1×10^{-4} cm^2 s^{-1} when $\psi = -1$ bar.

Given enough time, most crop plants will extract water from the soil occupied by their roots until ψ_{soil} falls to about -15 to -20 bar, at least in

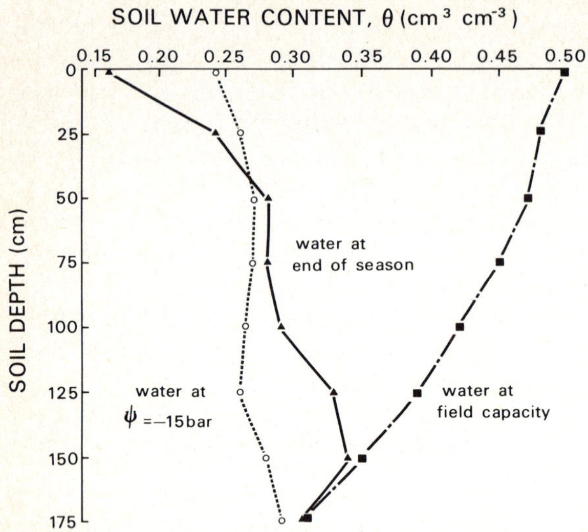

Fig. 3. Distribution of soil water content with depth under a grain sorghum crop growing on Houston Black Clay at Temple, TX. Redrawn from Jordan and Miller (1980).

the top half of the rooting zone, and even in the bottom half many crops will reduce ψ to such a level, during a terminal drought, that there is negligible water left in accessible soil at a ψ greater than −15 to −20 bar (Power et al., 1961; Fischer and Kohn, 1966). Yet there are many other cases of crops in similar circumstances failing to extract substantial amounts of apparently available water from the bottom halves of their rooting zones (Schultz, 1971; Hurd, 1974; Walter and Barley, 1974; Jordan and Miller, 1980). Figure 3, adapted from fig. 25.2 of Jordan and Miller (1980) clearly illustrates the problem; a total of 50 mm of water at a potential greater than −15 bar remained in the soil to the depth of the rooting zone at the end of the season despite a prolonged drought. Why did the roots fail to extract the water? Did they arrive too late? The water content in excess of θ at $\psi = -15$ bar varied from 0.01 to 0.07 $cm^3\ cm^{-3}$ within the rooting zone. Assuming that D is $1.0 \times 10^{-5}\ cm^2\ s^{-1}$ and that the roots can maintain $\psi = -15$ bar at their surface, one can show, using an equation similar to eq. (6) of Tinker (1976), that:

$$\theta_a = \theta'_a \exp(-1.5\, L\, t) \qquad (1)$$

where θ_a = water content in excess of θ at $\psi = -15$ bar;

$$\theta'_a = \theta_a \quad \text{at} \quad t = 0 \qquad (2)$$

and t = elapsed time (day).

Figure 4 shows θ_a as a function of time for $L = 0.5$ and 0.1, and $\theta'_a = 0.05$ and 0.02, and assuming that the roots take up water on a square wave basis for 12 h each day. Given $L = 0.5\ cm^{-2}$ and $\theta'_a = 0.05$, this simple physical

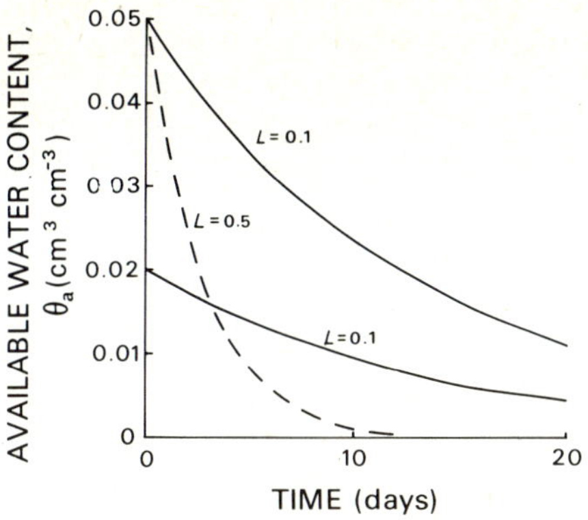

Fig. 4. Available water content, θ_a, as a function of time for rooting densities, L, of 0.1 and 0.5 cm cm^{-3}, and initial θ_a of 0.05 and 0.02, and assuming a soil water potential at the root surface of −15 bar, and a diffusivity of 1×10^{-5} cm^2 s^{-1} (see Eq. 1).

model predicts that 80% of the water will be used in only 4 days. If these conditions obtain in a layer of soil 50 cm thick the transpiration rate would average 5 mm day^{-1} during this period, which is not far below typical potential rates during grain filling. For L = 0.1 cm^{-2} the rate would be 1 mm day^{-1}, and would ensure that most of the water would be used within 10 days. Jordan and Miller's (1980) data suggest that if we define the bottom of the rooting zone as the depth at which L has reached 0.1 cm^{-2}, then the zone had reached about 125 cm by 76 days after planting, and 170 cm by 88 days, by which time the crop was probably just past anthesis. At the same times, L had reached 0.5 cm^{-2} at depths of about 100 and 140 cm, respectively. Jordan and Miller (1980) suggested that L in the bottom half of the rooting zone was too low for the roots to extract all the available water by maturity, and possibly they were right. But if they were right, then the roots must have been much less effective in taking up water than Fig. 4 implies. Possible reasons for the roots being ineffective include: (a) clumping of the roots in, for example, fracture planes in the soil, so that the path length for the flow of water to the roots would be much greater than if the roots were uniformly distributed; and (b) impermeability of a large proportion of the total root length. Little is known about what fraction of the total root length is involved in taking up water from fairly dry soils in the field, but I have estimated from laboratory experiments with wheat seedlings growing in soil of medium texture at ψ = −5 bar, that the effective L was about one-third of the actual (Passioura, 1980). If a similar ratio applied in the field, one would have to increase L accordingly to get the time-courses described in Fig. 4.

A high resistance to axial flow in the roots could conceivably be another reason for the failure of some crops to fully extract available water from the bottoms of their rooting zones. Such a resistance seems unlikely in dicots, for one would expect that, with their facility for secondary growth, they would have an ample supply of xylem vessels, as seems to be the case with cotton (Taylor and Klepper, 1978; Meyer and Ritchie, 1980). In monocots, however, and particularly in spring cereals, inadequate xylem may be a problem, especially if drought prevents the growth of nodal roots so that the plants are relying entirely on their seminal roots to extract the water from the subsoil (Passioura, 1974). The complete failure of sorghum crops that fail to develop nodal roots is a striking example of this problem. Sorghum, with its single seminal root containing typically only one main xylem vessel of 50 μm diameter would require (assuming Poiseuille flow) a pressure gradient of about 0.5 bar cm^{-1} along its seminal root for a transpiration rate of 1 mg s^{-1}, which would typically occur in a plant having a leaf area of about 100 cm^2 on a day of high evaporative demand. Even with this very small leaf area for a sorghum plant it would require a very large pressure drop along the seminal root to supply adequate water. If the root were carrying water at 1 mg s^{-1} through a layer of dry topsoil 20 cm thick the pressure drop along the root would be about 10 bar. Once the nodal root system becomes well established, however, with many root axes containing many large xylem vessels each, axial transport may not be a problem.

The development of effective deep roots depends on both environmental and genetic factors and no doubt there is a strong interaction between the two which makes the job of the breeder very difficult. To select for rooting depth in a breeding program is impractical. Perhaps the best we can do is search for parents with deep roots and inject them, without further explicit selection, into a breeding program, as Hurd (1974) appears to have done. But before embarking on such a breeding program, it is important to consider the long-term worth of putting a lot of effort into producing plants that can extract the last few centimeters of available water from the bottoms of their rooting zones. Ostensibly, the effort would seem well worthwhile. The additional water would not only increase the total water supply, and presumably the biomass, by say 20%, it might also increase the harvest index, because it would become available towards the end of the crop's life, during grain-filling. But the several centimeters of water that appears to be there for the taking may be partly an illusion. It may be the accumulation of several years of incomplete extraction by preceding crops and may represent only a few millimeters a year. The long-term benefit of complete extraction can only be judged by determining the local recharge of the water table, for this recharge reflects the average amount of water that escapes the roots each year. It is important to be sure that the recharge is sufficiently large before embarking on an expensive breeding program.

WATER-USE EFFICIENCY

The amount of above-ground biomass produced per unit of water transpired, the *T*-efficiency of Tanner and Sinclair (1983), varies little between plants of the same metabolic class (C_3, C_4, or CAM) growing in the same environment. Tanner and Sinclair (1983) have thoroughly discussed the factors influencing T-efficiency and conclude that there is little scope for improvement. Perhaps they are right, but given the lack of explicit selection pressure on root growth, and given that any assimilate spent in the roots is assimilate unavailable for the shoot, there may be a chance for a marginal improvement in T-efficiency if root/shoot ratios are reduced. To reduce this ratio in a droughted crop may seem counterproductive. After all, if water is limiting, it would seem better to increase the roots in order to get more water, as would seem appropriate for wild plant competing with neighbours. But as Donald (1968, 1981) has pointed out, we would expect to get the best performance, as a crop, from plants that are weak competitors as individuals. There is no point in a droughted crop investing a parcel of assimilate in its roots if the extra water thereby obtained does not allow the shoots to at least replace the assimilate so spent. Thus, in the instances mentioned earlier, in which the crops extracted all available water by maturity, there is a prima facie case that the root systems were too big; even in those crops that leave several centimeters of apparently available water in the subsoil, the amount of roots in the topsoil might be reduced with some benefit.

The root/shoot ratio in many crops at maturity is about 0.1 (Welbank et al., 1974), which is so low that there seems to be little scope for reducing it. Reducing the ratio by half would free, ostensibly, only 5% more assimilate for the shoot. But there are three other factors that must be taken into account. The first is that old roots tend to decay, and the dry weight that they once contained is often lost by maturity, so that the final harvest is an underestimate of the total dry weight; it is true that saprophytic organisms can also attack old leaves, but generally it is much easier to make a complete harvest of the shoot. The second is that drought usually increases the root/shoot ratio, which can exceed 0.3 in droughted crops (Struik and Bray, 1970; Schultz, 1974). And the third is that soil-grown roots appear to respire, or at least consume assimilate, much more rapidly per unit dry weight than does the shoot (Sauerbeck and Johnen, 1976; Martin and Kemp, 1980) so that the ratio of dry weight between the two may greatly underestimate the proportion of assimilate spent by the roots. The leakage of mucilage and metabolites from the roots, and the steady production and sloughing off of root-cap cells, are a drain on assimilate not paralleled in the shoot. Martin and Kemp (1980) found as much previously assimilated carbon in the soil as in the roots of young wheat plants.

It is hard to assess the cost in water of producing dry matter in the roots. It is probably at least double that for the shoot. A ratio in the range of 500 to 1 000 for transpiration/dry-matter is plausible. From the point of view only of water relations, the discussion centered on Fig. 4 would imply that a plant that achieves $L = 0.5\ \mathrm{cm}^{-2}$ throughout the wetted soil profile should face no limitations within the soil to the reasonably rapid extraction of water; Rickman et al. (1978) showed such behavior for their dryland winter wheat. If roots at $L = 0.5\ cm^{-2}$ extend to a depth of 2 m, the total root length per hectare would be 1×10^8 m. Jordan and Miller (1980) estimated that sorghum roots weigh about 6 $mg\,m^{-1}$. Estimates from the data of Merril and Rawlins (1979) for sorghum (5 $mg\ m^{-1}$), and of Gregory et al. (1978) for wheat (4 $mg\ m^{-1}$), are similar. Taking 5 $mg\ m^{-1}$ as typical, 1×10^8 m of root would weight 500 kg and would require 250 to 500 $t\,ha^{-1}$ (25 to 50 mm) of transpired water to produce. If the roots and wetted profile extended to 1 m, as they often do, instead of to 2 m, these amounts would be halved. The striking thing about these calculated weights is how small they are; the root weights of droughted crops are typically several times larger, being in the range of 1 000 to 2 000 $kg\,ha^{-1}$ (Schultz, 1974), and presumably requiring about 100 mm of water to produce.

Does this mean that we may in principle be able to breed plants with smaller but adequate root systems that will release dry matter equivalent to about the weight of current root systems for incorporation into the shoot? Perhaps, but we must first enquire if the apparently excessive root length in many of our current varieties is needed for other purposes. There are several conceivable benefits that could accompany a large rooting density in the plow layer, which is where most of the apparently excessive root length usually occurs. These include: (a) a rapid uptake of recent rain before it evaporates, as discussed earlier; (b) the provision of spare capacity in case of disease; (c) the ability to extract poorly mobile nutrients like potassium and phosphorus; and (d) the ability to compete with weeds for both water and nutrients. Each of these benefits is plausible, particularly the last two, but they are both amenable to management: banding the fertilizer would ensure that it remains concentrated during the life of the crop and therefore readily accessible to a few roots; good farming practice, or, looking ahead, the invention of better herbicides, will obviate the problem of weeds.

So perhaps drought resistance might be improved by, paradoxically, decreasing the size of the root system. Given that a large root/shoot ratio may be more appropriate for a wild plant than a cultivated one, we might expect that domestication has reduced the root/shoot ratio. No data appear to be available, but given that there has been little, if any, improvement in aboveground biomass production in modern cultivars (improvements in harvest index accounting for improved yields) (Gifford and Evans, 1981), there may have been no shift in root/shoot ratio.

HARVEST INDEX

Harvest index, HI, the ratio between harvestable yield and aboveground biomass, may vary from zero to more than 0.5 in a droughted crop. Its value depends on the pattern of water use (De Wit, 1958), especially in determinate crops like cereals. Early drought tends to give a large HI; late drought tends to give a low HI, and so does a sharp drought at or slightly before flowering. A uniformly distributed drought need not damage HI, as is illustrated by Figs. 1 and 2.

The basis of these responses is best discussed in relation to the two main components of traditional yield-component analysis, namely, the number of grains set per unit area and the average weight per grain. Rapidly falling water status about the time of flowering, resulting from sharply felt drought, can severely damage grain number (Salter and Goode, 1967; Fischer, 1973; Morgan, 1980). No matter how good are the following conditions, HI will always be low in these circumstances. A severe drought late in the life of the crop, after the grain number has been firmly set and is no longer susceptible to the plastic behavior of the plant, results in shrivelled grain and a low HI. The following discussion concerns the ways in which roots influence the pattern, or timing, of water use, and hence the harvest index, of determinate crops.

Plants that grow in the wild and compete with other plants for a limited supply of water often use the water fast until there is so little left that it becomes a means of survival rather than of production (Fischer and Turner, 1978). To conserve large amounts of water for later use does not make good evolutionary sense if a prodigal competitor is nearby. With crop plants, on the other hand, where it is the performance of the community rather than of the individual that matters, conservative behavior is likely to give the highest yields (Donald, 1968, 1981), or at least the highest HI if there is drought, through the way it buffers the water status of the plants. Buffering implies that the crop must drop its transpiration rate below the potential rate even though a substantial amount of available water remains in the soil. In practice this means that the crop must keep its leaf area index, LAI, below about 2.5 once the effects of a drought become anything more than mild: a low LAI enables the crop to maintain a high water status while using water slowly. To a good approximation the transpiration rate of a crop with open stomata is proportional to LAI up to about 2.5 after which it becomes independent of LAI (Ritchie, 1974). The transpiration rate at high LAI can only be controlled by the stomata, and in these circumstances the risk of a rapidly falling water status is great, with its concomitant rapid senescence of leaves, and the possible drop in fertility. Droughted crops do tend to vary their transpiration rates by varying leaf area rather than stomatal conductance (Legg et al., 1979; Leach, 1980).

Ideally when the LAI of the crop should be adjusted to the available water content in the soil well before the likely onset of severe water stress. What are the means by which roots influence the development of leaf area? The most straightforward effects are morphological ones associated with the rate of uptake of water, such as the slow exploration of a wet subsoil that was apparent with Jordan and Miller's (1980) sorghum, or a large resistance to axial flow in the roots which becomes apparent once the topsoil is dry (see below). If the roots extract water at say half the potential evaporation rate for many days then the crop will presumably develop and hold a green LAI of about 1.5.

The mechanism by which plants adjust their leaf areas to their water supplies is unknown. It presumably involves subtle effects on the water relations of the shoot and may be largely physical. But there are other, even more subtle, ways, apparently unrelated to the water status of the shoot, and probably involving growth regulators, in which roots can influence leaf area (Troughton, 1980).

The above discussion has been rather general. The remainder of this section treats of a specific case in which a simple and heritable anatomical feature of wheat roots can have a large influence on the pattern of water use of wheat plants.

In many wheat-growing areas, wheat crops are exposed to severe droughts during grain-filling, and in some areas it is common for the crops to receive so little rain during the growing season that they rely largely on water contained in the soil at the time of sowing that accumulated during a fallow. Grain yield in these circumstances may depend very much on the amount of usable water contained in the soil at about the time of flowering (Nix and Fitzpatrick, 1969). There is presumably an optimum ratio, given a limiting supply of water, between the amount of water used before flowering, and that used, or available for use, after. This optimum will no doubt depend on the pattern of evaporative demand, but for plants growing during a slowly increasing evaporative demand the optimum seems to be about 2 : 1. This ratio gives a high HI while ensuring that essentially all available water is used by maturity, in both the glasshouse (Passioura, 1976, 1977) and the field (Angus et al., 1980). Where drought brings about a low HI, then, it may often be because of a too-rapid use of water before flowering. How could we set about slowing the early water use where this is deemed necessary? In some areas, the farmers have evolved techniques such as: growing their crops on heavy soils, particularly cracking soils, that release water comparatively slowly to the plants growing on them; ensuring that the crops are not well endowed with nitrogen, which tends to produce a large LAI when in good supply; or using low seeding rates or wide rows. But there is, in principle, another way of getting the crops to buffer their use of water. This involves generating a large hydraulic resistance in the xylem connecting the transpiring shoots to the roots exploring a wet subsoil. Such a resistance would manifest itself once the topsoil became dry and would prevent the crops from rapidly using the water in the subsoil. The morphology of a wheat plant is well suited for inducing this resistance; the following account of it is based largely on the review by Troughton (1962).

Field-grown wheat plants typically have three to four seminal roots (those arising from the seed) that penetrate deeply into the subsoil. If the crowns of the plants are in moist soil they will produce many nodal roots which, in a long-season (winter) wheat crop, will also penetrate deeply into the subsoil. But in a short-season crop, the nodal roots are largely restricted to the top 30 cm of the soil and in prolonged early droughts may even not develop at all. Once the topsoil is dry, then, the water collected by the roots in the subsoil is transported through the dry top soil by only about three seminal roots. Furthermore, each of these roots typically contains only one major xylem vessel, whose diameter is usually within the range 55 to 70 μm in the basal parts of the roots, and which increases in diameter with increasing depth (Meyer and Alston, 1978). Assuming Poiseuille flow, large gradients of pressure would be needed to sustain flow through the smaller of these vessels at rates large enough to satisfy the evaporative demand in a typical wheat crop (Passioura, 1972, 1974; Greacen et al., 1976; Meyer and Alston, 1978). There is nevertheless, a case for making the vessels even smaller if there is evidence that the crop is using its water too quickly.

The average diameter of these vessels is influenced by several factors (Richards and Passioura, 1981a, b): high temperatures during germination and establishment decrease it; small seeds produce small vessels, and so do seeds, for a given seed size, that came from a droughted crop. Age does not influence vessel size, because once a vessel has matured it is comprised of a series of dead cells that have lost the metabolic machinery needed for change. The genetic component of variation in vessel size is substantial, with a narrow-sense heritability of about 50%. My colleague, R.A. Richards, has embarked on a back-cross breeding program to incorporate small xylem vessels into the seminal roots of some of the best Australian cultivars. He is currently testing some 40 BC3F5 lines for their ability to yield under drought.

Agronomically, one may be able to influence vessel size by selecting small seed or seed harvested from droughted crops, but doing this could lead to problems of establishment. Cornish (1981) has suggested that it may be possible to introduce a large hydraulic resistance between the shoot and the seminal roots by deep sowing. The intracoleoptile internode that forms between the seed and the crown of a wheat plant has a large hydraulic resistance per unit length owing to the vessels within it being very small — typically less than 20 μm in diameter. When a seed is sown deeply, it produces a long internode (for the crown usually forms near the surface of the soil irrespective of the sowing depth) and the hydraulic resistance within this internode is correspondingly large. Cornish's (1981) suggestion is an interesting one and may provide a simple way of agronomically improving drought resistance in certain circumstances. It may be more useful with the tall wheats than with the semidwarfs because of the much longer internodes that they can produce; there is a maximum length that the internode can attain, and if one sows more deeply than this length establishment will be poor.

Drought resistance is often correlated with an inability to grow well when water is ample. This correlation is particularly clear when we compare xeric species with mesic ones, but it is also common within a species. Laing and Fischer (1977) and Keim and Kronstad (1979) provide examples in wheat. One of the attractions of inducing a large hydraulic resistance in the seminal roots or intracoleoptile internode of wheat as described above is that it may improve the drought resistance without prejudicing the plant's ability to respond to a good season. For in a good season, when the topsoil is well-watered, it is the nodal root system that supports the plant, and the high resistance in the seminal system does not then manifest itself.

GENERAL DISCUSSION AND CONCLUSION

The yield of a crop is the outcome of myriad processes occurring at many time scales. When we set out to improve that yield we usually do so within a conceptual framework that helps us form ideas about what is an ideal plant. This framework and our notions of the ideal are often quite implicit and very rudimentary. In the context of roots and drought resistance the traditional and popular framework is the extremely simple one that more, and more vigorous, roots gives more drought resistance. I have tried to show that this is not necessarily true, and that a more useful framework is provided by the identity discussed in the Introduction in which yield is viewed as the product of water used, efficiency of water use, and harvest index. This framework is also very simple, but it does help us to focus on a wide range of physiological and morphological attributes of possible significance to drought resistance, and furthermore, it helps us assess their importance (Passioura, 1981).

An important though little-discussed feature of physiological processes that influence drought resistance is the time scale at which they operate. We have seen how harvest index is correlated with how well the water supply is buffered, or alternatively how damaging rapid changes in water status can be, particularly around the time of flowering. Very many papers have been written on the behavior of plants that have been subjected to sudden and severe water stress in glasshouse or laboratory. Leaf water potential of a large plant in a small pot will change by several bars per day if one stops watering the pot. This change is much faster than what normally happens in the field and may elicit responses in the plant that, despite their intrinsic physiological interest, may have little bearing on performance in the field (Begg and Turner, 1976).

Given the lack of major variation in transpiration ratio among plants of a given metabolic class growing in similar environments, and given the sensitivity of harvest index to both rapid changes in water status and the depletion of soil water prior to flowering, it is likely that we will make most rapid progress in improving drought resistance by concentrating on those processes that buffer the rate of use of soil water, or increase its effective supply. The injection of deep-rooted lines into a breeding program for areas in which substantial amounts of water are left in the subsoil at maturity, and the induction of a large resistance within the plant to the flow of water through a dry topsoil in areas where the plants are prone to use water too quickly when they are young, are two ways that provide a reasonable chance of improving drought resistance in the near future.

REFERENCES

Angus, J.F., Nix, H.A., Russell, J.S. and Kruizinga, J.E., 1980. Water use, growth and yield of wheat in a subtropical environment. Aust. J. Agric. Res., 31: 873—886.

Begg, J.E. and Turner, N.C., 1976. Crop water deficits. Adv. Agron., 28: 161—217.

Cornish, P.S., 1981. Resistance to water-flow in the intracoleoptile internode of wheat. Plant Soil, 59: 119—125.

De Wit, C.T., 1958. Transpiration and crop yields. Versl. Landbouwkd. Onderz., 64(6): 1—88.

Donald, C.M., 1968. The breeding of crop ideotypes. Euphytica, 17: 385—403.

Donald, C.M., 1981. Competitive plants, communal plants, and yield in wheat crops. In: L.T. Evans and W.J. Peacock (Editors), Wheat Science — Today and Tomorrow. Cambridge University Press, Cambridge, pp. 223—247.

Fischer, R.A., 1973. The effect of water stress at various stages of development on yield processes in wheat. In: R.O. Slatyer (Editor), Plant Response to Climatic Factors. Proc. Uppsala Symp., 1970. UNESCO, Paris, pp. 233—241.

Fischer, R.A. and Kohn, G.D., 1966. Soil water relations and relative turgidity of leaves in the wheat crop. Aust. J. Agric. Res., 17: 269—280.

Fischer, R.A. and Turner, N.C., 1978. Plant productivity in the arid and semiarid zones. Annu. Rev. Plant Physiol., 29: 277—317.

Gifford, R.M. and Evans, L.T., 1981. Photosynthesis, carbon partitioning and yield. Annu. Rev. Plant Physiol., 32: 485—509.

Greacen, E.L., Ponsana, P. and Barley, K.P., 1976. Resistance to water flow in the roots of cereals. In: O.L. Lange, L. Kappen and E.-D. Schulze (Editors), Water and Plant Life. Springer, Berlin, pp. 86—100.

Gregory, P.J., McGowan, M., Biscoe, P.V. and Hunter, B., 1978. Water relations of winter wheat. 1. Growth of the root system. J. Agric. Sci., Camb., 91: 91—102.

Hurd, E.A., 1974. Phenotype and drought tolerance in wheat. Agric. Meteorol., 14: 39—55.

Jordan, W.R. and Miller, F.R., 1980. Genetic variability in sorghum root systems: implications for drought tolerance. In: N.C. Turner and P.J. Kramer (Editors), Adaptation of Plants to Water and High Temperature Stresses. Wiley Interscience, New York, NY, pp. 383—399.

Keim, D.L. and Kronstad, W.E., 1979. Drought resistance and dryland adaptation in winter wheat. Crop Sci., 19: 574—576.

Kramer, P.J., 1969. Plant and Soil Water Relationships: A Modern Synthesis. McGraw-Hill, New York, NY, 482 pp.

Laing, D.R. and Fischer, R.A., 1977. Adaptation of semidwarf wheat cultivars to rainfed conditions. Euphytica, 26: 129—139.

Leach, J.E., 1980. Photosynthesis and growth of spring barley: some effects of drought. J. Agric. Sci., Camb., 94: 623—635.

Legg, B.J., Day, W., Lawlor, D.W. and Parkinson, K.J., 1979. The effect of drought on barley growth: models and measurements showing the relative importance of leaf area and photosynthetic rate. J. Agric. Sci., Camb., 92: 703—716.

Martin, J.K. and Kemp, J.R., 1980. Carbon loss from roots of wheat cultivars. Soil Biol. Biochem., 12: 551—554.

Merrill, S.D. and Rawlins, S.L., 1979. Distribution and growth of sorghum roots in response to irrigation frequency. Agron. J., 71: 738—745.

Meyer, W.S. and Alston, A.M., 1978. Wheat responses to seminal root geometry and subsoil water. Agron. J., 70: 981—986.

Meyer, W.S. and Ritchie, J.T., 1980. Water status of cotton as related to taproot length. Agron. J., 72: 577—580.

Morgan, J.M., 1980. Possible role of abscisic acid in reducing seed set in water-stressed wheat plants. Nature, 285: 655—657.

Nix, H.A. and Fitzpatrick, E.A., 1969. An index of crop water stress related to wheat and grain sorghum yields. Agric. Meteorol., 6: 321—337.

Passioura, J.B., 1972. The effect of root geometry on the yield of wheat growing on stored water. Aust. J. Agric. Res., 23: 745—752.

Passioura, J.B., 1974. The effect of root geometry on the water relations of temperate cereals (wheat, barley, oats). In: J. Kolek (Editor), Structure and Function of Primary Root Tissues. Veda, Bratislava, pp. 357—363.

Passioura, J.B., 1976. Physiology of grain yield on wheat growing on stored water. Aust. J. Plant Physiol., 3: 559—565.

Passioura, J.B., 1977. Grain yield, harvest index and water use of wheat. J. Aust. Inst. Agric. Sci., 43: 117—120.

Passioura, J.B., 1980. The transport of water from soil to shoot in wheat seedlings. J. Exp. Bot., 31: 333—345.

Passioura, J.B., 1981. The interaction between the physiology and the breeding of wheat. In: L.T. Evans and W.J. Peacock (Editors), Wheat Science — Today and Tomorrow. Cambridge University Press, Cambridge, pp. 191—201.

Power, J.F., Grunes, D.L. and Reichman, G.A., 1961. The influence of phosphorus fertilization and moisture on growth and nutrient absorption by spring wheat. I. Plant growth, N uptake, and moisture use. Soil Sci. Soc. Am. Proc., 25: 207—213.

Richards, R.A. and Passioura, J.B., 1981a. Seminal root morphology and water use of wheat. I. Environmental effects. Crop Sci., 21: 249—252.

Richards, R.A. and Passioura, J.B., 1981b. Seminal root morphology and water use of wheat. II. Genetic variation. Crop Sci., 21: 253—255.

Rickman, R.W., Allmaras, R.R. and Ramig, R.E., 1978. Root-sink descriptions of water supply to dryland wheat. Agron. J., 70: 723—727.

Ritchie, J.T., 1974. Atmospheric and soil water influences on the plant water balance. Agric. Meteorol., 14: 183—198.

Salter, P.J. and Goode, J.E., 1967. Crop responses to water at different stages of growth. Res. Rev. 2, Commonwealth Bureau of Horticultural and Plantation Crops, Commonwealth Agricultural Bureau, Farnham Royal, Buks., 246 pp.

Sauerbeck, D. and Johnen, B., 1976. Der Umsatz von Pflanzenwurzeln in Laufe der Vegetationsperiode und dessen Beitrag zur "Bodenatmung". Z. Pflanzenernähr. Bodenkd., 3: 315—328.

Schultz, J.E., 1971. Soil water changes under fallow-crop treatments in relation to soil type, rainfall and yield of wheat. Aust. J. Exp. Agric. Anim. Husb., 11: 236—242.

Schultz, J.E., 1974. Root development of wheat at the flowering stage under different cultural practices. Agric. Rec., 1: 12—17.

Struik, G.T. and Bray, J.R., 1970. Root-shoot ratios of native forest herbs and *Zea mays* at different soil moisture levels. Ecology, 51: 892—893.

Tanner, C.B. and Sinclair, T.R., 1983. Efficient water use in crop production: research or re-search. In: H.M. Taylor, W.R. Jordan and T.R. Sinclair (Editors), Limitations to Efficient Water Use in Crop Production. American Society of Agronomy (In press).

Taylor, H.M. and Klepper, B., 1978. The role of rooting characteristics in the supply of water to plants. Adv. Agron., 30: 99—128.

Tinker, P.B., 1976. Transport of water to plant roots in soil. Phil. Trans. R. Soc. London B, 273: 445—461.

Troughton, A., 1962. The roots of temperate cereals (wheat, barley, oats and rye). Publ. No. 2/1962, Commonwealth Bur. Pastures and Field Crops, Hurley, Berks., 91 pp.

Troughton, A., 1980. Production of root axes and leaf elongation in perennial ryegrass in relation to dryness of the upper soil layer. J. Agric. Sci., Camb., 95: 533—538.

Vartanian, N., 1981. Some aspects of structural and functional modifications induced by drought in root systems. Plant Soil, 63: 83—92.

Walter, C.J. and Barley, K.P., 1974. The depletion of soil water by wheat at low, intermediate and high rates of seeding. Proc. 10th Int. Congr. Soil Sci., Moscow, Vol. 1. Nauka, Moscow, pp. 150—158.

Welbank, P.J., Gibb, M.J., Taylor, P.J. and Williams, E.D., 1974. Rep. 1973, Rothamsted Exp. Stn., Part 2, pp. 26—66.

Agricultural Water Management, 7 (1983) 281—299
Elsevier Science Publishers B.V., Amsterdam — Printed in The Netherlands

STRATEGIES FOR CROP IMPROVEMENT FOR DROUGHT-PRONE REGIONS

W.R. JORDAN, W.A. DUGAS, Jr. and P.J. SHOUSE

Texas Agricultural Experiment Station, Blackland Research Center, P.O. Box 748, Temple, TX 76503 (U.S.A.)

(Accepted 22 November 1982)

ABSTRACT

Jordan, W.R., Dugas, W.A., Jr. and Shouse, P.J., 1983. Strategies for crop improvement for drought-prone regions. *Agric. Water Manage.*, 7: 281—299.

Simulation models for sorghum and wheat were used to predict grain yield resulting from specific modifications to the plant or in how the plant reacts to its environment. Model predictions were made for Lubbock, TX, Temple, TX and Manhattan, KS using daily weather data for the 30-year period between 1 January 1949 and 31 December 1978. Sorghum modifications included maturity, deep rooting, and enhanced osmoregulation, but only deep rooting was tested for wheat.

Simulated yields for an early maturing sorghum hybrid, assumed to reach anthesis 6 to 7 days earlier than the standard hybrids, were in general lower than the standard. A yield advantage due to early maturity was predicted only in years when yields were reduced at least 40% by low water supply. At Manhattan, the early hybrid outyielded the standard in 17% of the years and at Temple in about 33% of the years, but earliness was not favored at Lubbock.

Substantial increases in grain yield were simulated for sorghum and wheat crops capable of deeper rooting. Yields of the deeper rooted crops exceeded those of the normal rooted crop by at least 20% in 20 to 38% of the years across locations. The water use efficiency of deep rooted sorghum, expressed as grain yield/total crop water use, was significantly greater at Temple and Manhattan than the normal rooted crop, because the additional water was available later in the season when a high fraction of the total evapotranspiration occurs as transpiration. As simulated, enhanced osmoregulation resulted in 20% higher yields only 1 year in ten at Lubbock, but additional quantitative data are needed to accurately simulate benefits from this mechanism.

INTRODUCTION

Annual crop plants possess a variety of anatomical, morphological and physiological features that enable them to survive and reproduce in water-limited environments. A partial list of some common mechanisms for drought resistance of wheat and sorghum is shown in Table I. The degree to which these mechanisms can be expressed places limits on the extent to which the harmful consequences of drought can be avoided or tolerated. Because

0378-3774/83/$03.00

TABLE I

Mechanisms that may contribute to drought resistance of wheat and sorghum growing under water-limited conditions

Mechanism of drought resistance	Dry matter or yield reduced?	Reversible on relief of stress?	Reference	
			Wheat	Sorghum
(A) Drought escape				
(1) Rapid phenological development	no?	no	Fischer (1981)	Blum (1970)
(2) Developmental plasticity	no	yes	Angus and Moncur (1977), Fischer (1981)	Blum (1979)
(B) Drought tolerance at high water potential				
(1) Reduction of water loss				
(a) Increase in stomatal resistance	yes	yes	Jones (1977), Blum (1981)	Henzell et al. (1975)
(b) Reduction in leaf area (senescence)	yes	no	Fischer and Kohn (1966)	Jordan (1983)
(c) Reduction in leaf area (orientation)	yes	yes	—	Begg (1980)
(d) Increase in epicuticular wax	no	no	Johnson et al. (1983)	Blum (1975), Jordan et al. (1983)
(e) Increase in cuticular resistance	no	no	—	Blum (1979)
(2) Maintenance of water uptake				
(a) Increased root depth/density	no?	yes?	Hurd (1974), Ansana and Singh (1968)	Jordan and Miller (1980)
(b) Increased liquid phase conductance	no	no	Passioura (1972, 1977)	—
(C) Drought tolerance at low water potential				
(1) Maintenance of turgor				
(a) Osmotic adjustment	no?	yes	Morgan (1980)	Jones and Turner (1978)
(b) Increase in cellular elasticity	no	unknown	—	Turner and Jones (1980)
(c) Decrease in cell size	yes	limited	—	Turner and Jones (1980)
(2) Desiccation tolerance				
(a) Protoplasmic tolerance	no?	unknown	Blum and Ebercon (1981)	Sullivan and Blum (1970) Jordan and Monk (1980)
(b) Maintenance of chloroplast integrity	no?	unknown	—	Sullivan and Blum (1970)

After Turner (1979) as modified by Jordan and Monk (1980) and Turner (1982)

several genes may be involved with each mechanism listed in Table I, the genotype × environment interaction for each may be complex.

The more obvious mechanisms for drought resistance, often exhibited by plants growing in desert environments, are strongly associated with survival. In an agronomic sense, drought resistance implies maintenance of productivity at levels that provide an economically acceptable yield for the producer. Since, by experience, man has discovered the general environmental limits for successful production, most crops are rarely exposed to conditions where survival mechanisms would be important. It follows then that greatest emphasis in crop improvement programs should be directed toward crop modifications that contribute to maintenance of dry matter production or yield.

The challenges to crop improvement programs dealing with drought resistance are threefold: (a) traits that promote maintenance of productivity must be identified; (b) variability for the traits must be assessed and superior genetic sources identified within the species or closely related species; (c) the gene or genes governing the traits must be incorporated into good agronomic cultivars. While these goals appear superficially simple, they have been extremely difficult to achieve in practice. Despite a body of literature replete with examples of specific responses to water deficits, very few traits can be shown to be causally related to maintenance of productivity during drought. While a few breeding programs deal with individual traits to enhance drought resistance (Blum, 1979; O'Toole and Chang, 1979; Townley-Smith and Hurd, 1979; Richards and Passioura, 1981b; Jordan and Miller, 1980), most rely on traditional approaches using selection under favorable agronomic conditions and testing for yield under water-limited conditions. Some programs also utilize selections made during drought, but because the timing, duration and intensity of droughts are seldom repeatable in the field, genetic gains from one cycle of selection are often lost in succeeding cycles. For this reason, a thorough understanding of the seasonal water supply is basic to the formulation of a strategy for crop improvement.

In the following sections, we develop a conceptual framework for crop improvement in the water-limited Great Plains region of the United States. Crop response will be estimated with deterministic, process-based simulation models using historical weather data.

EXPERIMENTAL PROCEDURES

The yield gains from specific modifications to crop maturity, rooting pattern and capacity for osmoregulation were predicted for grain sorghum and winter wheat crops grown at Temple and Lubbock, TX and Manhattan, KS. General features of the climate for these three locations are illustrated in Fig. 1. Even though the Great Plains is generally a region of summer rainfall maximums, potential evapotranspiration exceeds precipitation throughout much of the growing season for summer crops.

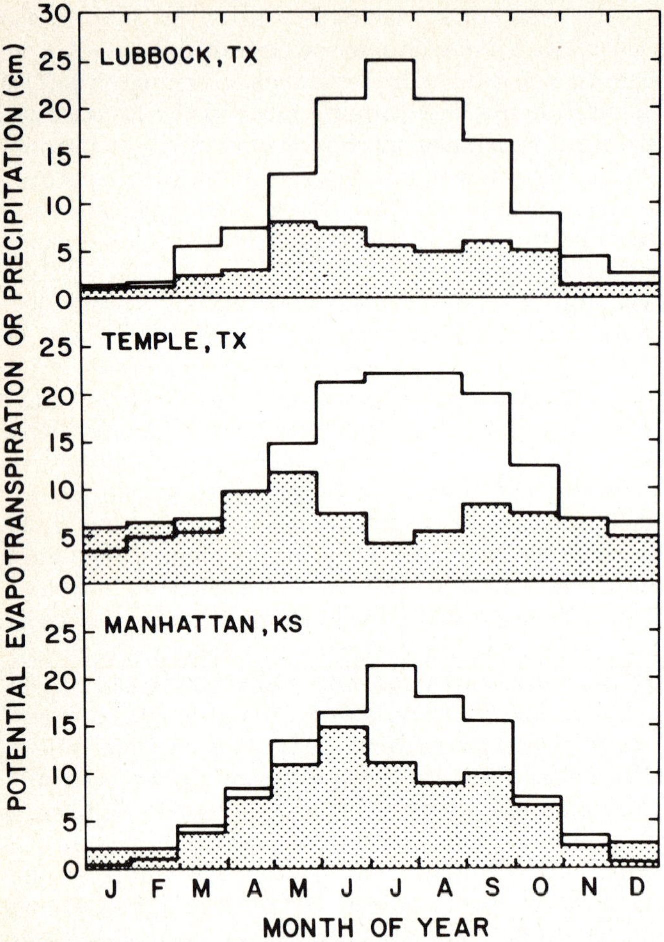

Fig. 1. General features of the climate at Lubbock and Temple, TX and Manhattan, KS as represented by the 30-year average monthly distribution of precipitation (shaded bars) and potential evapotranspiration calculated by the method of Priestley and Taylor (1972). Adapted from Jordan and Sullivan (1982).

Both crop simulation models, CSM's, used in these analyses require daily weather information as inputs to predict the daily accumulation of dry matter and its distribution into various plant parts. The sorghum model, SORGF, utilizes inputs of maximum and minimum air temperatures, solar radiation and rainfall in addition to plant, location and management-specific inputs (Table II). Details of SORGF have been presented (Arkin et al., 1976) along with a validation for the central and southern Great Plains (Vanderlip and Arkin, 1977).

The wheat model, TAMW, was validated with data from ten locations ranging from South Dakota to Texas (Maas and Arkin, 1980). To produce a

TABLE II

Inputs for for crop simulation models for wheat (TAMW) and sorghum (SORGF) at Temple and Lubbock, TX and Manhattan, KS

Input	Temple	Lubbock	Manhattan
Crop-independent			
latitude (°N)	31	33.5	39.2
soil albedo (%)	9	9	9
U (cm)[a]	1.5	1.3	1.0
CONA[b] (mm day$^{1/2}$)	3.5	4.0	4.5
soil	Frio silty clay loam (fine, montmorillonitic, thermic Cumulic Haplustolls)	Pullman silty clay (fine, mixed thermic Torrertic Paleustolls)	Grundy silty clay loam (fine, montmorillonitic, mesic Vertic Argiudolls)
Wheat-specific (TAMW)			
sowing date	25 October	15 October	6 October
row spacing (cm)	20	20	20
inter-plant spacing (cm)	1.9	1.9	1.9
sowing depth (cm)	4	4	4
maximum depth of soil water extraction (m)			
'normal' cultivar	2.2	2.2	1.2
'deep rooting' cv	2.8	2.6	1.8
max. ASW[c] (cm)			
'normal' cultivar	18.8	22.2	17.2
'deep rooting' cv	25.4	26.7	25.8
Sorghum-specific (SORGF)			
sowing date	15 March	24 May	29 May
row spacing (cm)	69	69	69
plant population (10^3 plants ha^{-1})	148	148	148
sowing depth (cm)	5	5	5
number of leaves	17	17	17
maximum depth of soil water extraction (m)			
'normal' hybrid	2.0	2.0	1.2
'deep rooting' hybrid	2.8	2.4	1.6
max. ASW[c] (cm)			
'normal' hybrid	18.8	21.8	17.4
'deep rooting' hybrid	26.2	26.6	23.7

[a]Upper limit of stage 1 soil evaporation.
[b]Slope of cumulative stage 2 soil evaporation with time.
[c]Maximum plant available soil water and rooting depth for the normal cultivar were measured during a terminal drought.

simulation of wheat growth, development and yield, TAMW requires the input of daily values of radiation, rainfall, snow depth, and maximum and minimum air temperatures. Additional crop and location specific inputs are given in Table II.

The drained upper and lower limits of water availability for the given soils were obtained from field measurements of wheat and sorghum crops (Ritchie and Ratliff, 1981). The lower limit of soil water availability was measured when the crops were near death from soil water deficit, i.e. a terminal drought. The difference between the drained upper and lower limits for extraction was termed the maximum plant available soil water,

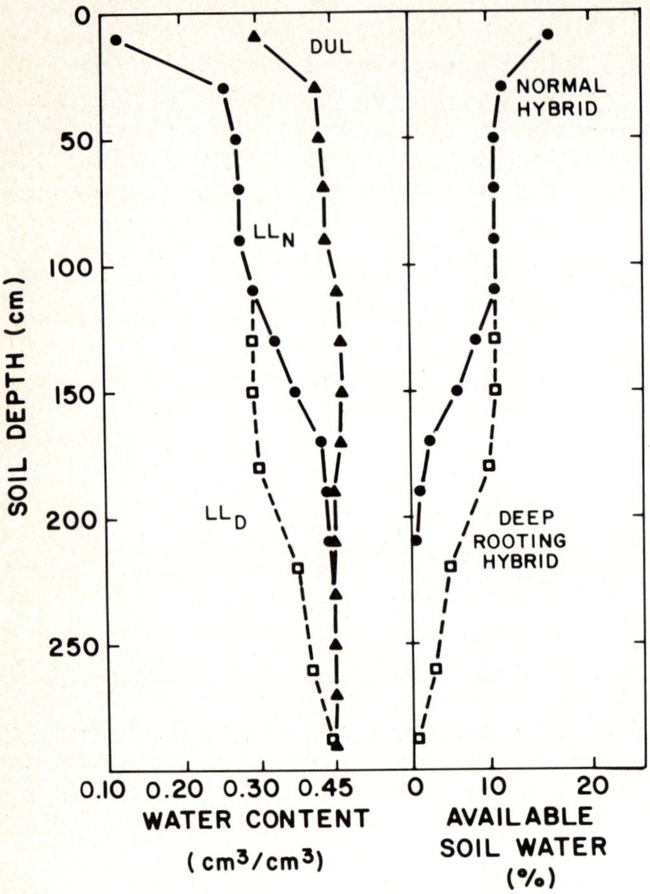

Fig. 2. An illustration of the distribution of the upper and lower limits of soil water availability (left) and plant available soil water (right) for a grain sorghum hybrid grown on a Frio silty clay loam soil near Temple, TX. The solid and dashed curves represent the limits for normal and deep rooting hybrids. DUL, drained upper limit; LL_N and LL_D, lower limit for water extraction by normal and deep rooting sorghum crops during a terminal drought (L.F. Ratliff, Blackland Research Center, Temple, TX, personal communication, 1982, for DUL and LL_N).

ASW (Table II). Fig. 2 illustrates the available soil water distribution and amount for a normal and deep rooted (assumed) grain sorghum crop growing on a Frio soil near Temple, TX. Actual daily ASW was determined in the model with a soil water balance subroutine. To initialize the simulations for both crops, we assumed a soil water content that was 50% maximum ASW on 1 January 1949. Once initialized, the model was run daily for the 30-year period between 1 January 1949 and 31 December 1978. Thus, crop water use during the preceding season affected the initial soil water conditions at the beginning of each successive cropping year.

To test the effects of specific crop modifications on predicted grain yield during the 30-year period, we altered one or more of the following crop-specific inputs or basic functional relationships within the CSMs:

(1) Crop maturity (sorghum only) — The number of days between planting and anthesis is taken as a measure of maturity. In SORGF, maturity is determined by specifying the number of leaves that develop. For comparison, we assumed the early hybrid formed 15 leaves while the late hybrid formed 17 leaves. The corresponding difference in anthesis date is about 6 to 7 days, depending upon location.

(2) Root systems — In the original version of SORGF the soil profile was treated as a single compartment from which water was removed to match the evaporative demand until some lower limit of available soil water was reached. We modified the water balance subroutine of SORGF to match that included in TAMW so that the soil was layered and soil water was extracted sequentially from the surface layer downward. We did not modify the functional relationships that determined final rooting depth or dry matter partitioning of either wheat or sorghum. We assumed that more water is extracted from the deeper soil layers (Fig.2) in accordance with root density profiles illustrated by Hurd (1974) and Jordan and Monk (1980). For the deep rooted cultivar, water was extracted in greater amounts from intermediate depths around 1 m and from greater depths than the normal cultivar to yield an additional 6.6 and 7.4 cm of ASW for the whole profile at Temple, TX (Table II). The actual plant available soil water in the deep layers depends on both preseason precipitation and total water use by the preceding crop.

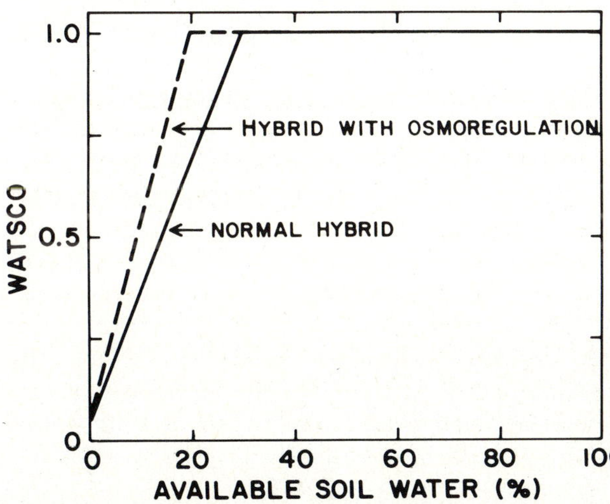

Fig. 3. Relation of available soil water to activity of turgor-dependent processes (WATSCO) for a normal grain sorghum hybrid and a hybrid with enhanced osmoregulation (modified from Arkin et al., 1976).

(3) Osmoregulation (sorghum only) — Osmoregulation is one form of acclimation to soil water deficit and is manifested as solute accumulation in plant tissues. One consequence of osmoregulation is the maintenance of cellular turgor when plant water potential falls. Turgor-dependent processes such as photosynthesis appear to be maintained at high rates to lower levels of available soil water when osmoregulation occurs, although other ramifications of this mechanism are poorly understood. We modeled enhanced osmoregulation by lowering the threshold value of available soil water at which water (turgor-)dependent processes (WATSCO) are reduced below potential values (Fig. 3).

PREDICTED BENEFITS FROM SPECIFIC CROP MODIFICATIONS

Crop maturity

Previous work. The importance of properly matching crop maturity to seasonal patterns of water availability, especially in Mediterranean climates, is established for sorghum (Blum, 1970) and wheat (Fischer and Wood, 1979). Much less well known, however, are effects of maturity in regions with unpredictable drought typical of the Great Plains. Following an analysis of phenological and yield data from hard red winter wheat trials between 1937 and 1965, Johnson et al. (1968) concluded that maturity played an important part in yield improvement, particularly under drought conditions. For locations in the Central Great Plains, Reitz (1974) cites yield increases of from 57 to 97 kg ha^{-1} for each day varieties matured earlier than the standard, 'Kharkof'.

The optimum flowering date for a location where yield is water-limited is determined by the pattern of water use and dry matter production during the growing season. When dry matter of an Australian wheat crop was varied by plant population, planting date and fertility, Fischer (1979) and Doyle and Fischer (1979) concluded that dry matter at anthesis, DM_a, was highly correlated with yield under terminal drought. This correlation results from a close association of DM_a with kernel number and with total dry matter available for remobilization to grain after flowering (source strength). Total dry matter available for grain filling arises from photosynthate produced during the period combined with the fraction of DM_a that can be remobilized to the grain. The fraction of DM_a available for grain filling appears to be about 10% (Fischer, 1981, based on data of Bidinger et al., 1977), but this value remains open to question (Gallagher et al., 1976). The fraction of grain dry matter derived from DM_a increases with post-flowering drought severity and the absolute amounts of DM_a remobilized to grain may be limited by the length of time the crop remains alive (Passioura, 1977).

The optimum ratio of pre- to post-flowering water use is estimated at about 2 : 1 by Passioura (1983) for spring wheat subjected to a slowly increasing evaporative demand. Greater water use before flowering is usually associated with greater DM_a, but soil evaporation may comprise 20 to 40% of the total water use depending on cultural practices, evaporative demand and frequency of rainfall (Ritchie, 1974; Fischer and Turner, 1978). Because transpiration is coupled with photosynthesis and dry matter accumulation and soil evaporation is not, one strategy for more efficient water use is to achieve full ground cover earlier in the season thus reducing soil evaporation. However, if planting and flowering dates are not altered, earlier attainment of ground cover implies greater preflowering water use, because water use is roughly proportional to LAI values below about 3 (Ritchie, 1974). In this situation, the increased risk of severe water shortage after flowering may negate any potential benefit from reducing soil evaporation. Serious water deficits that develop suddenly after flowering may lead to incomplete grain filling and low harvest index (Passioura, 1977; Angus et al., 1980), but if water deficits develop slowly, yield may be reduced up to 50% without altering the harvest index (Boyer and McPherson, 1975; Garrity et al., 1982).

Crop maturity and simulated yields. Because the literature supports the view that late maturity is undesirable for crops grown in the Great Plains, a comparison of maturity dates presents an excellent situation for testing the predictive powers of CSM's. Two such CSM's have already been used to test the effects of earlier maturity on yields of wheat (Ritchie, 1980) and grain sorghum (Jordan and Sullivan, 1982). For wheat grown at Altus, OK, the analysis presented by Ritchie (1980) affirmed the benefit of earlier floral initiation (2 March versus 6 March). In 50% of the years, the predicted yield advantage of the earlier 'variety' was about 240 kg ha^{-1} or 20%. Expressed in a manner similar to Reitz (1974), early floral initiation was worth 60 kg ha^{-1} day^{-1} compared with the later variety.

The effect of earlier maturity on sorghum yields at three Great plains locations is shown in Fig. 4. When water supplied allowed high yields, the later maturing hybrid was fovored at all three locations, a finding in agreement with the negative correlation between early maturity and yield under optimum conditions. As expected, the advantage of early maturity was manifest only in years where yields were greatly reduced by lower water supply. At Manhattan, the early hybrid was favored in 17% of the years and at Temple in about 33% of the years, but earliness was not favored at Lubbock. Based on these simulations, the risks associated with production of the later maturing hybrid are more than compensated for by their higher yield potential.

These results should be regarded with some degree of caution because we assumed identical planting dates each year. At Lubbock, planting is normally delayed until sufficient rainfall has occurred to insure establishment of adequate stands. Should the planting date be greatly delayed, an earlier maturing hybrid may be preferred to reduce risks of yield loss from low temperatures near seasons end. There may indeed have been several instances

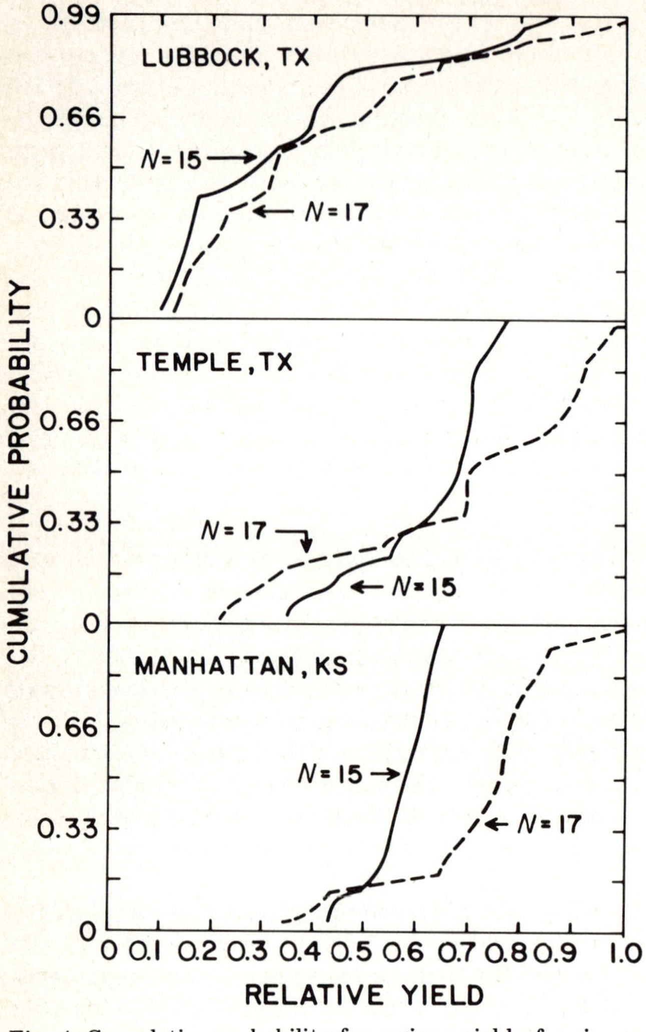

Fig. 4. Cumulative probability for a given yield of grain sorghum at Lubbock and Temple, TX and Manhattan, KS for an early (N = 15 leaves) and late (N = 17 leaves) maturity genotype (see text for details). Yields were simulated with SORGF using daily weather inputs for a 30-year period and are expressed relative to the highest yield at each location. Adapted from Jordan and Sullivan (1982).

during the 30 year simulation that early maturity would have promoted higher yields, but these situations must be regarded as exceptions and would play a small role in determining the optimum maturity from a crop improvement standpoint.

Root systems

Previous work. In the conceptual framework developed by Passioura (1977) and expanded by Fischer (1981), the grain yield of winter wheat is closely related to both the total seasonal crop water use and the distribution of water use before and after anthesis. The literature contains examples of at least two philosophies of root modifications which may alter either the rate of soil water extraction by the crop or the total amount water extracted under water-limited conditions. One philosophy, developed by Passioura (1977), centers about the conservation of water early in the season so greater amounts are available for grain filling. A second philosophy proposes that increased supplies of water may be made available by deeper rooting under certain circumstances.

For spring wheat grown in a terminal drought, Passioura (1977; Richards and Passioura, 1981a, b) suggests that the resistance of seminal roots to water flow could control the rate of water use by the crop. Under these specific conditions the crop must often grow to maturity on stored soil water, but total water supplies are limited because deep soil layers are not recharged annually. The nodal root system does not develop unless the soil surface is wetted by rainfall, therefore in dry years the plants depend wholly on water collected and transported by the seminal roots. If the root resistance is high, a lower leaf water potential will be required to transport water to the shoot. Because low leaf water potentials are aften symptomatic of drought stress, and vegetative growth is reduced by such stress, the end result is reduced DM_a, not as a result of low soil water availability, but due to high root resistance. A greater fraction of the ASW is thus conserved for use during grain filling and high harvest index is maintained. This view is more fully developed by Passioura in this volume.

The risk of yield reduction associated with high root resistance seems minimal although the water savings by restricing DM_a unnecessarily may not be as large as envisioned. Soil water evaporation is greater for crops with incomplete canopies and some of the potential water savings from reduced DMa are not realized (Fischer, 1981). Above average rainfall early in the season would not only contribute to soil water storage, but it would also stimulate nodal root development and increase DM_a. One could envision a situation where above average rainfall near anthesis would not be efficiently utilized if DM_a (hence, potential grain numbers and source size) were too low, but the probabilities for this to occur in the Australian situation are low.

Even though terminal drought is a feature of portions of the Great Plains, production in substantial areas is often limited not by a deficiency of stored soil water but by an inability of the crop to either fully extract water stored deep in the profile or extract it fast enough or soon enough to contribute to yield maintenance. In these situations, crop improvement to enhance either rooting depth and/or rooting density at greater soil depths would appear

Osmoregulation

Previous work. When a plant cannot maintain water uptake rates sufficient to replace transpiration loss, leaf water potential falls. Osmotic adjustment is one means of maintaining turgor and preserving turgor-dependent processes at low water potentials (Table I). Full or partial turgor maintenance during a soil drying cycle via this mechanism seems widespread among crop species (Turner and Jones, 1980). Both sorghum and wheat are capable of osmoregulation when exposed to soil water deficit (Jones and Turner, 1978; Stout and Simpson, 1978; Turner et al., 1978; Morgan, 1980) and genetic variability in osmoregulatory capacity has been reported for both crops.

The benefits derived from turgor maintenance via osmoregulation are less well studied than the phenomenon itself, but most seem to be associated with maintenance of high stomatal conductances during drought (Turner et al., 1978). Presumably, maintenance of high stomatal conductance permits maintenance of photosynthesis as leaf water potentials decline (Jones and Rawson, 1979). If this photosynthate were used to maintain root growth, as is often observed in droughted crops, larger volumes of soil may be explored and harmful dehydration may be postponed or avoided if rainfall is not delayed too long. There is little definitive information on how much additional soil water may be made available or how effective the additional water is in yield maintenance. If rooting depth is not promoted, the additional water available to the crop in lowering the leaf water potential a few bars is not large and would support transpiration only a few days (Jordan and Miller, 1980). However, if water deficits occur after anthesis, maintenance of high photosynthetic rates for even a few additional days during grain filling may have a significant impact on grain yield.

Yield predictions from SORGF. The predicted yield advantage associated with enhanced osmoregulation is shown in Table III. Maximum benefits were predicted for Lubbock where the modification resulted in a simulated yield increase of 20% only about 1 year in 10. No substantial yield advantage was predicted for Manhattan or Temple.

A yield benefit due to osmoregulation is not supported by experimental evidence for any crop, therefore it is difficult to comment on the validity or value of these results. We have been conservative, perhaps too much so, in assuming the only effect of osmoregulation is to lower the ASW threshold from 30—20% for the WATSCO function (Fig. 3). For these three soils this modification allowed use of only about 2 cm of additional water at potential rates, enough to satisfy unrestricted transpiration of a fully developed crop for about 3 to 4 days. Because the differences in process activities (WATSCO) were not large at any time, and the differences were apparent for only a few days when ASW was low, the benefits in terms of yield were small. This same conclusion may be reached from results of simulations in which osmoregulation and deep rooting were combined (Table III). The yield advantage from the two mechanisms combied was only slightly superior to deep rooting alone with the effect most apparent at Lubbock.

DISCUSSION AND CONCLUSIONS

Commercial seedsmen and state and federal agricultural research agencies have, for the large part, effectively utilized most of the existing genetic variability for maturity in our major cereal grains. In certain regions, however, there may be room for further improvement. Fischer (1979) and Nix (1975) conclude that over much of southern Australia the flowering date for winter wheat should be as early as permitted by the risk of spring frost. Earlier maturity in these regions must be accompanied by greater tolerance to spring frost and greater dry matter production at low temperatures (Fischer, 1981). No doubt the same is true for portions of the Great Plains winter wheat belt.

Results from CSM's (Fig. 4 and Ritchie, 1980) suggest that earlier anthesis dates may enhance yields of winter wheat and grain sorghum under certain conditions. However, it may be more advantageous to alter flowering date by early planting rather than by shortening the vegetative growth period. Recent trends toward earlier planting are apparent for the southern Great Plains where rainfall is sufficient to insure satisfactory stand establishment. Although early sorghum planting is often an attempt to avoid midseason insect pests, the water requirement for crop growth is reduced because of the lower evaporative demand associated with cooler temperatures. Fischer (1979) estimated the water requirement to produce dry matter in a winter wheat crop increased from 11 mm t^{-1} ha^{-1} in winter months to 40 mm t^{-1} ha^{-1} in summer months. Similar differences would be expected for early versus late planted summer crops. For summer crops, germinability (Thomas and Miller, 1979) and establishment of adequate stands at low temperatures are major limitations to early planting, but recent reports suggest genetic gains may be possible in both areas.

Potential benefits from deep rooting are two-fold: total soil water available to the crop may be greater and the additional water is available later in the season when crops are generally more sensitive to water stress and evaporative demands are higher. The CSM's for both crops predicted a substantial yield advantage occurred in response to additional water made available from deep rooting for all three locations. Analyses of predicted ET and grain yields for all years and locations also suggest a significantly greater water use efficiency is associated with deep rooting at Temple and Manhattan, probably because most of the additional water is used in transpiration and less is lost in evaporation from the soil.

Deep rooting has been a difficult character to incorporate into breeding programs. No easy screening procedures are available that can accurately predict deep rooting under field conditions. Genetic variability in root and shoot characteristics of sorghum measured on plants grown in solution culture (Jordan et al., 1979) are often not expressed to the same degree in the field during drought (Jordan and Miller, 1980). Moreover, field screening for deep rooting may be complicated by untimely rainfall which stimulates root development near the soil surface. Our experience suggests deep rooting cultivars can be identified only in those years when the crop must grow for long periods (i.e. 30 days) on stored soil water. It would appear that, at best, we will be successful in identifying a few parental lines to be used in breeding programs without further evaluation until elite lines emerge.

The question as to whether deep rooting can be incorporated without loss of grain yield has been discussed by Jordan and Miller (1980) and Passioura (1983). There is no simple answer, but Passioura's (1983, this issue) analysis suggests the dry matter gain associated with the increased water supply will more than offset the dry matter investment in forming and maintaining new roots.

Osmoregulation, as simulated by extraction of a slightly greater fraction of the ASW at potential rates, did not result in a detectable yield difference in most years. If osmoregulation is to contribute to yield maintenance during drought, it appears that it does so by providing additional photosynthate required for continued root growth and subsequent water uptake from unexplored regions of the profile. This implies that active root growth must occur deep in the soil profile during grain filling. Deep root growth does occur with sorghum during grain filling (Jordan and Miller, 1980), but this phenomenon is not well documented for other crops.

The examination of potential benefits from the specific crop modifications discussed above was made possible by the existence of documented, validated CSM's. While not perfect, CSM's appear to provide a suitable means to interpret weather information in terms of growth, development and yield. We believe CSM's have an important role to play in formulating strategies for crop improvement for drought-prone regions.

REFERENCES

Angus, J.F. and Moncur, M.W., 1977. Water stress and phenology in wheat. Aust. J. Agric. Res., 28: 177—181.

Angus, J.F., Nix, H.A., Russel, J.S. and Kruizinga, J.E., 1980. Water use, growth and yield of wheat in a subtropical environment. Aust. J. Agric. Res., 31: 873—886.

Ansana, R.D. and Singh, D.N., 1968. On the relation between flowering time, root-growth and soil-moisture extraction in wheat under non-irrigated cultivation. Indian J. Plant Physiol., 10: 154—169.

Arkin, G.F., Vanderlip, R.L. and Ritchie, J.T., 1976. A dynamic grain sorghum growth model. Trans. ASAE, 19: 622—630.

Begg, J.E., 1980. Morphological adaptations of leaves to water stress. In: N.C. Turner and P.J. Kramer (Editors), Adaptation of Plants to Water and High Temperature Stresses. Wiley Interscience, New York, NY, pp. 33—42.

Bidinger, F., Musgrave, R.B. and Fischer, R.A., 1977. Contribution of stored pre-anthesis assimilate to grain yield in wheat and barley. Nature, 270: 431—433.

Blum, A., 1970. Effect of plant density and growth duration on grain sorghum yield under limited water supply. Agron. J., 62: 333—336.

Blum, A., 1975. Effect of BM gene on epicuticular wax deposition and the spectral characteristics of sorghum leaves. Sabrao J., 7: 45—52.

Blum, A., 1979. Genetic improvement of drought resistance in crop plants: a case for sorghum. In: H. Mussell and R.C. Staples (Editors), Stress Physiology in Crop Plants. Wiley Interscience, New York, NY, pp. 429—445.

Blum, A., 1981. Drought avoidance in wheat and its rapid estimation by remote infrared thermal leaf canopy measurements. In: Proc. 3rd International Wheat Conference 22 May—3 June, Madrid. Nebr. Agric. Exp. Stn. Misc. Publ. 41, p. 643.

Blum, A. and Ebercon, A., 1981. Cell membrane stability as a measure of drought and heat tolerance in wheat. Crop. Sci., 21: 43—47.

Boyer, J.S. and McPherson, H.G., 1975. Physiology of water deficits in cereal crops. Adv. Agron., 27: 1—23.

Doyle, A.D. and Fischer, R.A., 1979. Dry matter accumulation and water use relationships in wheat crops. Aust. J. Agric. Res., 30: 815—829.

Fischer, R.A., 1979. Growth and water limitation to dryland wheat yield in Australia: A physiological framework. J. Aust. Inst. Agric. Sci., 45: 83—94.

Fischer, R.A., 1981. Optimizing the use of water and nitrogen through breeding of crops. In: J. Monteith and C. Webb (Editors), Soil Water and Nitrogen. Junk, The Hague, The Netherlands, pp. 249—278.

Fischer, R.A. and Kohn, G.F., 1966. The relationship of grain yield to vegetative growth and post-flowering leaf area in the wheat crop under conditions of limited soil moisture. Aust. J. Agric. Res., 17: 281—295.

Fischer, R.A. and Turner, N.C., 1978. Plant productivity in the arid and semiarid zones. Annu. Rev. Plant. Physiol., 29: 277—317.

Fischer, R.A. and Wood, J.T., 1979. Drought resistance in spring wheat cultivars. III. Yield associations with morpho-physiological traits. Aust. J. Agric. Res., 30: 1001—1020.

Gallagher, J.N., Biscoe, P.V. and Hunter, B., 1976. Effect of drought on grain growth. Nature, 264: 541—542.

Garrity, D.P., Watts, D.G., Sullivan, C.Y. and Gilley, J.R., 1982. Moisture deficits and grain sorghum performance: Evapotranspiration-yield relationships. Agron. J., 74: 815—820.

Henzell, R.G., McCree, K.J., Van Bavel, C.H.M. and Schertz, K.F., 1975. Method for screening sorghum genotypes for stomatal sensitivity to water deficits. Crop Sci., 15: 516—518.

Hurd, E.A., 1968. Growth of roots of seven varieties of spring wheat at high and low moisture levels. Agron. J., 60: 201—205.

Hurd, E.A., 1974. Phenotype and drought tolerance in wheat. Agric. Meteorol., 14: 39—55.

Johnson, D.A., Richards, R.A. and Turner, N.C., 1983. Yield, water relations, gas exchange, and surface reflectances of nearisogenic wheat lines differing in glaucousness. Crop Sci., 23: 318—325.

Johnson, V.A., Shafer, S.L. and Schmidt, J.W., 1968. Regression analysis of general adaptation in hard red winter wheat (*Triticum aestivum* L.). Crop Sci., 8: 187—191.

Jones, H.G., 1977. Aspects of the water relations of spring wheat (*Triticum aestivum* L.) in response to induced drought. J. Agric. Sci., 88: 267—282.

Jones, M.M. and Rawson, H.M., 1979. Influence of the rate of development of leaf water deficits upon photosynthesis, leaf conductance, water use efficiency, and osmotic potential in sorghum. Physiol. Plant., 45: 103—111.

Jones, M.M. and Turner, N.C., 1978. Osmotic adjustment in leaves of sorghum in response to water deficits. Plant Physiol., 61: 122—126.

Jordan, W.R., 1983. Whole plant responses to water deficits: An overview. In: H.M. Taylor, W.R. Jordan and T.R. Sinclair (Editors), Limitations to Efficient Water Use in Crop Production, American Society of Agronomy, Madison, WI (In press).

Jordan, W.R. and Miller, F.R., 1980. Genetic variability in sorghum root systems: implications for drought tolerance. In: N.C. Turner and P.J. Kramer (Editors), Adaptation of Plants to Water and High Temperature Stresses. Wiley Interscience, New York, NY, pp. 383—399.

Jordan, W.R. and Monk, R.L., 1980. Enhancement of drought resistance of sorghum: progress and limitations. Proc. Annu. Corn Sorghum Res. Conf., 35: 185—204.

Jordan, W.R. and Sullivan, C.Y., 1981. Reaction and resistance of grain sorghum to heat and drought. In: Sorghum in the Eighties. Proc. International Symposium on Sorghum, Vol. 1, 2—7 November 1981, Patancheru, India, ICRISAT, Patancheru, pp. 131—142.

Jordan, W.R., Miller, F.R. and Morris, D.E., 1979. Genetic variation in root and shoot growth of sorghum in hydroponics. Crop Sci., 19: 468—472.

Jordan, W.R., Monk, R.L., Miller, F.R., Rosenow, D.T. and Clark, L.E., 1983. Environmental physiology of sorghum. I. Environmental and genetic control of epicuticular wax load. Crop Sci. (In press).

Ketring, D.L., Jordan, W.R., Smith, O.D. and Simpson, C.E., in press. Genetic variability in root and shoot growth characteristics of peanuts. Peanut Sci., 9: 68—72.

Maas, S.J. and Arkin, G.F., 1980. TAMW: a wheat growth and development simulation model. Res. Cent. Progr. Model Doc. 80-3, Texas Agricultural Experiment Station, College Station, TX.

Morgan, J.M., 1980. Differences in adaptation to water stress within crop species. In: N.C. Turner and P.J. Kramer (Editors), Adaption of Plants to Water and High Temperatures Stresses. Wiley Interscience, New York, NY, pp. 369—382.

Nix, H.A., 1975. The Australian climate and its effects on grain yield and quality. In: A. Lazenby and E.M. Matheson (Editors), Australian Field Crops. Vol. 1, Wheat and Other Temperature Cereals. Angus and Robertson, Sydney, pp.183—226.

O'Toole, J.C. and Chang, T.T., 1979. Drought resistance in cereals — rice: a case study. In: H. Mussell and R.C. Staples (Editors), Stress Physiology in Crop Plants. Wiley Interscience, New York, NY, pp. 373—405.

Passioura, J.B., 1972. The effect of root geometry on the yield of wheat growing on stored water. Aust. J. Agric. Res., 23: 745—752.

Passioura, J.B., 1977. Grain yield, harvest index, and water use of wheat. J. Aust. Inst. Agric. Sci., 43: 117—120.

Passioura, J.B., 1983. Roots and drought resistance. Agric. Water Manage., 7: 265—280.

Priestly, C.H.B. and Taylor, R.J., 1972. On the assessment of surface heat flux and evaporation using large-scale parameters. Mon. Weather Rev., 100: 81—92.

Quisenberry, J.E., Jordan, W.R., Roark, B.A. and Fryrear, D.W., 1981. Exotic cottons as genetic sources for drought resistance. Crop Sci., 21: 889—895.

Raper, C.D., Jr. and Barber, S.A., 1970. Rooting systems of soybeans. I. Differences in root morphology among varieties. Agron. J., 62: 581—584.

Reitz, L.P., 1974. Breeding for more efficient water use — is it real or a mirage? Agric. Meteorol., 14: 3—11.

Richards, R.A. and Passioura, J.B., 1981a. Seminal root morphology and water use of wheat. I. Environmental effects. Crop Sci., 21: 249—252.

Richards, R.A. and Passioura, J.B., 1981b. Seminal root morphology and water use of wheat. II. Genetic variation. Crop Sci., 21: 253—255.

Ritchie, J.T., 1974. Atmospheric and soil water influences on plant water balance. Agric. Meteorol., 14: 183—198.

Ritchie, J.T., 1980. Water management and water efficiencies for American agriculture. In: Proc. Weather and Agriculture Symposium, 1—2 October 1979, Kansas City, MO. Mo. Agric. Exp. Stn. Publ. MX-245, University of Missouri, Columbia, MO, pp. 15—42.

Ritchie, J.T. and Ratliff, L.F., 1981. Estimating the limits of soil water availability to plants. I. Comparison of field and laboratory measurements. Agron. Abstr., p. 218.

Stout, D.G. and Simpson, G.M., 1978. Drought resistance of *Sorghum bicolor*. I. Drought avoidance mechanisms related to leaf water status. Can. J. Plant Sci., 58: 213—224.

Sullivan, C.Y. and Blum, A., 1970. Drought and heat resistance of sorghum and corn. Proc. Annu. Corn Sorghum Res. Conf., 25: 55—66.

Taylor, H.M., Burnett, E. and Booth, G.D., 1978. Taproot elongation rates of soybeans. Z. Acker Pflanzenbau, 146, 33—39.

Thomas, G.L. and Miller, F.R., 1979. Base temperatures for germination for temperate and tropically adapted sorghums. In: Proc. 11th Biennial Grain Sorghum Res. and Util Conf., Grain Sorghum Prod. Assoc., Wichita, KS, p. 61.

Townley-Smith, T.F. and Hurd, E.A., 1979. Testing and selecting for drought resistance in wheat. In: H. Mussell and R.C. Staples (Editors), Stress Physiology in Crop Plants. Wiley Interscience, New York, NY, pp. 447—464.

Turner, N.C., 1979. Drought resistance and adaptation to water deficits in crop plants. In: H. Mussell and R.C. Staples (Editors), Stress Physiology in Crop Plants. Wiley Interscience, New York, NY, pp. 343—372.

Turner, N.C., 1982. The role of shoot characteristics in drought resistance of crop plants. In: Symp. Principles and Methods in Crop Improvement for Drought Resistance: With Emphasis on Rice. International Rice Research Institue, Los Baños, Philippines, pp. 115—134.

Turner, N.C. and Jones, M.M., 1980. Turgor maintenance by osmotic adjustment: A review and evaluation. In: N.C. Turner and P.J. Kramer (Editors), Adaptation of Plants to Water and High Temperature Stresses. Wiley Interscience, New York, NY, pp. 87—103.

Turner, N.C., Begg, J.E. and Tonnet, M.L., 1978. Osmotic adjustment of sorghum and sunflower crops in response to water deficits and its influence on the water potential at which stomata close. Aust. J. Plant Physiol., 5: 597—608.

Vanderlip, R.L. and Arkin, G.F., 1977. Simulating accumulation and distribution of dry matter in grain sorghum. Agron. J., 69: 917—923.

Introduction to Part 5

REMOTE SENSING OF PLANT STRESS AND WATER USE

Remote sensing involves measurement of electromagnetic radiation that is reflected or emitted from the earth's surface. Water stress affects the spectral characteristics of plants and plant canopies and leaf/canopy temperature. Thus, remote sensing at reflective and thermal infrared wavelengths can be used to monitor plant response to drought. In addition, because water has a significant impact on thermal and dielectric properties of soil, remote sensing at thermal infrared and microwave wavelengths can be used to monitor precipitation and soil water, the input side of the drought problem. The following papers describe recent advances and applications in remote sensing of stress and water use.

JAMES L. HEILMAN
Texas A&M University
College Station, TX, U.S.A.

Agricultural Water Management, 7 (1983) 303—321
Elsevier Science Publishers B.V., Amsterdam — Printed in The Netherlands

DROUGHT DETECTION AND QUANTIFICATION BY REFLECTANCE AND THERMAL RESPONSES

C.L. WIEGAND[1], P.R. NIXON[1] and R.D. JACKSON[2]

[1] *Agricultural Research Service, U.S. Department of Agriculture, P.O. Box 267, Weslaco, TX 78596 (U.S.A.)*

[2] *U.S. Water Conservation Laboratory, Agricultural Research Service, U.S. Department of Agriculture, Phoenix, AZ (U.S.A.)*

(Accepted 23 November 1982)

ABSTRACT

Wiegand, C.L., Nixon, P.R. and Jackson, R.D., 1983. Drought detection and quantification by reflectance and thermal responses. *Agric. Water Manage.*, 7: 303—321.

Drought can be detected by its effect on plant canopy development as indicated by measurements of biomass production, leaf area index (LAI), and ground cover in the reflective wavelengths (0.4—2.5 μm) and by canopy minus air temperature in the thermal infrared (8—14 μm) wavelengths. Both measures are quantitative, but quantification in terms of drought's economic impact requires that relationships between canopy 'appearance' as revealed by the measurements and effect on salable plant parts be established. The measurements themselves are becoming routine and their quantification in terms of drought's economic impact on grain sorghum and wheat grain yields have been demonstrated, but are not yet routine. Seven considerations in a system for drought quantification are discussed, and results of semi-operational applications to drought assessment that incorporate most of the seven considerations are described. Future drought assessment systems will likely combine physiologically based crop growth and yield models with soil characteristics, weather data, and Earth satellite spectral scanner observations.

INTRODUCTION

Polcyn et al. (1969) discussed a nine-step procedure for planning and executing spectral investigations in ecology. Their procedure, reduced to seven steps, is:

(1) define clear objectives;

(2) list known manifestations of the parameters of interest;

(3) define sensors needed to gather the data;

(4) obtain use of sensors with the desired characteristics;

(5) define data collection programs (aerospace and ground-based);

(6) define data reduction steps (may require development of new or modified procedures);

(7) analyze results in terms of biological consequences and management decisions.

0378-3774/83/$03.00

The steps fit equally well to research and application in any field. For example, if the person conducting the study has been unable to isolate measurable responses to the stresses imposed (step 2), or has not developed procedures to extract the meaningful information from the observations made (step 6) the study or project has little prospect of success. The objective of this paper is to illustrate these seven steps using drought as the situation of interest. In so doing we will emphasize the foliage characteristics of the individual plants or plant communities because, except during senescence or dormancy, leaves dominate the responses observed. The wavelengths to be considered are the reflective (0.4—2.5 μm) and thermal infrared (8—14 μm).

MANIFESTATIONS OF DROUGHT

Drought stressed versus nonstressed plants and plant canopies manifest one or more of the following:

decreased growth rate and cumulative green biomass that results in a lower leaf area index (LAI) and vegetative ground cover; shortened internodes;

lower leaf relative water content and water potential; cellular adaptations;

leaf color change;

leaf droop and roll; leaf orientation change;

elevated canopy and soil temperature;

shortened green leaf area duration;

lower near surface and profile soil water content.

Of these manifestations, the reduction in leaf area and vegetative ground cover are the most consequential economically. They result from the sensitivity to plant water status of cell enlargement and the transport of nutrients and water to various plant parts (Kramer, 1969). Assuming other factors are optimal, photosynthesis and growth are proportional to amount of light intercepted by the canopy and, therefore, to LAI (Ritchie, 1974). Consequently, drought may be detected and quantified by measuring its effect on plant canopy development as indicated by biomass production, LAI, or ground cover. Such measures use the plants or plant canopies as integrators of the cumulative effect of drought during the year or growing season up to the time of observation. Intensity or severity of drought on a particular date can also be quantified through canopy minus air temperature measurements (Jackson et al., 1981).

Since we want to emphasize the plant canopy manifestations of drought using the reflective and thermal infrared wavelengths, we need to consider plant leaf and canopy interactions with electromagnetic radiation in these wavelengths.

USEFUL WAVELENGTHS AND NEEDED SENSORS FOR CANOPY OBSERVATION

Within the reflective wavelength interval (WLI), 0.4—2.5 μm, three regions have been recognized: the visible (0.4—0.75 μm), affected primarily by chlorophyll and other pigments; the reflective near-infrared (0.75—1.35 μm) where leaf structure is important; and, the water absorption (1.35—2.5 μm) region.

In the reflective near-infrared, the amount of air space in the leaves is important because light is refracted and scattered in passing from hydrated cell walls to mesophyll air spaces (Gausman, 1974; Gausman et al., 1974a). Air spaces form in dicots such as soybeans and sunflower when the leaves are one-fourth to one-third full size. Leaves that develop under stress such as drought or soil salinity have fewer air spaces than nonstressed leaves and consequently are less reflective in the 0.75—1.35 μm WLI than nonstressed leaves. Also, water-stressed leaves are less hydrated than nonstressed leaves and absorb less of the energy in wavelengths dominated by water absorption than do nonstressed leaves. Reviews on physiological and anatomical factors are given by Gates et al., 1965; Knipling, 1970; Woolley, 1971; Wiegand et al., 1972; Myers et al., 1975; and Gausman et al., 1978.

The general plant leaf spectrum and the regions mentioned previously are illustrated in Fig. 1 (after Gausman et al., 1975, 1978) by the laboratory spectra of single leaves (A) and field spectra of cotton canopies (*Gossypium hirsutum* L.) from nematode-stressed and noninfested plots (B). The nematodes impaired root functioning and, therefore, like drought-affected plants, they were shorter and had fewer, smaller, and darker green leaves than the nonstressed plants. Field-grown, nonstressed versus stressed leaf average values were: area, 25.5 versus 17.3 cm^2 (P = 0.01); water content, 77.5 versus 78.3% (P = 0.01); and, thickness, 0.16 versus 0.15 mm (NS). Ground cover was complete in the instantaneous field of view (IFOV) for both the nematode-infested and nematode-treated field areas.

Fig. 1. indicates the general magnitude of the reflectance response, and shows that differences are observable for both individual leaves and crop canopies. Laboratory data indicate that a response may be observable under field conditions, but do not guarantee it (Myers et al., 1966). Amount of sunlit soil and shadows in the IFOV, planting configuration, soil wetness, condition of the atmosphere, sun and observer angles and plant architecture (leaf angle, size, leaf arrangement, compactness) are variables that affect field spectra (Colwell, 1974; Jackson et al., 1979). Consequently, Park et al. (1977) suggest that a change in reflectance of about 10% is a significant change in terms of plant stress.

Fig. 2 presents the linear correlation coefficients between percent vegetative cover of Milam and Penjamo wheat cultivars and reflectance at seven wavelengths (0.55, 0.65, 0.73, 0.90, 1.10, 1.65 and 2.20 μm) measured with a field spectroradiometer on various days during the growing season (Leamer et al., 1978). The wheat emerged 1 December and by 31 December ground cover averaged 25%. Vegetative cover and LAI increased into February as tillering and plant development occurred, then the plants began to senesce. The magnitude and sign of the correlation coefficients depend, respectively, on the reflectance contrast between the plants and soils and whether the plants or soil are the more reflective.

In the visible wavelengths, the soil is generally more reflective than the plants so that information about plant cover is obtained from the extent to which the plants obscure the soil background (Wiegand et al., 1974). In the 0.75—1.35 μm interval, leaves are nearly transparent. The diffuse light trans-

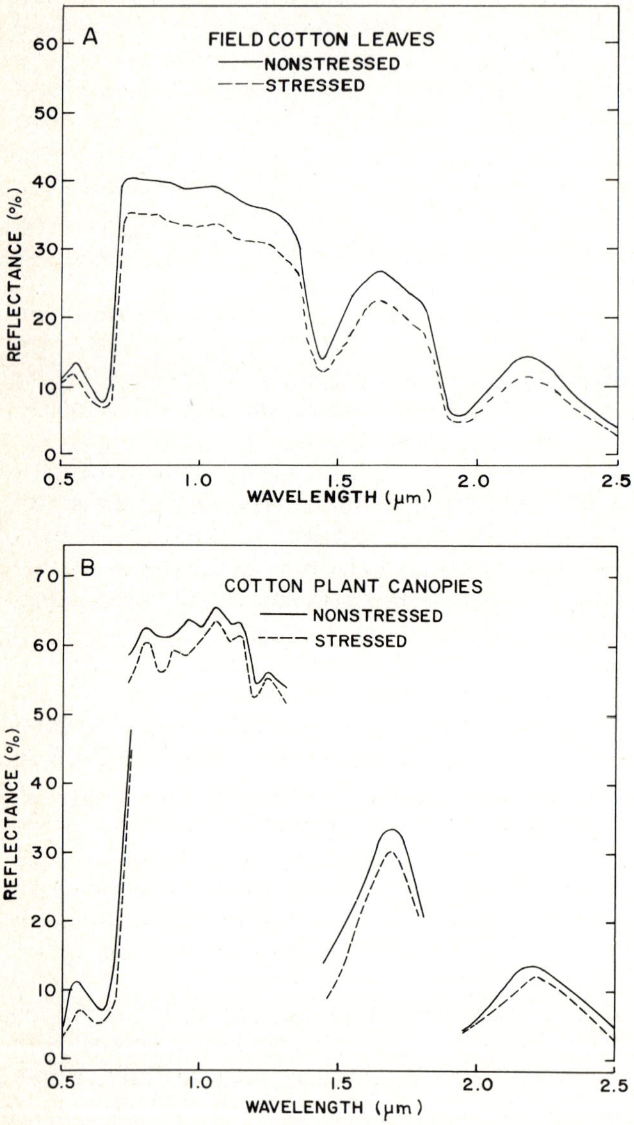

Fig. 1. Total diffuse reflectance spectra over the 0.5- to 2.5-μm waveband for cotton leaves from plants grown in the field under nematode-infected and noninfested conditions (A), and spectroradiometric reflectance spectra of the plant canopies in the same plots (B). After Gausman et al. (1975, 1978).

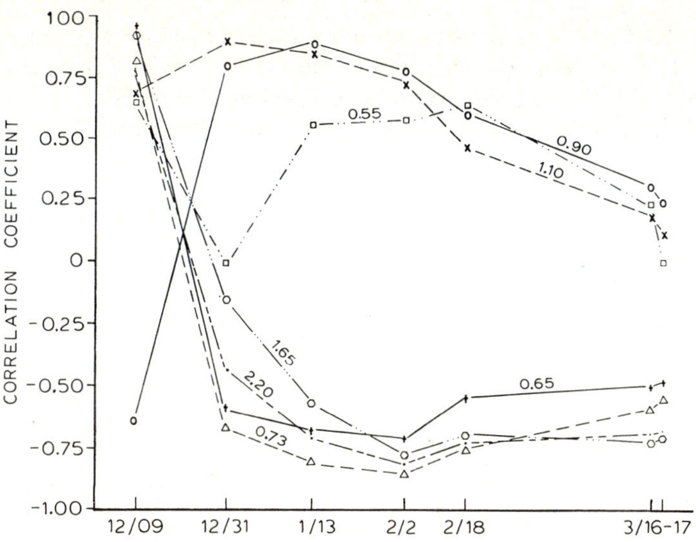

Fig. 2. Linear correlation coefficients between vegetative cover and reflectance for seven wavelengths (0.55, 0.65, 0.75, 0.90, 1.10, 1.65, and 2.20 μm) on seven dates during the growing season. After Leamer et al. (1978).

mitted by upper leaves as well as the specular light that penetrates between the leaves is scattered (upward and other directions) by lower leaves, if present, and is not fully attenuated until the LAI reaches 8. Consequently, this wavelength interval is especially useful for LAI and biomass determinations (Wiegand et al., 1979).

As seen in Fig. 2, after 31 December correlations in the WLI affected by water absorption (1.65 and 2.20 μm) are the mirror image of the visible green (0.55 μm) so that reflectance or radiance in these intervals are often correlated. Significant correlations have been shown between relative water content at wilting and below and reflectance at the two strong water absorption bands, 1.45 and 1.95 μm, in laboratory studies (Carlson et al., 1971; Thomas et al., 1971). However, in field studies it would be difficult to separate differences due to leaf water content from leaf structure changes associated with availability of water during leaf development and changes in the soil background and shadow components of the field of view as the plants wilted.

The strongest and most durable correlations are for the visible red (0.65 μm) and far red (0.75 μm), and reflective infrared plateau wavelengths (0.90 and 1.10 μm). Therefore, sensors — be they handheld, truck lift-, airplane-, or satellite-mounted — that operate in these wavelengths will be the most useful reflective wavelengths for measuring drought's impact on vegetation.

Another useful WLI for drought analysis is the emitted thermal infrared (8—14 μm) where canopy temperature can be observed. Canopy temperature's usefulness is supported by the observation that as profile soil water is depleted, the partitioning of net radiation, R_n, for a canopy shifts with more of it going into sensible heat flux from the canopy to the air, H, and less of it into latent heat flux to the air λE. The relationships are expressed by:

$$R_n = G + H + \lambda E \tag{1}$$

where the soil heat flux, G, below the canopy is small and ignored; λ is the heat of vaporization of water (and all terms have the units W m^{-2}); H and λE can be expressed as:

$$H = \rho c_p \, (T_c - T_A)/r_a \tag{2}$$

and

$$\lambda E = \rho c_p \, (e_c^* - e_A)/[\gamma(r_a + r_c)] \tag{3}$$

where ρ is the density of air (kg m^{-3}); c_p the heat capacity of air (J kg^{-1} $°C^{-1}$); T_c the surface temperature (°C); T_A the air temperature (°C); e_c^* is the saturated vapor pressure (Pa) at T_c; e_A the vapor pressure of the air (Pa); γ is the psychrometric constant (Pa $°C^{-1}$), r_a is the aerodynamic resistance (s m^{-1}), and r_c is the canopy resistance (s m^{-1}) to vapor transport (Jackson et al., 1981; Jackson, 1982). Monteith (1973) discussed the procedures and assumptions leading to Eqs. (2) and (3).

Combining Eqs. (1), (2), and (3), assuming that G is negligible, and defining Δ as the slope of the saturated vapor pressure-temperature relation $(e_c^* - e_A^*)/(T_c - T_A)$ (in units of Pa $°C^{-1}$), we obtain:

$$T_c - T_A = \frac{r_a R_n}{\rho c_p} \frac{\gamma(1 + r_c/r_a)}{\Delta + \gamma(1 + r_c/r_a)} - \frac{e_A^* - e_A}{\Delta + \gamma(1 + r_c/r_a)} \tag{4}$$

which relates the difference between the canopy and the air temperatures to the vapor pressure deficit of the air $(e_A^* - e_A)$, the net radiation, and the aerodynamic and crop resistances. The upper limit of $T_c - T_A$ can be found from Eq. (4) by allowing the crop resistance r_c to increase without bound, i.e., as $r_c \to \infty$:

$$T_c - T_A = \frac{r_a R_n}{\rho c_p} \tag{5}$$

Experimentally, $r_a \to \infty$ can be approximated for a standing crop canopy that has senesced or be calculated from theory (Monteith, 1973). The lower bound, found by setting $r_c = 0$ in Eq. (4) is the case of wet plants acting as a free water surface.

The senior author and co-workers obtained direct experimental support for the variables included in Eq. (4) from the thermal band (10.5—12.5 μm) data of the heat capacity mapping mission (HCMM) satellite for 8 summer dates in 1978 (27 May; 6, 12, 17, and 23 June; 3 and 8 July; and, 15 August). The ground sites had been preselected, using high altitude photographs and Landsat images, as representative and were either cropland (grain sorghum, citrus, cotton, improved pasture) or rangeland. HCMM overpassed the test area between 14.00 and 14.16 CST during the test period. At a central weather station within the 100 km × 200 km test area, saturation deficit of the air, SDEF (mbar), insolation were monitored continuously and integrated hourly and daily to yield insolation the hour of the pass, IHROP (ly h^{-1}; ly, langley = cal cm^{-2} = 42.186 8 kJ m^{-2}) and day of the overpass, IDOP (ly day^{-1}). For each ground site we determined the distance from the Gulf coast, DIST (km), and water holding capacity, WCAP (cm), of the root zone for the dominant soil type. From the reporting weather station nearest each site we obtained maximum air temperature, T_{max} (°C), the day of the overpass and from weather station data (dryland) or direct observation (irrigated sites) we determined days since rainfall or irrigation, DSR (days). There were 168 site-date observations in the data set.

The HCMM-measured surface temperature, T_s (where $T_s = T_c$) for sites with adequate vegetative ground cover), was regressed on the mentioned independent variables. The first five variables included were SDEF, DIST, IHROP, IDOP, and WCAP and each contributed significantly to the relation. SDEF alone accounted for 55% of the variation (Table I) in surface temperature in agreement with its importance in Eq. (4). DIST is a geographic variable but is related in this case to the strong gradient in rainfall away from the coast, 1 cm per 4.5 km, for the test area. The insolation variables IHROP

TABLE I

Means and standard deviations of each variable included in a synoptic analysis of HCMM-measured canopy temperature (T_c) and the linear correlation coefficients for each with T_c for HCMM sample sizes of one (36 ha) and nine pixels (T_{c1}, T_{c3})

Variable[a]	Units	Mean	Standard deviation	Correlation coefficient, r, versus	
				T_{c1}	T_{c3}
SDEF	mbar	28.8	7.1	0.738**	0.758**
DIST	km	101.4	33.1	0.721**	0.729**
IHROP	1 y h^{-1}	74.3	24.0	0.600**	0.613**
IDOP	1 y day^{-1}	408.6	46.6	0.128	0.126
WCAP	cm	16.2	1.4	−0.170*	−0.171
Tmax	°C	36.4	2.1	0.686**	0.712**
DSR	days	17.9	17.5	0.472**	0.481**
			Mean	33.00	32.75
			Standard deviation	4.60	4.48

*Significant at $P = 0.05$.
**Significant at $P = 0.01$.
[a]Saturation deficit of the air, distance inland, insolation the hour and day of the overpass, water holding capacity of the rootzone, daily air temperature maximum, and days since rainfall or irrigation, respectively.

and IDOP are the source of R_n in Eq. (4) and the driver of the energy balance to which T_c is responding.

Table I summarizes the means, standard deviations, and linear correlation coefficients for each variable of the analysis. T_{max} was highly correlated with the HCMM-observed temperatures for the 36 ha area central pixel T_{c1} and the 3-by-3 matrix of pixels T_{c3} surrounding the central pixel. T_{max} was highly significantly correlated with SDEF, DIST, and IHROP, and they rather than T_{max} were included in the multiple regression.

Equation (4) can be modified and made relevant to potential evapotranspiration rate by defining a canopy resistance at potential evapotranspiration r_{cp}. (Note that r_{cp} probably differs among crops and that it may change with environmental conditions during the day.) Setting $r_c = r_{cp}$ in Eq. (4):

$$T_c - T_A = \frac{r_a R_n}{\rho c_p} \frac{\gamma^*}{\Delta + \gamma^*} - \frac{e_A^* - e_A}{\Delta + \gamma^*} \tag{6}$$

where

$$\gamma^* = \gamma(1 + r_{cp}/r_a) \tag{7}$$

Eq. (4) and (6) describe a linear relation between $T_c - T_A$ and the vapor pressure deficit, $e_A^* - e_A$. Thus, for a particular temperature the lower bound is a line extending from the intercept at $e_A^* - e_A = 0$ (saturated air) to a value of $e_A^* - e_A = e_A^*$ (completely dry air). The upper bound is independent of vapor pressure and dependent only on r_a and R_n. The bounds are shown by the lines labeled 0 (lower) and ∞ (upper) in Fig. 3. Lines for intermediate values of r_c (5, 50, and 500 $s\,m^{-1}$) are also shown. All lines in this figure were calculated for a temperature of 30°C, and net radiation of 600 $W\,m^{-2}$.

The ratio of actual, E, to potential evapotranspiration, E_p, is an index of crop water status. Combining Eqs. (1), (2), and (3) and solving for actual λE (for any r_c) and ratioing to λE_p (for $r_c = r_{cp}$) gives:

$$E/E_p = \frac{\Delta + \gamma^*}{\Delta + \gamma(1 + r_c/r_a)} \tag{8}$$

where γ^* is defined as in Eq. (7), and the net radiation difference for the E and E_p cases is assumed to be small.

The ratio E/E_p ranges from 1 (ample water, $r_c = r_{cp}$) to 0 (no available water, $r_c \to \infty$). A crop water stress index that ranges from 0 to 1 is:

$$\text{CWSI} = 1 - E/E_p = \frac{\gamma(1 + r_c/r_a) - \gamma^*}{\Delta + \gamma(1 + r_c/r_a)} \tag{9}$$

The ratio r_c/r_a needed to solve Eq. (9) is obtained from Eq. (4) expressed in terms of r_c/r_a. The CWSI is a function of net radiation, canopy and air temperature, vapor pressure deficit, and crop and aerodynamic resistances.

Fig. 4 (after Jackson, 1982) relates the CWSI and the extractable water (Ritchie, 1981) used to time for two differentially irrigated wheat plots

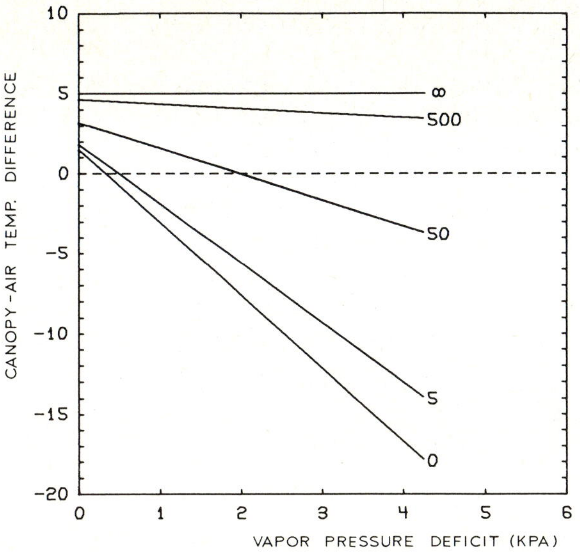

Fig. 3. Theoretical relation between the canopy — air temperature difference and the vapor pressure deficit of the air. Numbers at the end of the lines indicate the value of the canopy resistance r_c used for the calculations. All calculations were for an air temperature T_A of 30°C, net radiation R_n of 600 W m^{-2}, and an aerodynamic resistance r_a of 10 s m^{-1}. After Jackson et al. (1981) and Jackson (1982).

designated treatments A (upper) and B (lower). The wheat was planted on day of year (DOY) 37 and irrigated on DOY 39, 1980, to achieve emergence. Thereafter treatment A was irrigated only on DOY 100, whereas B was irrigated on days 93 and 114. Soil water contents were measured at 0.2 m intervals to a depth of 1.6 m 2 or 3 times per week, and were smoothed and interpolated to obtain the continuous lines shown in Fig. 4. Canopy temperature was measured daily between 13.40 and 14.00 h by a handheld infrared thermometer aimed at 30° from the horizontal. Wet- and dry-bulb temperatures were measured with a psychrometer at the 1.5 m height. Net radiation was taken as 0.75 of the measured insolation. For Fig. 4 a value of r_{cp} = 5 s m^{-1} was used.

Treatment A plants were stressed prior to the irrigation on day 100 (CWSI = 0.67, and fraction extractable water used from the surface 1.1 m of soil= 0.73). The CWSI did not reach a minimum until 6 days after irrigation, indicating that the plants required that much time to recover from the water deficit, i.e., to rehydrate tissue and grow new root hairs. Treatment B plots were not as stressed when irrigated on day 93 (CWSI = 0.55, extractable water used = 0.68) as on day 114 (CWSI = 0.66, extractable water used = 0.87) but again several days were required for the CWSI to reach a minimum following irrigation. There is also a tendency for the minimum CWSI to increase with plant maturity. Generally the extractable water used and the

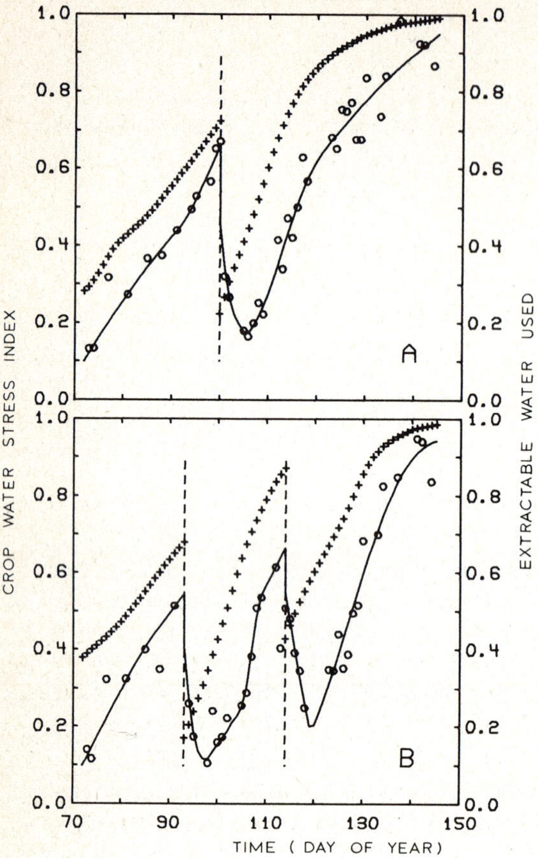

Fig. 4. The CWSI (circles represent data points) and the relative extractable water used (plus symbols) as a function of day of year for two wheat plots. Dashed vertical lines indicate irrigations. From Jackson (1982).

CWSI curves roughly parallel each other until extractable water used reaches 0.85.

The equations presented demonstrate that several interacting environmental variables and species specific plant characteristics determine plant water status and canopy temperature. Nonetheless, the difference between canopy and air temperature can be an excellent characterizer of intensity of drought so that infrared thermometry, which utilizes the 8– 14 μm WLI can be a valuable tool in drought analysis. The CWSI reduces the variables to a single numerical value for relating the physical measurements to plant responses.

DATA COLLECTION PROGRAMS

Data collection is highly specific for the study objectives. For drought- or water stress-related investigations, the useful wavelengths are the visible, the reflective infrared, the thermal, and the microwave (1—30 cm). (Microwaves are covered in this issue by other contributions). The sensors can be hand-held, truck-mounted, or deployed in helicopters, airplanes, or spacecraft, depending on the scope of the study. In any case the use of the sensors has to be arranged for — by ownership, by cooperating with others who have access to the necessary sensors, or by using data in the public domain such as that from the NOAA operational meteorological satellites.

If one is planning to use an equation such as (4) or (9), the variables that measurements must address are well defined. However, questions remain about the timing of observations during the day, the frequency of sampling during the season, and the adequacy of sampling. If one is using a relatively new approach such as canopy temperature, then he also usually needs to measure the accepted traditional characterizers of plant water stress or drought such as plant water potential, soil profile water content, plant biomass or grain yield, and the weather variables, precipitation, air temperature, and insolation. Consequently, the problem is usually one of keeping the studies down to a manageable size with the resources available.

In a later section we highlight past and probable future approaches to drought assessment in which data collection programs are necessarily alluded to.

Observed biomass, LAI, and percent cover are increasingly being characterized (as documented in Wiegand et al., 1979, e.g.) through the use of vegetation indices (Kauth and Thomas, 1976; Richardson and Wiegand, 1977; Tucker, 1979). They are ratios, differences, or linear combinations of band responses. The indices can be calculated from the instrument responses expressed as digital count, bidirectional reflectance, radiance, or voltage. There is currently a proliferation of measuring instruments available which differ in calibration and WLI. Preprocessing is often done to bring all observations to an equivalent sun angle or to reduce atmospheric effects. Thus, the pattern of response of vegetation indices among instruments during a growing season is qualitatively similar. However, quantification of the spectral indices against plant canopy parameters depends on in-situ calibration.

The most extensively analyzed data have come from the LANDSAT series of earth observation satellites. Since the same sensor system views the whole globe, data are comparable from widely separated sites. The multispectral scanner on board has two visible (0.5—0.6 and 0.6—0.7 μm) bands and two near-infrared (0.7—0.8 and 0.8—1.1 μm) bands. It has been shown that virtually all soils (Kauth and Thomas, 1976) as well as cloud shadows and cloud tops (Richardson and Wiegand, 1977) fall on a line when a visible and a reflective near-infrared band are plotted against each other, and that the greater the density of green vegetation, the greater the displacement of paired visible-near infrared band observations from the soil line. The 'greenness' or green vegetation index (GVI) of Kauth and Thomas (1976) and the perpendicular vegetation index (PVI) of Richardson and Wiegand (1977) express this orthogonal displacement of observations containing vegetation from the soil line. PVI is zero for bare soil, whereas GVI ranges from -8 to

−11 for bare soil. Both indices track canopy development and are a quantitative expression of green vegetation present, if atmospheric path radiance (Kauth et al., 1979) and viewing angle are adequately corrected for.

Since the launch of Landsat-1 in 1972, the ARS group at Weslaco, TX has been obtaining Landsat acquisitions during the growing season. In 1973, 1975, 1976, and 1977 ground truth was also obtained for a selection of irrigated and dryland fields. Fig. 5 displays the combined data for the 4 years for four variables: plant population, POP (No. plants × $10^{-3}\,ha^{-1}$), plant cover, PC (%), leaf area index, LAI (unitless), and grain yield ($kg\,ha^{-1}$) versus the vegetation index, PVI, observed during the grain filling stage. The symbol T on the figure designates irrigated fields planted in double row (DR) configuration whereas S and D designate single row (SR) planting configurations for the irrigated and dryland fields, respectively. The POP and PC relations tend to segregate between irrigated and dryland fields because of the

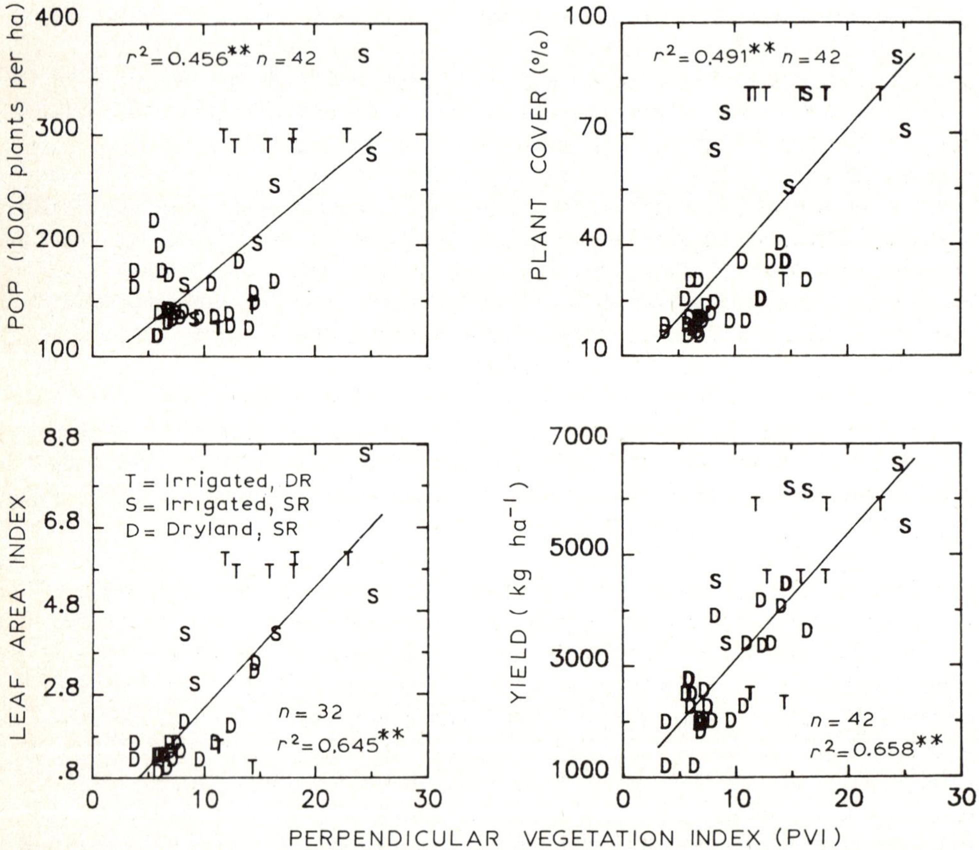

Fig. 5. Scatter diagrams of the vegetation index PVI versus plant population, plant cover, leaf area index, and grain yield of sorghum for fields ground truthed and observed by Landsat during grain filling in 1973, 1975, 1976 and 1977.

large difference in seeding rate. Although all relations are significant at the 1% level, the highest coefficient of determination, r^2, is for grain yield versus PVI. This is a rather remarkable finding considering the diversity in weather, soils, grower practices, and genetics represented in the four years of data. We take the findings as strong evidence that plant canopy development responses to their soil and aerial environments are meaningfully expressed by the spectral vegetation indices. To the extent that the canopy condition and vigor relate to yield, the vegetation indices should relate to yield.

In Fig. 5, the LAI is expressed by:

$$\text{LAI} = -0.41 + 0.29\ (\text{PVI})$$

and grain yield is expressed by:

$$\text{Yield}\ (\text{kg ha}^{-1}) = 840 + 227\ (\text{PVI})$$

For the production area where the relation was developed on average PVI for the irrigated sorghum is 16, corresponding to a yield of 4 470 kg ha^{-1}, whereas it is 10 for dryland sorghum (3 100 kg ha^{-1}). If one has the spectral data to determine PVI in other or future years, then it is known whether the crop is better than or worse than average and the yield deviation from the average can be calculated from the regression equation. In this sense, the information is quantitative.

Such relationships could be developed for major crops for many production areas from historical Landsat, and Crops and Livestock Reporting Service, Statistical Reporting Service, USDA, county statistics. Inclusion of unusually wet and dry years on the Great Plains, such as 1982 and 1980, respectively, in the data set help establish the limits of applicability of the relation, and, again to express it numerically.

Another way to use the vegetation indices is illustrated in Fig. 6. In it the mean PVI for irrigated and dryland fields, for each year are expressed as spectral crop development profiles (Crist and Malila, 1980). The spectral profiles are defined by equations of the form:

$$\text{PVI} = a\ (t - t_0)^b e^{c(t-t_0)^n}$$

in which t = day of the year; t_0 = emergence day; a, b, and c, constants with values close to unity; and n = approximately 5 for sorghum.

The left side of spectral profiles represent greenup or development of the crop and the right side the brownout or senescence of the crop. The slopes at the flex points on each side of the curve express the rates of greenup an and senescence in the units PVI day^{-1}, and the area under such curves between the flex points is an integral crop greenness analogous to leaf area duration. The curve fitting procedure enables one to assign a maximum PVI (PVI_{max}) for a particular field or site and to date its occurrence for each crop season. If the emergence dates for particular fields of interest or spring greenup dates of winter cereals and pastures are unknown the curves help estimate them (Badhwar, 1980). Sorghum reaches PVI_{max} in such curves in

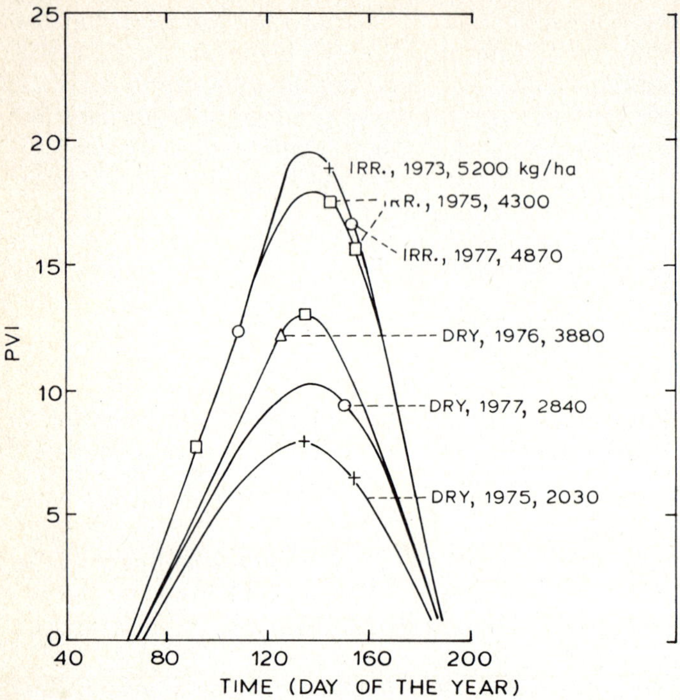

Fig. 6. Spectral canopy development profiles for irrigated and dryland sorghum fields in each of 3 years.

extreme south Texas on about day 140 (20 May) or about 10 days after heading. This is about midway between the time the flag leaf is fully developed and physiological maturity of the grain.

The curves for irrigated fields in Fig. 6 are quite stable; however, canopy development observed for the dryland fields was limited by available soil water. The rainfall data given in Table II for McCook, TX, show that the

TABLE II

Precipitation (mm) for the July through December profile replenishment and the January through June 1975, 1976, and 1977 crop seasons at McCook, TX

Year	Replenishment July–December (mm)	Year	Crop season J	F	M	A	M	J	Annual total
1974	312	1975	32	14	9	13	59	48	487
1975	496	1976	16	1	38	65	65	11	692
1977	492	1977	50	41	5	8	37	133	766
Long-term	255	Long-term	37	34	25	36	61	68	516

annual totals from July of the previous year through June of the 1975, 1976, and 1977 crop years was close to or exceeded the long term amounts. However, in terms of crop performance, 1975 was an extreme drought year, 1977 was a below average year, and yields in 1976 were above average. The rainfall was low in March and April of both 1975 and 1977, but it is not apparent from the rainfall data that the grain yields would differ as they did. The direct look at the canopies expressed by the vegetation index sensed crop condition that related to grain yield. We advocate the use of spectral vegetation indices in future analyses of drought.

DROUGHT ANALYSES, PAST AND FUTURE

The quantification of drought in terms of biological consequences has been elusive. Fig. 4 suggests that an extractable soil water balance would be useful. Heilman et al. (1977) and Kanemasu et al. (1977) used an evapotranspiration model with solar radiation, air temperature, precipitation, and leaf area index estimates from Landsat data as inputs to follow the water content of the surface 150 cm of soil for a five-state region. Their soil moisture estimates compared favorably with the traditional crop moisture index (CMI) of Palmer (1968) and could be associated with yield. However, a water balance has two shortcomings: over the growing season, the volume of soil from which roots are actively extracting water at a given time is poorly known; and, when the evaporative demand exceeds the rate the roots can extract water, the relation between the soil profile water balance change and plant stress becomes less predictable. The latter deficiency can be overcome if enough information is available to invoke Eq. (9) in which the ratio of actual evapotranspiration, E, to potential evapotranspiration, E_p, appears. Even though Eq. (9) is primarily a research tool, interpretation of the biological consequences of soil water balances will likely be improved if the CWSI, or $(1 - E/E_p)$, can be calibrated in terms of the more directly observable $(T_c - T_A)$ for geographic areas of interest.

In a different approach, Thompson (1976) compared spectral monitoring of winter wheat moisture deficiencies using LANDSAT digital and image data with the meteorological CMI. The LANDSAT digital tapes were obtained for portions of five states (Colorado, New Mexico, Kansas, Oklahoma, and Texas) during the 'normal' 1975 winter wheat growing season and the droughty 1976 growing season. A vegetation index, the green index number (GIN) was calculated from the tasseled cap transformation of Kauth and Thomas (1976) for fields of sample segments in the study area for the two seasons. The GIN was compared with the weekly CMI published by NOAA and with gauged rainfall for the study segments. The approach has also been applied to the crop reporting districts of South Dakota by Thompson (1976) and Thompson and Wehmanen (1977). Later the technique was applied to the entire US Great Plains wheat producing area, to Russia's spring wheat region and to Australia for the 1977 season (Thompson and

Wehmanen, 1979). They also found that their procedure worked in the corn-soybean growing areas of Iowa, Illinois and Indiana (Thompson and Wehmanen, 1980), an area with different crops and higher rainfall than the Great Plains.

Differences in condition predicted by the GIN and the CMI in the above studies were most frequently caused by differences in water holding capacity between the sample segment locations and the values used in the CMI, or by different rainfall recorded at the segment location than at the weather stations used to compute the CMI.

Thompson (1976) outlined an operational procedure as follows:

(1) "Use meteorological data to signal potential drought area.

(2) Determine vegetation greenness of segments within and outside potential area and evaluate against a normal year.

(3) If potential damage is indicated, use full-frame imagery in conjunction with segment greenness to outline areal extent" ...

(4) "Using full-frame imagery, monitor precipitation for areal extent and effectiveness" ...

(5) "Generate specific yield results for affected area. Crop vigor comparison may give indication of yield reduction and a need to adjust yield results.

(6) After harvest, continue to monitor meteorological data to determine if soil water is recharged" — as indicator of soil water conditions for the next crop.

Further developments include those of Kanemasu and co-workers (1982) who incorporated reflectance and thermal observations into the physiologically-based Kansas State University wheat (KSWHT) model using data from the 1978—1979 and 1979—1980 season, serial wheat (var. Produra) sowings of the U.S. Water Conservation Laboratory, Phoenix, AZ. For 37 planting date-irrigation treatment plots, the mean difference between physiological model predicted and observed yields was 15 bu ac^{-1} (1 010 kg ha^{-1}). When LAI as estimated from the GVI and band ratios, measured by an Exotech Model 100-A handheld radiometer, were input to the KSWHT model the mean difference between observed and predicted plot means was 6 bu ac^{-1} (400 kg ha^{-1}). When both LAI and soil water were estimated from reflectance and canopy temperature, the model predicted mean yields within 2 bu ac^{-1} (135 kg ha^{-1}) of the mean observed yields. The canopy temperature was entered into the model by knowing that, for a given vapor pressure deficit, VPD (mbar), if $T_c - T_A$ between a given plot and that predicted for well-watered plots by: $T_c - T_A = 3.38 - 0.325$ (VPD) (preheading), and $T_C - T_A = 2.88 - 0.211$ (VPD) (postheading) (Idso et al., 1981) was >2 C, the soil moisture, Θ (mm), was set by:

$$\Theta = 212.5 + 3.37 \text{ (VPD)}$$

(Kanemasu et al., 1982) within the plant available water content range

10—50%. Unless the canopy minus air temperature difference exceeded 2°C, the evapotranspiration subroutine in the crop model estimated the daily water use for the water balance.

Increased use of canopy temperature in drought analysis will build on the modest amount of experience to date. Continued progress hinges on developing relationships similar to those in Fig. 4 for crops and atmospheric environments of interest, additional relationships including $T_0 - T_A$ versus $(1 - E/E_p)$, and learning how to introduce T_c into the crop development and yield response model subroutines for respiration, photosynthesis, evapotranspiration, stress, dry matter accumulation, and phenology. The prospects are exciting in that new interpretation tools are becoming available that directly sense crop canopy response in ways that indicate how crop yields will be affected. Future systems for drought analysis will likely use the plant available water and rooting depth information implicit in soil surveys, weather data, physiologically-based crop development and yield models, and Earth satellite spectral scanner observations of the plant canopies.

REFERENCES

Badhwar, G.D., 1980. Crop emergence date determination from spectral data. Photogramm. Eng. Remote Sensing, 46: 369—377.

Carlson, R.E., Yarger, D.N. and Shaw, R.H., 1971. Factors affecting the spectral properties of leaves with special emphasis on leaf water status. Agron. J., 63: 486—489.

Colwell, J.E., 1974. Vegetation canopy reflectance. Remote Sensing Environ., 3: 175—183.

Crist, E.P. and Malila, W.A., 1980. A temporal-spectral analysis technique for vegetation applications of Landsat. In: Proc. 14th Int. Symp. Remote Sensing of Environment, II. Environ. Res. Inst. Michigan, Ann Arbor, MI, pp. 1031—1040.

Gates, D.M., Keegan, H.J., Schleter, T.C. and Wiedner, V.R., 1965. Spectral properties of plants. Appl. Opt., 4: 11—20.

Gausman, H.W., 1974. Leaf reflectance of near-infrared. Photogramm. Eng. Remote Sensing, 40: 183—191.

Gausman, H.W., Allen, W.A. and Escobar, D.E., 1974. Refractive index of plant cell walls. Appl. Opt., 13: 109—111.

Gausman, H.W., Heald, C.M., Jr. and Escobar, D.E., 1975. Effect of *Rotylenchulus reniformis* on reflectance of cotton plant leaves. J. Nematol., 7: 368—374.

Gausman, H.W., Escobar, D.E. and Rodriguez, R.R., 1978. Effects of stress and pubescence on plant leaf and canopy reflectance. In: G. Hildebrandt and H.-J. Boehnel (Editors), Proc. Int. Symp., Freiburg, Germany. Int. Arch. Photogramm., 22: 719—749.

Heilman, J.L., Kanemasu, E.T., Bagley, J.O. and Rasmussen, V.P., 1977. Evaluating soil moisture and yield of winter wheat in the Great Plains using Landsat data. Remote Sensing Environ., 6: 315—326.

Idso, S.B., Jackson, R.D., Pinter, P.J., Jr., Reginato, R.J. and Hatfield, J.L., 1981. Normalizing the stress-degree day parameter for environmental variability. Agric. Meteorol., 24: 45—55.

Jackson, R.D., 1982. Canopy temperature and crop water stress. In: D. Hillel (Editor), Advances in Irrigation. Academic Press, New York, NY, pp. 43—85.

Jackson, R.D., Reginato, R.J., Pinter, P.J., Jr., and Idso, S.B., 1979. Plant canopy information extraction from composite scene reflectance of row crops. Appl. Opt., 18: 3775—3782.

Jackson, R.D., Idso, S.B., Reginato, R.J. and Pinter, P.J., Jr., 1981. Canopy temperature as a crop water stress indicator. Water Resour. Res., 17: 1133—1138.

Kanemasu, E.T., Heilman, J.L., Bagley, J.O. and Powers, W.L., 1977. Using LANDSAT data to estimate evapotranspiration of winter wheat. Environ. Manage., 1: 515—520.

Kanemasu, E.T., Feyerherm, A., Latchford, W., Killian, T. and Wagner, J., 1982. Crop model development using remotely sensed thermal and optical data. Development Planning and Research Associates, Inc., Manhattan, KS, 231 pp.

Kauth, R.J. and Thomas, G.S., 1976. The tasselled cap — a graphic description of the spectral-temporal development of agricultural crops as seen by LANDSAT. In: Proc. Symp. Machine Proc. Remotely Sensed Data. IEEE Cat. 76, CH1103-1-MPRSD, Institute Electrical and Electronic Engineers, New York, NY, pp. 41—51.

Kauth, R.J., Lambeck, P.F., Richardson, W., Thomas, G.S. and Pentland, A.P., 1979. Feature extraction applied to agricultural crops as seen by Landsat. In: Proc. LACIE Symp. II. JSC-16015, NASA, Johnson Space Center, Houston, TX, pp. 705—721.

Knipling, E.B., 1970. Physical and physiological basis for the reflectance of visible and near-infrared radiation from vegetation. Remote Sensing Environ., 1: 155—159.

Kramer, P.J., 1969. Plant and Soil Water Relationships: A Modern Synthesis. McGraw-Hill, New York, NY, 482 pp.

Leamer, R.W., Noriega, J.R. and Wiegand, C.L., 1978. Seasonal changes in reflectance of two wheat cultivars. Agron. J., 70: 113—118.

Monteith, J.L., 1973. Principles of Environmental Physics. Edward Arnold, London, 241 pp.

Myers, V.I., Wiegand, C.L., Heilman, M.D. and Thomas, J.R., 1966. Remote sensing in soil and water conservation research. In: Proc. 4th Symp. Remote Sensing of Environment. Inst. Sci. and Tech., Univ. Michigan, Ann Arbor, MI, pp. 801—813.

Myers, V.I. (Author-Editor), Heilman, M.D., Lyon, R.J.P., Namken, L.N., Simonett, D., Thomas, J.R., Wiegand, C.L. and Woolley, J.T., 1975. Crops and soils. Chapter 22 in: R.G. Reeves (Editor-in-Chief), Manual of Remote Sensing. American Society of Photogrammetry, Falls Church, VA, pp. 1715—1813.

Palmer, W.C., 1968. Keeping track of crop moisture conditions nationwide. The new crop moisture index. Weatherwise, 21: 156—161.

Park, A., Kanemasu, E., Boatwright, G., Whitman, R., Cook, P. and Hardy, J., 1977. Episodic events and economic yield. In: Proc. Crop Spectra Workshop, February 1—3, 1977, Sterling, VA. Ecosystems Inst., Gambrills, MD, pp. 39—43.

Polcyn, F.C., Spansail, N.A. and Malila, W.A., 1969. How multispectral sensing can help the ecologist. In: P.L. Johnson (Editor), Remote Sensing in Ecology. Univ. Georgia Press, Athens, GA, pp. 194—218.

Richardson, A.J. and Wiegand, C.L., 1977. Distinguishing vegetation from soil background information. Photogramm. Eng. Remote Sensing, 43: 1541—1552.

Ritchie, J.T., 1974. Atmospheric and soil water influences on the plant water balance. Agric. Meteorol., 14: 183—198.

Ritchie, J.T., 1981. Soil water availability. Plant Soil. 58: 327—338.

Thomas, J.R., Namken, L.N., Oerther, G.F. and Brown, R.G., 1971. Estimating leaf water content by reflectance measurements. Agron. J., 63: 845—847.

Thompson, D.R., 1976. Results of LACIE integrated drought analysis (Southern U.S. Great Plains Drought 1975—76). Rep. LACIE-00424, JSC-11336, NASA, Johsnon Space Center, Houston, TX, 69 pp.

Thompson, D.R. and Wehmanen, O.A., 1977. The use of Landsat digital data to detect and monitor vegetation water deficiencies. In: Proc. 11th Int. Symp. Remote Sensing of Environment, II. Environ. Res. Inst. Michigan, Ann Arbor, MI, pp. 925—931.

Thompson, D.R. and Wehmanen, O.A., 1979. Using Landsat digital data to detect moisture stress. Photogramm. Eng. Remote Sensing, 45: 201—207.

Thompson, D.R. and Wehmanen, O.A., 1980. Using Landsat digital data to detect moisture stress in corn-soybean growing regions. Photogramm. Eng. Remote Sensing, 46: 1087—1093.

Tucker, C.J., 1979. Red and photographic infrared linear combinations for monitoring vegetation. Remote Sensing Environ., 8: 127–150.

Wiegand, C.L., Gausman, H.W. and Allen, W.A., 1972. Physiological factors and optical parameters as bases of vegetation discrimination and stress analysis. In: Proc. Seminar Operational Remote Sensing. American Society Photogrammetry, Falls Church, VA, pp. 82—100.

Wiegand, C.L., Gausman, H.W., Cuellar, J.A., Gerbermann, A.H. and Richardson, A.J., 1974. Vegetation density as deduced from ERTS-1 MSS response. In: 3rd Earth Resources Tech. Satellite-1 Symp. I, Section A. U.S. Government Printing Office, Washington, DC, pp. 93—116.

Wiegand, C.L., Richardson, A.J. and Kanemasu, E.T., 1979. Leaf area index estimates for wheat from Landsat and their implications for evapotranspiration and crop modeling. Agron. J., 71: 336—342.

Woolley, J.T., 1971. Reflectance and transmittance of light by leaves. Plant Physiol., 47: 656—662.

Agricultural Water Management, 7 (1983) 323—340
Elsevier Science Publishers B.V., Amsterdam — Printed in The Netherlands

REMOTE MONITORING OF RANGELAND PRODUCTION

EUGENE L. MAXWELL

Solar Energy Research Institute, 1617 Cole Blvd., Golden, CO 80401 (U.S.A.)

(Accepted 1 November 1982)

ABSTRACT

Maxwell, E.L., 1983. Remote monitoring of rangeland production. *Agric. Water Manage.*, 7: 323—340.

Use of Landsat vegetation indices (combinations of spectral bands) to measure standing crop biomass production in rangelands is addressed. The theoretical basis for vegetation indices is briefly reviewed for ratios, differences and normalizations. Both optical and digital indices are discussed.

The reliability of indices as indicators of rangeland primary production is evaluated in relation to drought impact. Seasonal green-up and senescence cycles are shown to be important factors when using indices as stress or drought indicators. Experimental results for the short grass prairies of Colorado show the onset and recovery from drought from 1973 to 1978.

The use of vegetation indices to monitor the changing quantity and condition of the vegetative cover of the earth's surface is one of the most promising applications of Landsat data. This application appears to be reliable and inexpensive and uses the attributes of multispectral information, rather than suffering from the loss of spatial information. By responding to the primary productive functioning of plants, the indices provide a measure that is difficult to make in any other way. Rather than being a substitute for more conventional measurements, the vegetation index provides a unique measure of vegetation that will be of value to many users for many applications, including the monitoring of drought impact.

INTRODUCTION

Some of the most widespread and universal problems facing the people of this earth have to do with the condition and availability of the vegetation covering the surface. Food, fiber and fuel are provided by the vegetation, as well as products for constructing shelters. Vegetation is also vital for the maintenance of a habitable environment because it has a dominant effect on the oxygen and carbon dioxide in our environment.

The measurement of standing crop biomass (the quantity of vegetative biomass in-place at a given point in time) and the annual production of biomass has therefore taken on worldwide importance. Because of the global nature of the problem, satellite remote sensing is an obvious candidate for the measurement system. The current Landsat system has the potential for

0378-3774/83/$03.00

such global monitoring, using vegetation indices derived from multispectral scanner data. This paper addresses the use of vegetation indices to monitor primary production on rangelands. Because most rangelands are uncultivated, changes in biomass production will be directly related to changes in precipitation, grazing pressure and range condition. Under stable and known management conditions, satellite multispectral images can provide an effective means of monitoring drought.

By definition, a vegetation index should be a variable which is deterministically related to the vegetation characteristic or characteristics of interest. In this instance, our concern is the measurement of standing crop biomass and/or the measurement of biomass production.

Some of the more available references to indices (including some not discussed here) include Jordan (1969); Maxwell (1976); Pearson et al., (1976); Thompson and Wehmanen (1979); Tucker (1979) and Wiegand and Richardson (1982). Recent research reported by Maxwell et al. (1980) evaluated the use of Landsat to monitor drought impact. Key results of that research project are briefly summarized in this paper.

THEORETICAL DISCUSSION

EM interactions with vegetation

The density variations observed in a photograph or the variations in the digital numbers recorded by Landsat result from the varying interactions which take place between surface materials and the incident electromagnetic (EM) energy. The use of EM remote sensing methods to measure vegetation characteristics is unique because the measurement of the reflected energy from vegetation is directly related to its biological functioning. This is one of the few instances in which the extraction of information from remotely sensed data is not dependent on some secondary or tertiary relationship.

The interactive mechanisms of a green leaf can be explained with reference to the leaf cross section shown in Fig. 1 (after Salisbury and Parke,

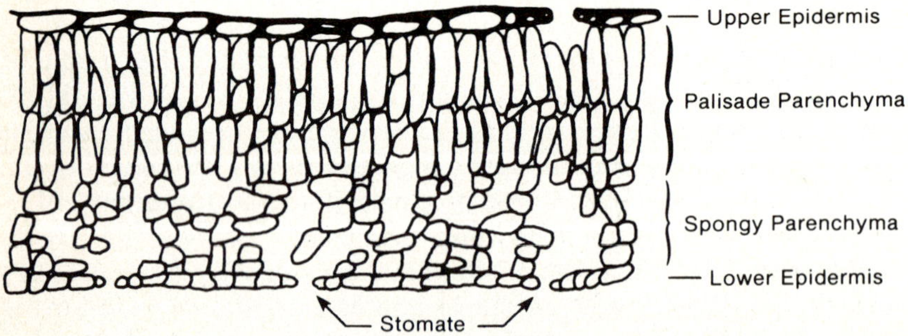

Fig. 1. Typical crossection of a leaf indicating internal structure.

1970). The light entering the leaf passes through the epidermal cells into the palisade region where most of the chlorophyll content of the leaf exists. The allowed energy states of chlorophyll results in strong absorption at red and blue wavelengths. Light which is not absorbed will reach the spongy parenchyma section of the leaf which is characterized by irregular cells separated by air spaces or voids partially filled with water. The different refractive indices of air, water, and cellulose results in refraction and reflection of the light rays passing through the leaf. This causes a considerable increase in the effective path length of the light rays, thereby increasing the probability that the light will be absorbed. For near IR wavelengths (not absorbed by leaf pigments) the multiple reflections within the leaf increases the probability that the rays will ultimately exit from the top of the leaf as reflected energy. These phenomena result in spectroreflectance characteristics for green vegetation such as that shown in Fig. 2. The strong pigment absorption processes result in very low reflectance in the visible portion of the spectrum, whereas near IR wavelengths exhibit very strong reflectance. The reflectance characteristics of a typical soil (brown silty loam) are shown for comparison.

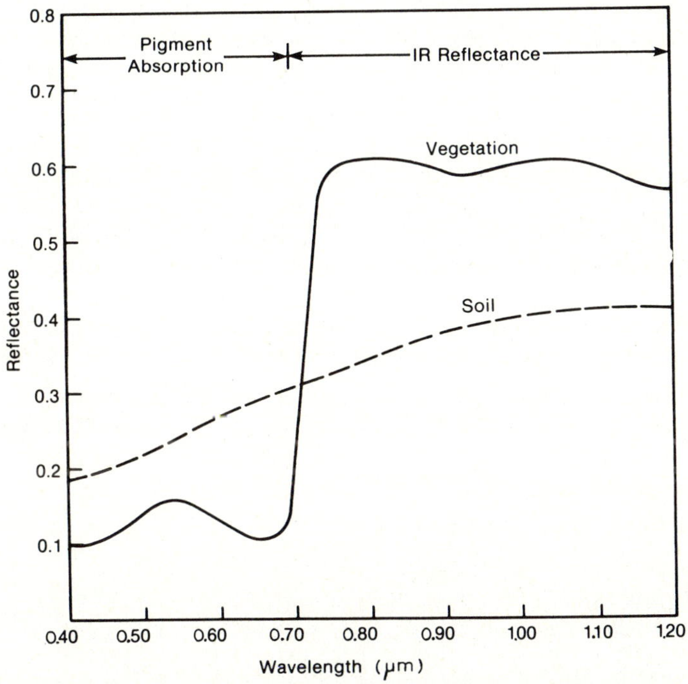

Fig. 2. Typical spectroreflectance characteristics of green vegetation and brown silty loam soil. Note the chlorophyll absorption bands in the blue (0.40—0.50 μm) and red (0.60—0.70 μm) portions of the spectrum.

Forming indices

From Fig.2 one can deduce that as the vegetation covering the soil increases from 0 to 100% canopy cover the reflectance in the visible part of the spectrum will decrease whereas the reflectance in the near IR will increase. Therefore, an IR/Red ratio or an IR—Red difference will form an index which is closely related to the quantity of photosynthetically active material on the earth's surface. In order to reduce sensitivity to changes in background (soil) reflectivity and atmospheric transmittance, the difference is often divided by the sum of the two bands, (IR—Red)/(IR+Red), which forms the normalized-difference (ND) index.

The Landsat bands used for forming indices are Band 5 (0.60—0.70 μm). Band 6 (0.70—0.80 μm) and Band 7 (0.80—1.00 μm). Band 5 encompasses the red portion of the spectrum and bands 6 and 7 are in the near IR.

Evaluating indices (Landsat MSS and TM)

Some of the most common indices generated from Landsat bands are listed and defined in Table I. The relative usefulness of these indices for measuring biomass is indicated by the *F*-values (ratios of variances) given in Table II. Data sets from rangeland, crops, crops plus range, and irrigated hay were used by Maxwell et al. (1980) to evaluate the indices. Although the normalized difference index using bands 7 and 5 has a considerably higher mean ranking than the other indices, the most significant observation is the similarity of the *F*-values. The variations in ranking for the different data sets may in part be an artifact of those sets as are *F*-value differences between sets. All in all, one must conclude that any of these six indices will provide a useful measurement of vegetative biomass, which is consistent with the results obtained by Tucker (1979).

From the reflectance curves shown in Fig. 2 one might conclude that the use of almost any part of the near infrared spectrum would be equally useful

TABLE I

Selected vegetation indices

Symbol	Name	Computation[a,b]
7/5	7/5 Ratio	$(DN_7/DN_5) \times 30$
ND	Normalized difference	$[((DN_7-DN_5)/(DN_7+DN_5))+1] \times 125$
RND	Root normalized difference	$[((DN_7-DN_5)/(DN_7+DN_5))+1] \times 30$
6/5	6/5 Ratio	$(DN_6/DN_5) \times 30$
ND-6	Normalized difference-6	$[((DN_6-DN_5)/(DN_6+DN_5))+1] \times 125$
RND-6	Root normalized difference-6	$[((DN_6-DN_5)/(DN_6+DN_5))+1] \times 30$

[a] DN, digital number recorded by Landsat.
[b] Band 5, 0.60—0.70 μm; Band 6, 0.70—0.80 μm; Band 7, 0.80—1.00 μm.

TABLE II

Comparison of initial F-values for the vegetation indices for four data sets (rank for each set shown in parentheses)

Index	F-values rangeland	F-values crops	F-values crops & range	F-values irrigated hay	Mean rank
7/5	614 (2)	1 272 (5)	1 930 (4)	220 (6)	4.25
ND	627 (1)	2 314 (1)	1 961 (3)	495 (2)	1.75
RND	540 (4)	1 877 (3)	1 998 (2)	368 (4)	3.25
ND-6	525 (5)	1 769 (4)	1 807 (5)	385 (3)	4.25
RND-6	606 (3)	2 026 (2)	1 699 (6)	554 (1)	3.00
6/5	517 (6)	1 238 (6)	2 036 (1)	204 (5)	4.50

in forming a vegetation index. Remote sensing systems, however, do not measure reflectance. Rather, they measure the radiance from the scene which is a function of the irradiance, surface reflectance, atmospheric transmittance, and atmospheric radiance parameters. This functional relationship can be expressed in a form similar to that shown in Eq. (1), which also shows the functional dependence on wavelength, solar elevation angle and the observation or look angle:

$$L = \frac{1}{\pi} \int_{\lambda_3}^{\lambda_2} E(\lambda)\, [\rho(\lambda,\theta,\phi)\, T_0(\lambda,\theta)\, T_u(\lambda,\phi) \sin\theta + \rho_A(\lambda)]\, d\lambda \qquad (1)$$

where L = surface radiance in the band from λ_1 to λ_2; $E(\lambda)$ = solar spectral irradiance at the top of the atmosphere; $\rho(\lambda, \theta, \phi)$ = surface reflectance; $T_0(\lambda, \theta)$ = downward atmospheric transmittance; $T_u(\lambda, \phi)$ = upward atmospheric transmittance; $\rho_A(\lambda)$ = atmospheric upward scattering coefficient; λ = wavelength; θ = solar elevation; and, ϕ = downward look angle from the aircraft or satellite.

The effect of the atmosphere on radiance measurements is observed in the narrow band radiance data from Ungar et al (1977), an example of which is shown in Fig. 3. Several major absorption bands are noted. The Landsat-D Thematic Mapper (TM) system has been designed with bands from 0.63 to 0.69 μm and 0.76 to 0.90 μm so as to reduce the effect of the oxygen and water absorption bands noted on Fig.3. These new bands should significantly improve the performance of vegetation indices used for the measurement of biomass.

Normalizing effects of indices

The vegetation indices defined in Table I are not only sensitive to biomass variations, but they are also relatively insensitive to variations in soil background reflectance and atmospheric absorption and scattering effects. In other words, the ratios and differences used in forming the indices tend to

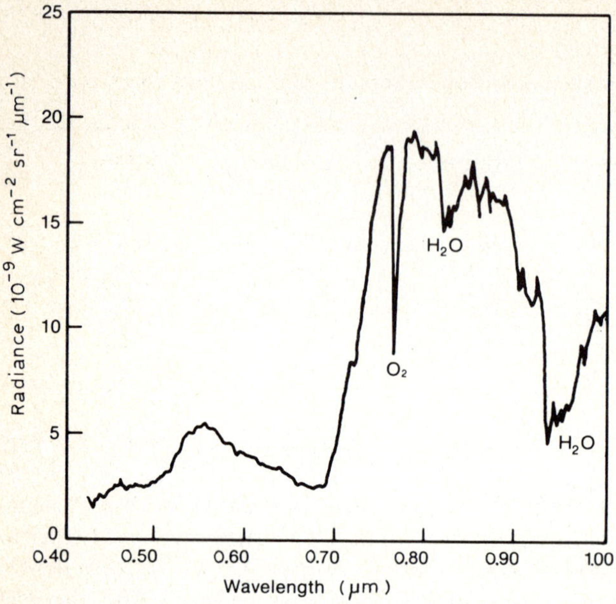

Fig. 3. Spectral radiance characteristics of a sugar beet field in the Imperial Valley, CA on 15 May 1975. From Ungar et al. 1977, annotations added.)

TABLE III

F-statistic evaluation of the sensitivity of the original landsat bands and the vegetation indices to variations in soil color

Varable	*F*-statistic
Band 4	291.0
Band 7	136.0
RND	15.6
RND-6	15.6
Band 5	14.0
ND	13.9
7/5 ratio	13.5
6/5 ratio	9.4
ND-6	9.2
Band 6	7.0

normalize radiance changes lacking spectral structure. For instance, the raising or lowering of the soil curve in Fig. 2, such as might result from soil moisture changes, will be normalized by ratioing any two bands. Complete normalization is not expected because most soil and atmospheric changes exhibit some spectral variations.

Several bare fields having different soil color were used to evaluate the original Landsat bands and the vegetation indices as soil discriminators. The *F*-statistics for each of these variables, when used to discriminate among the soil types, are shown in Table III. These statistics indicate that Landsat bands 4 and 7 are most sensitive to soil color variations, whereas the vegetation indices are relatively insensitive. Malila (1982) has shown that vegetation indices also effectively normalize for atmospheric changes.

DROUGHT APPLICATION

A project directed at the development of methods for monitoring drought impact using Landsat data was undertaken with NASA sponsorship during 1978 and 1979 (Maxwell et al., 1980). The project included a systematic analysis of digital and optical products, image availability, and procedures for assessing drought impact.

Using a digital vegetation index

Biomass values from ground measurements are needed to calibrate digital vegetation index values derived from Landsat. Field data obtained from many locations in Colorado were used to establish the graphical relationship between the normalized difference index and biomass shown in Fig. 4. The difference between the row crops and rangeland data is probably related to plant structure. The initial growth of row crops does not produce much canopy cover. Therefore, even though this predominantly vertical growth produces a significant quantity of biomass in the fields, the field reflectance is still dominated by the soil.

Ultimately, when the leaf area index becomes so high that all of the incident light is either absorbed or reflected from the vegetation canopy the vegetation index will no longer be sensitive to changes in biomass. It is likely the rangeland data will asymptote around 10 000 kg ha^{-1}. The asymptote for row crops, forbs and shrubs will occur at a much higher biomass. It is important to recognize that once these asymptotic values are reached, or once maximum growth of any given vegetation type has been achieved, further variations in vegetation index values will be related only to changes in crop vigor. This also serves to remind us that the vegetation indices will always be a function of both biomass and vegetation vigor.

A decision was made to define a 'universal' nonasymptotic vegetation index/biomass relationship as shown in Fig. 4, spanning the rangeland and row crop data. The dashed lines were visually fit around the mean so as to encompass 90% of the data points (excluding the row crop data below 10 000 kg ha^{-1}). These lines were used to bracket non-overlapping biomass ranges. The results indicate that 2 to 1 changes in biomass can be accurately monitored.

A regression equation could have been fit to the greater portion of the curve in Fig.4 such that a biomass value could be computed for each pixel. However, a simpler analytical technique making use of the entire curve was selected. The index values for each of the biomass ranges were used in con-

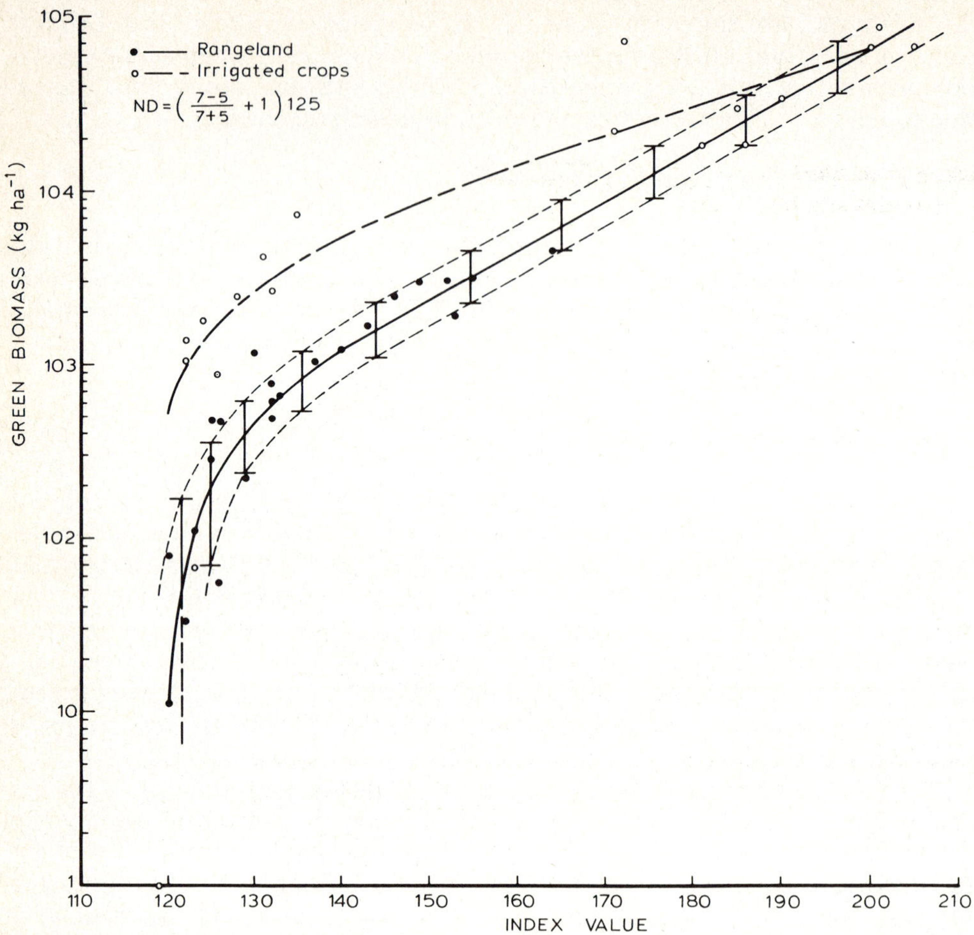

Fig. 4. Green biomass vs. the normalized difference (ND) index. Data are from numerous fields in eastern Colorado and several image dates in 1973 and 1978. This composite 'universal' curve ignores known differences of vegetation types. This will limit the absolute accuracy of biomass measurements.

nection with a level slicing or graymapping routine. This procedure is very inexpensive and produces biomass maps as well as statistical summaries (by summing the number of pixels falling within each of the biomass ranges). The method assigns index ranges and printer symbols to more or less equally spaced biomass values. Pixels in a scene are then assigned biomass values and symbols according to the ranges within which their index values fall.

Fig. 5. Computer generated graymaps of nine levels of biomass in the Earl quadrangle near Trinidad, CO. Biomass ranges were based on universal curve shown in Fig. 7. Highest biomass areas were irrigated hay fields.

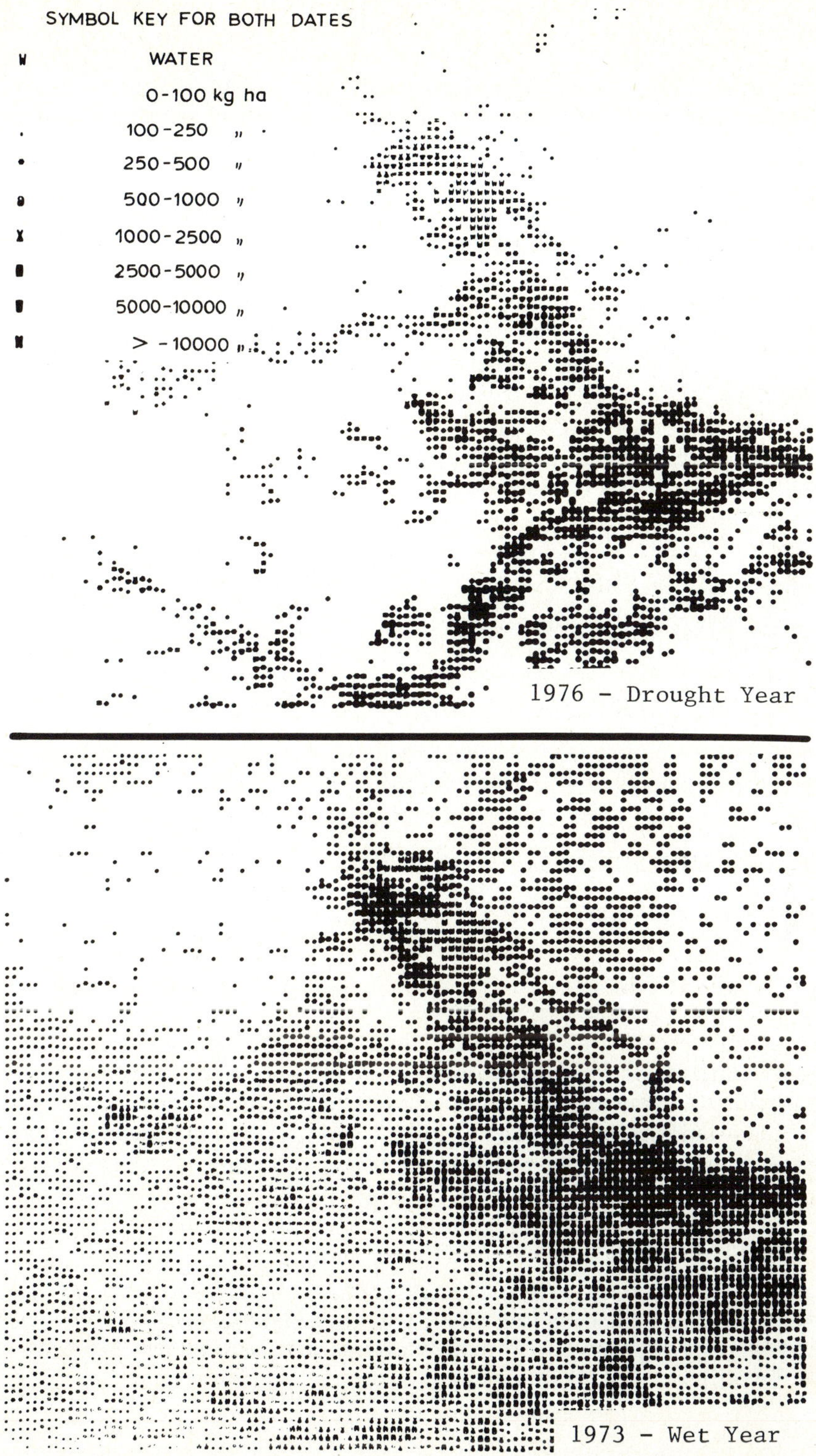
SYMBOL KEY FOR BOTH DATES
W WATER
0-100 kg ha
100-250 "
250-500 "
500-1000 "
1000-2500 "
2500-5000 "
5000-10000 "
> -10000 "
1976 - Drought Year
1973 - Wet Year

Typical results

Fig. 5 compares the results obtained using the graymapping procedure for 1973 (a wet year) and 1976 (a drought year) for a portion of the Earl quadrangle in southcentral Colorado. The effect of the drought on the rangeland and irrigated hay fields of this region is apparent. A more quantitative evaluation of the drought impact is available from the biomass totals for 1973, 1976 and 1978 which are given in Table IV. The accuracy of these total biomass numbers was not evaluated. However, because the image dates in each of the three years corresponded to the same point in the growing season, the relative magnitude of change for each of these three regions should be valid. It was especially encouraging to find such comparable results for three regions having distinctly different vegetative cover. The Julesburg region in northeast Colorado was almost all short grass prairie, the Earl region was dominated by irrigated hay fields and Weston was in the mountains of southcentral Colorado with deciduous shrubs dominating the ground cover. All three areas were experiencing severe drought in 1976.

TABLE IV

Estimates of biomass (kg) using the ND index, for portions of three quadrangles in Colorado

Year	Earl [a]	Julesburg [b]	Weston [c]
1973	57 100 000	6 340 000	4 480 000
1976	16 500 000	1 720 000	2 520 000
1978	43 500 000	5 910 000	6 610 000

[a] 5 420 ha of rangeland and irrigated hay.
[b] 1 960 ha of rangeland.
[c] 1 020 ha of mountain shrubland.

OPERATIONAL CONSIDERATIONS

There are many factors which could limit the usefulness of Landsat for monitoring the impact of drought on rangeland. The most significant of these are the availability of images and uncertainty about which factor has caused a change in rangeland condition and primary production. Furthermore, the cost of digital processing may be a deterrent for many users.

Fig. 6. The green-up and senescent cycle for rangeland in several Colorado counties. Data from ocular estimates of field observers. The upper curve for Conejos County is from an irrigated pasture. All data from 1978; rainfall events were regional, not measured at observation fields.

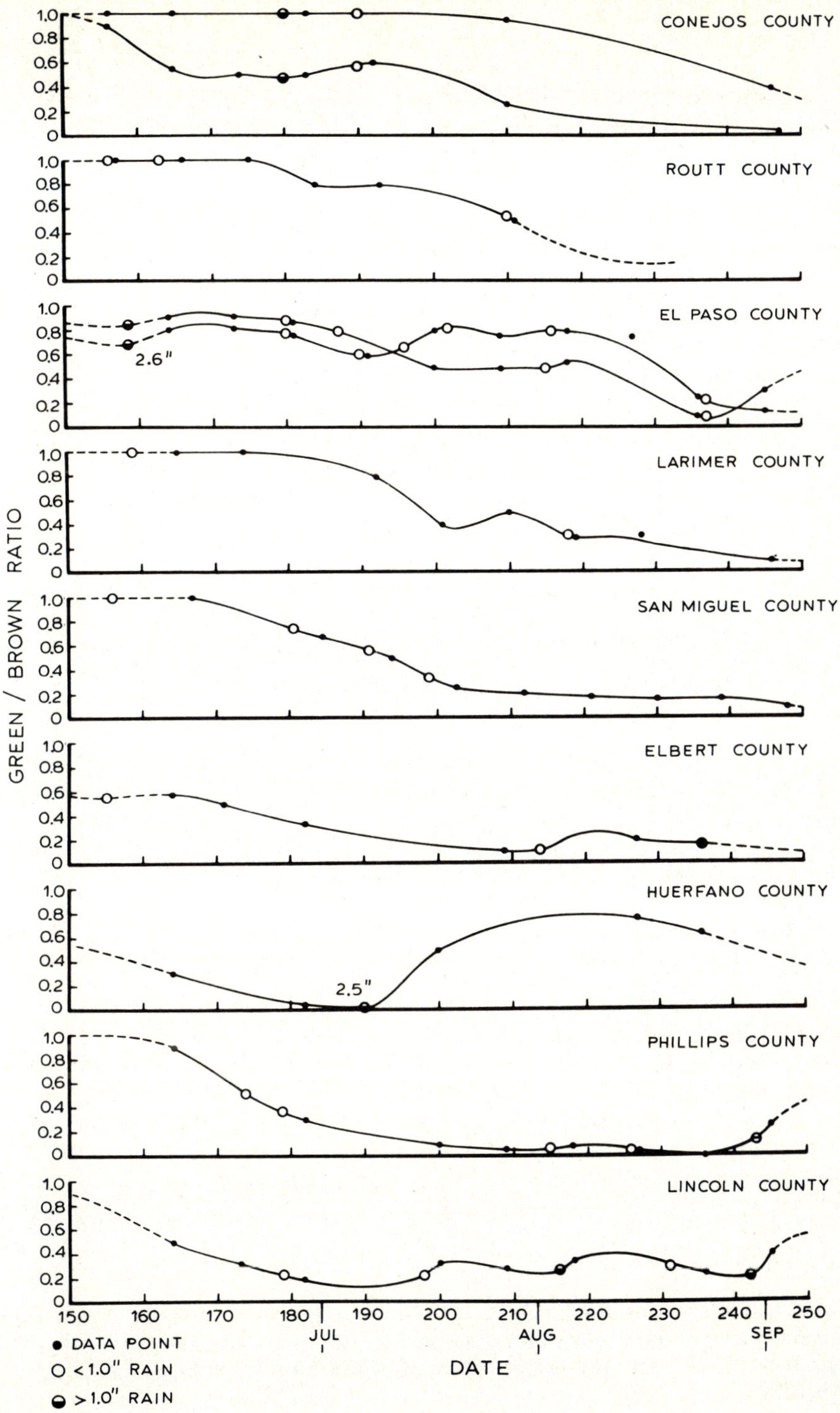
CONEJOS COUNTY
ROUTT COUNTY
EL PASO COUNTY
2.6"
LARIMER COUNTY
SAN MIGUEL COUNTY
ELBERT COUNTY
HUERFANO COUNTY
2.5"
PHILLIPS COUNTY
LINCOLN COUNTY
GREEN / BROWN RATIO
DATE
JUL
AUG
SEP
• DATA POINT
○ < 1.0" RAIN
◓ > 1.0" RAIN

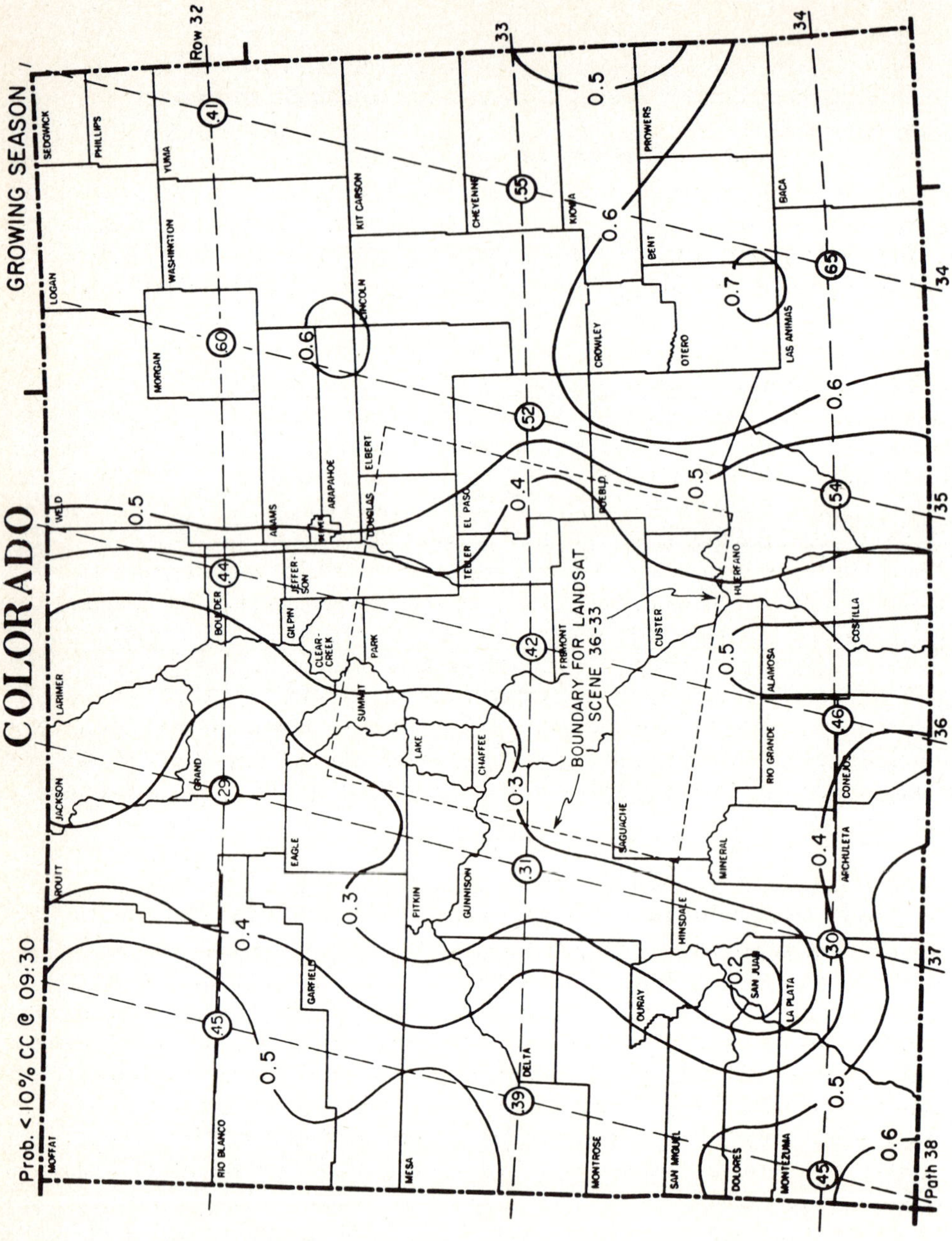

Fig. 7. Colorado map showing the probability of having less than 10% cloud cover at 09:30 MST on any day from 15 May to 15 September. Contours prepared from NOAA-VHRR images from 1974 to 1978; circled values are from Landsat cloud cover statistics from 1973 to 1978.

Temporal needs and image availability.

The need for several measurements during the growing season is indicated by the green-up and senescence cycle for rangeland in several Colorado counties during 1978 (see Fig.6). One should note that during the spring the rangelands tend to green up at most locations, even under drought conditons. Certainly there would be differences in total biomass in one area versus another, but these differences exist in part as a natural consequence of different climate and soil conditions. Supporting evidence of drought impact is obtained, therefore, by monitoring the onset of senescence. For instance, note that the upper curve for Conejos county was taken from an irrigated pasture which was not suffering from drought. Compare this curve with those counties experiencing the most severe drought conditions (Albert, Huerfano, Phillips and Lincoln). Larimer county was in relatively good condition and the sites measured in El Paso county benefited from the 6.6-cm (2.6-in) rain noted on Fig. 6. It is also interesting to note that Routt county was experiencing drought conditions, but at the higher elevations of that mountainous county senescence develops at a later date.

The plots in Fig. 6 show the need to monitor rangelands several times during the growing season to obtain a good measure of primary production and/or drought impact. Therefore, the availability of Landsat images several times during the growing season is of vital importance.

The probability of obtaining a cloud-free image on any given date for any location in Colorado is shown in Fig. 7. The contours for this figure were obtained from interpretations of four years of NOAA weather satellite images. Cloud cover was interpreted for each $1/2° \times 1/2°$ Lat.-Long. cell for mid-morning images, obtained on alternate days from 15 May to 15 September. The circled numbers, located at the nominal center of Landsat frames, were obtained from cloud cover statistics for Landsat images for the 1973 to 1978 growing seasons. These two sources of cloud cover information are in very good agreement and were verified by Weather Service observations at Denver, Alamosa and Grand Junction.

From Fig. 7 one sees that the probability of obtaining an essentially cloudfree image on any given overpass is about 60% for much of the eastern plains of Colorado and about 30% for the central Rocky Mountains. These results show that the 9-day repeat coverage provided by two Landsat satellites would likely provide images during time periods needed to monitor the plains of eastern Colorado. In other words, with two satellites there is a good chance of obtaining cloud-free coverage within 9 days of the optimum date. With only one satellite there is only a 50% probability of obtaining a cloud-free image during or close to any 15-day time period (e.g., late May). In the mountains the probability is less than 30%. For regions having more frequent cloud cover than Colorado the availability of Landsat images will be a more serious problem.

Interpreting change

Many factors can affect rangeland production from year to year and place to place. These include management practices, annual weather variations, species differences and soil differences. Assessment of drought impact will require, therefore, an in-depth knowledge of the region affected. In part, this will require satellite monitoring over a period of several years such that 'normal' conditions and 'normal' geographical variations are well understood. Governments cannot wait until a drought emergency arises and expect to use satellite images to accurately measure drought impact. With the understanding gained from several years of monitoring rangeland production, however, satellite multispectral scanners can be effective tools for measuring drought impact.

Using an optical vegetation index (OVI)

The cost of digital processing and the need for data at several times during the growing season prompted consideration of optical products for detecting change. Initially, it was assumed that false color composite prints (Landsat) would be used to provide an inexpensive drought monitoring capability. After a great deal of research and evaluation, however, it was concluded that quality control problems of standard color products could not be practically solved and that contrast enhanced products of the quality required would be almost as expensive as the cost of performing a full computer analysis of the digital data. The use of color products was therefore rejected for applications where comparisons from one date to the next are required, e.g., for drought assessment.

As an alternative optical product, an optical vegetation index (OVI) was generated by forming a sandwich of a band 7 positive transparency and a band 5 negative transparency. Because vegetated areas are bright on both of these products and nonvegetated areas are generally darker (more dense), the density (gray level) of a print made from the sandwich is strongly correlated with standing crop biomass. Such a print is designated 1/(7-5) to indicate a positive band 7, a negative band 5 and the inversion accomplished by printing from the sandwich.

An example of the OVI is shown in Fig. 8 where a comparison with band 7 and band 5 images can be made. Note for example, the difference in vegetation conditions in the San Luis Valley (west of the mountains) and the eastern plains. This difference is much more evident on the OVI product than on either band 5 or 7 and reflects the difference between an annual precipitation of 20 cm in that part of the San Luis Valley versus 30 to 36

Fig. 8. Positive prints from Landsat bands 5 and 7 and an optical vegetation index (OVI) printed from a sandwich of a band seven positive and a band five negative. The image is centered on Trinidad in southcentral Colorado on 6 August, 1978.

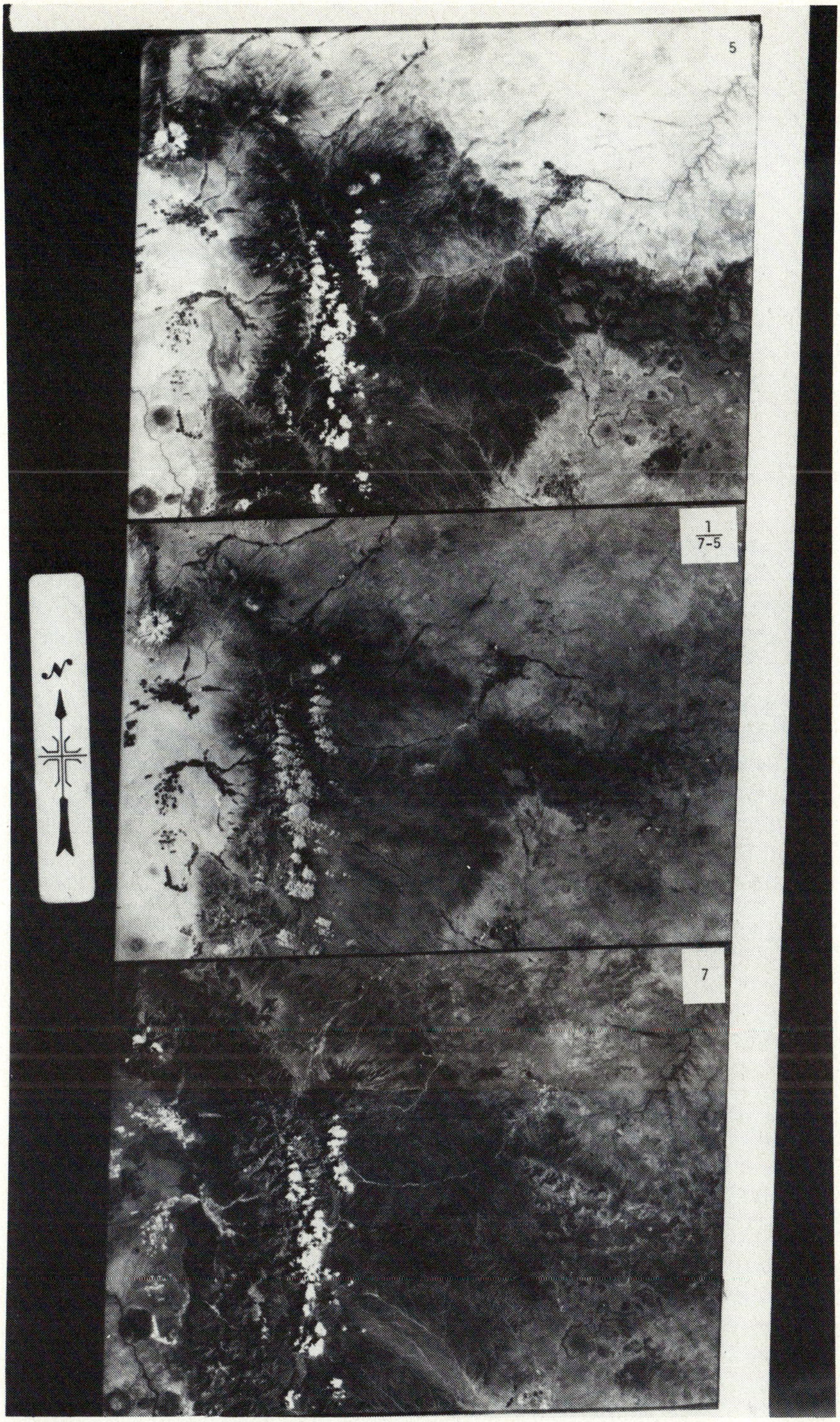
5
1/7-5
7
N

cm east of the mountains. Note also the dark volcanic cones in the lower left hand corner of each image. These cones are very sparsely vegetated and their dark appearance on both bands 5 and 7 results from the dark basaltic rocks of which they are comprised. The heavy vegetation in the irrigated agricultural fields in the San Luis Valley and in the bottom of the drainages from the mountains is indicated by the dark tone on the OVI product.

Although these initial results for the OVI product were very encouraging, subsequent analysis has shown that variations in the density of the standard black and white products produced by the EROS Data Center can produce a false indication of changes in vegetation condition. Therefore, caution must be exercised when using these products. In many instances such problems can be identified by examining locations of known land cover conditions, such as irrigated croplands, barren lands, and lakes and reservoirs.

NOAA and GOES images

Recent work by Duggin et al. (1982) and others has shown the feasibility of using the broad visible and near IR bands of weather satellites to form a vegetation index. These satellites have the distinct advantage of providing daily coverage (every 30 min from GOES). However, their broad spectral bands make their data more subject to atmospheric effects, i.e., they were not designed to monitor the earth's surface.

The spatial resolution (1 km) of these systems further limits their use for rangeland monitoring. Each pixel covers a 100 ha area, compared to Landsat's 1/2 ha resolution. Actually, the modulation transfer function of multispectral scanners limits the quantitative analysis of their data to areas at least 5 × 5 pixels in size (Maxwell et al., 1980). The radiance of smaller fields will be significantly affected by surrounding areas. Expressed in another way, pixels adjacent to a boundary between fields having different reflectances are not representative of either field. Hence, quantitative monitoring of rangeland production using NOAA satellite or GOES images would be limited to more or less homogeneous regions of 2 500 ha or larger. This is not unreasonable for many rangelands; furthermore, such data could be used to monitor average conditions even in heterogeneous regions.

CONCLUSIONS

Based on the current knowledge summarized in this paper, remote monitoring of rangeland production using satellite multispectral images and the application of such monitoring to drought impact assessment appears promising. Recommendations for using existing satellite systems and the limitations of such use are itemized below.

Most of the common vegetation indices (including those not evaluated in this paper) will perform satisfactorily to monitor rangeland production and as indicators of drought impact. The ratio and normalized difference indices are recommended because of their simplicity and their easy adaptation to most multispectral systems (ground, aircraft, land satellites and weather satellites).

Images should be obtained two or three times during the growing season (late spring, mid summer, late summer).

A repeat cycle of 16 to 18 days will limit reliable use of Landsat to semiarid and arid climates. Cloud cover problems will restrict the availability of images (at the right times during the growing season) in other climates.

NOAA and GOES satellites provide frequent image collection but their poor spectral and spatial resolution limit their use for this application. Their nominal 1 km resolution limits quantitative evaluations to areas of 2 500 ha and larger.

Use of an optical vegetation index (Black and White) formed from a near IR positive transparency sandwiched with a red band negative is recommended as an inexpensive method for monitoring large regions. The OVI should be used primarily to detect change (geographical and time).

The EROS Data Center and/or NOAA should be encouraged to provide an inexpensive digital and/or optical vegetation index product. The universal importance of vegetation monitoring dictates this need.

Digital indices can be used with a simple level slicing/graymapping routine to prepare biomass maps and forage/production statistics. Levels should be set to detect 3 dB or 2 to 1 changes in biomass, beginning at 50 to 100 kg ha^{-1}.

ACKNOWLEDGEMENTS

This paper has been compiled from the results of several research projects covering the period from 1973 to 1981. Many graduate students and others have been involved in the research, the majority of which was conducted at Colorado State University. Sponsors have included NASA, USGS and the Colorado Department of Agriculture. Support and encouragement for the research was received from many people including Drs. Tucker, Schnetzler and Cressy (NASA); Governor Lamm and Dr. David Carlson (Colorado) and Dr. Lee Miller (now at the University of Nebraska).

REFERENCES

Duggin, M.J., Piwinski, D., Whitehead, V. and Ryland, G., 1982. Evaluation of NOAA-AVHRR data for crop assessment. Appl. Opt., 21(11): 1873—1875.

Jordan, C.F., 1969. Derivation of leaf area index from quality of light on the forest floor. Ecology, 50: 663—666.

Malila, W., 1982. Tasseled cap transformation applications in arid/semi-arid lands. Presented ACSM-ASP Convention, March 14—20, 1982, Denver, CO.

Maxwell, E.L., 1976. A remote rangeland analysis system. J. Range Manage., 29: 66—73.

Maxwell, E.L., Aherron, R.M., Fits, D., Gross, G., Grunblatt, J. and Morse, A., 1980. Monitoring drought impact from Landsat. Rep. 1951, NASA Contract NAS 5-25081, Colorado State University, Fort Collins, CO, 214 pp.

Pearson, R.L., Tucker, C.J. and Miller, L.D., 1976. Spectral mapping of shortgrass prairie biomass. Photogramm. Eng. Remote Sensing, 42: 317—323.

Salisbury, F.B. and Parke, R.V., 1970. Vascular Plants: Form and Function, 2nd Edition. Wadsworth, Belmont, CA.

Thompson, D.R. and Wehmanen, O.A., 1979. Using Landsat digital data to detect moisture stress. Photogramm. Eng. Remote Sensing, 45: 201—207.

Tucker, C.J., 1979. Red and photographic infrared linear combinations for monitoring vegetation. Remote Sensing Environ., 8: 127—150.

Ungar, S.G., Collins, W., Coiner, J. et al., 1977. Atlas of Selected Crop Spectra. Imperial Valley, CA. NASA/GSFC, Greenbelt, MD.

Wiegand, C.L. and Richardson, A.J., 1982. Comparisons among a new soil index and other two- and four-dimensional vegetation indices. In: Proc. ACSM-ASP Convention, March 14—20, 1982, Denver, CO, pp. 210—227.

Agricultural Water Management, 7 (1983) 341—350
Elsevier Science Publishers B.V., Amsterdam — Printed in The Netherlands

ESTIMATION OF EVAPOTRANSPIRATION AT ONE TIME-OF-DAY USING REMOTELY SENSED SURFACE TEMPERATURES

J.L. HATFIELD[1], A. PERRIER[2] and R.D. JACKSON[3]

[1] *Department of Land, Air, and Water Resources, University of California, Davis, CA 95616 (U.S.A.)*
[2] *Station de Bioclimatologie, Institut National de la Recherche Agronomique, 78000 Versailles (France)*
[3] *U.S. Water Conservation Laboratory, Agricultural Research Service, U.S. Department of Agriculture, Phoenix, AZ 85040 (U.S.A.)*

Contribution from the California Agricultural Experiment Station, Institut de la Recherche Agronomique, and the Agricultural Research Service, U.S. Department of Agriculture

(Accepted 1 November 1982)

ABSTRACT

Hatfield, J.L., Perrier, A. and Jackson, R.D., 1983. Estimation of evapotranspiration at one time-of-day using remotely sensed surface temperatures. *Agric. Water Manage.*, 7: 341—350.

The estimation of evapotranspiration on a regional scale may be possible using remotely sensed inputs to surface energy balance models. Energy balance considerations lead to a relation that includes net radiation, surface and air temperatures, and an aerodynamic resistance, as inputs. The resistance term was examined as to its behavior under both stable and unstable temperature conditions, several surface roughness conditions, and at various windspeeds. The model shows that the evapotranspiration is higher than net radiation when the surface is cooler than the air and lower when the surface is warmer than the air. The aerodynamic resistance changes due to surface-air temperature differences play a substantial role in determining evapotranspiration.

To test the model, evapotranspiration was calculated using remotely sensed temperatures, with the remaining inputs conventionally assessed. The calculations were made for a one-time-of-day period near midday, as would be required for a remote sensing technique, and were compared to lysimetrically determined evapotranspiration. The measured data were obtained at several locations in the Western United States, and were for a variety of crops. The good agreement between calculated and measured values indicates that the goal of developing techniques that produce accurate evapotranspiration estimates over large areas is attainable.

INTRODUCTION

Remote sensing offers the possibility of evaluating evapotranspiration over large areas. Currently such estimations are made using meteorological variables that do not account for the spatial distribution of different cropped and bare soil surfaces, nor to the water status of any particular field. Remotely sensed surface temperatures provide a measure of the surface at a resolution of a few cm^2 for hand-held radiation thermometers to several km^2 for certain satellite sensors. Surface temperatures, combined with meteorological factors, can be used to provide improved estimates of areal evapotranspiration.

The concept of using remotely sensed surface temperatures in evapotranspiration estimations was espoused by Bartholic et al. (1972), Brown and Rosenberg (1973), and Stone and Horton (1974). The physical basis for the concept was outlined some years earlier by Montieth and Szeicz (1962), and Monteith (1963). Perrier (1975a,b,c) presented a detailed account of evapotranspiration under natural conditions, including the role of surface temperatures in the evaporation process. Jackson et al. (1977) combined several variables into a constant that was empirically obtained by regressing one time of day surface-air temperature differences and daily net radiation to the total daily evapotranspiration, in an attempt to evaluate evapotranspiration from predominately remotely sensed data. Seguin and Petit (1980) compared four methods for calculating evapotranspiration, two of which used remotely snesed surface temperatures as an input, one being the simplified approach of Jackson et al. (1977). They concluded that the temperature methods gave satisfactory results, with a precision of 10—15% compared to an energy balance method. Later, Seguin and Itier (1983) showed that stability and surface roughness should be included when evaluating the empirical factor suggested by Jackson et al. (1977).

Heilman et al. (1976), and Soer (1980) used thermal data derived from airborne sensors to estimate regional evapotranspiration. Heilman et al. adjusted the measured temperature for atmospheric attenuation and reported good agreement with measured values. They noted that the calculated values were overestimated at high rates of evapotranspiration. Soer reported computed values to be within 30% of values measured over a watershed.

The majority of investigations used several methods for calculating evapotranspiration to compare with those calculated using surface temperature inputs. Also, many investigations were conducted under conditions of reasonably high humidity, conditions that minimize the difference in temperature between the surface and the air (Jackson, 1982). In this report we examine a surface temperature based technique for estimating evapotranspiration under both stable and unstable conditions, and compare the results with evapotranspiration rates measured by weighing lysimeters.

ENERGY BALANCE EQUATIONS

The energy balance for a surface can be expressed as:

$$R_n + G = H + \lambda E \qquad (1)$$

where R_n is the net radiation, G the soil heat flux, H the sensible heat flux, and E the latent heat flux (evapotranspiration), all in units of J m^{-2} s^{-1}. The sensible heat flux (H) can be written as:

$$H = \rho c_p \, (T_c - T_A)/r_a \tag{2}$$

where ρc_p is the volumetric heat capacity (J m^{-3} °C^{-1}), T_c and T_A are the surface and the air temperatures (°C), respectively, and r_a is the aerodynamic resistance (s m^{-1}).

Substituting Eq. (2) into Eq. (1) and solving for γE yields:

$$\lambda E = R_n + G - \rho c_p \, (T_c - T_A)/r_a \tag{3}$$

which allows the calculation of evapotranspiration with remotely sensed surface temperatures as an input.

For neutral conditions, the aerodynamic resistance can be calculated from the expression:

$$r_a = \{\ln[(z-d)/z_0]\}^2/(k^2 u) \tag{4}$$

where z is the reference height (m), d is the zero plane displacement (m), z_0 is the surface roughness height (m), k is Von Karman's constant (0.38, unitless), and u is the windspeed at the reference height (m s^{-1}). Neutral conditions imply that $T_c \approx T_A$, a condition of little interest for using remotely sensed surface temperatures in evapotranspiration calculations. Under conditions of medium to low atmospheric humidity, $T_c - T_A$ may range from −10 to + 5°C, depending on the water status of the crop (Jackson et al., 1981). Thus, stability corrections must be applied to Eq. (4) if actual conditions are to be approximated.

Monteith (1963) showed that stability could be accounted for (at least approximately) by inclusion of the Richardson number. Expressing this number in terms of temperature differences and combining it with Eq. (4) results in the relation:

$$r_{a_c} = r_a \, [1 - n(z-d)g \, (T_c - T_A)/(T_0 u^2)] \tag{5}$$

where r_{a_c} is the stability corrected aerodynamic resistance; n is an empirical constant — Monteith (1973) suggested a value of 5; g is the acceleration due to gravity (9.8 m s^{-2}); T_0 (K) the average temperature (usually taken as the air temperature). A detailed discussion of the development of, and the assumptions underlying, Eqs. (1—5) is given by Monteith (1963, 1973).

Of the terms in Eq. (3), the aerodynamic resistance (as given by Eq. (4) or (5)) is the most difficult to evaluate, largely because of its dependence on windspeed and surface roughness. The effect of these factors on the calculation of r_{a_c}, and hence, evapotranspiration, are examined in the following section.

WINDSPEED AND SURFACE ROUGHNESS EFFECTS

A surface that is cooler than air will have a vertical transfer of sensible heat toward the canopy. This effect partially counters the momentum-induced aerodynamic resistance, and depends upon windspeed, canopy roughness, and the temperature gradient. Conversely, if the surface temperature is warmer than air, the aerodynamic resistance will decrease, because of buoyancy effects. Values of r_a for neutral conditions (horizontal dashed lines), and stability corrected values (solid lines) at several windspeeds and temperature gradients for a rough canopy (z_0 = 0.13 m) are shown in Fig. 1, and for a smooth canopy (z_0 = 0.05 m) in Fig. 2. The value of T_0 was taken as 303 K (this value was used in all calculations of r_{a_c}. The calculations show that, as the windspeed increases, the stability corrected r_{a_c} decreases, regardless of whether the canopy is warmer or cooler than the air. Also, as the surface becomes smoother (smaller z_0), values of r_{a_c} become larger, at the same value of windspeed and temperature difference. The roughness height (z_0) used in Fig. 1 is typical for canopies such as corn and sorghum, whereas the smaller z_0 as used in Fig. 2 is typical for crops such as alfalfa, beans, wheat, and cotton.

Equation (3) was used to evaluate the sensitivity of predicted evapotranspiration to changing canopy roughness and windspeed over a range of canopy-air temperature differences. The results are given in Fig. 3 for three roughness heights and two windspeeds. The net radiation was taken as 600 W m^{-2}. As the roughness decreases for a given windspeed, r_{a_c} increases and the evapotranspiration decreases. In other words, when the windspeed decreases, evapotranspiration becomes less sensitive to canopy-air temperature differences. This result is in agreement with the data of Verma et al. (1976).

The effects of roughness and windspeed on evapotranspiration are larger when the canopy is above air temperature. This suggests that morphological changes of crops (e.g. as a result of water stress) may increase the roughness and cause greater heat dissipation by convection. These calculations indicate that crops with rough canopies such as sorghum and corn would exhibit canopy temperatures near the air temperature for windspeeds above 1 m s^{-1}.

The calculations shown in Fig. 3 indicate that, whenever the canopy is cooler than the air, evapotranspiration will be more than the net radiation, because the direction of sensible heat transfer will be toward the canopy. These calculations agree with the experimental results of Sumayao et al. (1980).

Calculated values of evapotranspiration as a function of the canopy-air temperature difference for four levels of net radiation are shown in Fig. 4. The windspeed was taken as 1 m s^{-1} and the surface roughness as 0.13 m. The solid lines represent the stability corrected aerodynamic resistances and the dashed lines indicate the uncorrected resistances. Correcting for stability has only a small effect for $T_c - T_A$ values near zero, becoming larger as the canopy becomes warmer or cooler than the air. By not correcting for stability, evapotranspiration rates would be overpredicted at canopy-air temperature differences greater than about ±2°C from neutral conditions. This implies that at conditions near-neutral, the stability correction is minor.

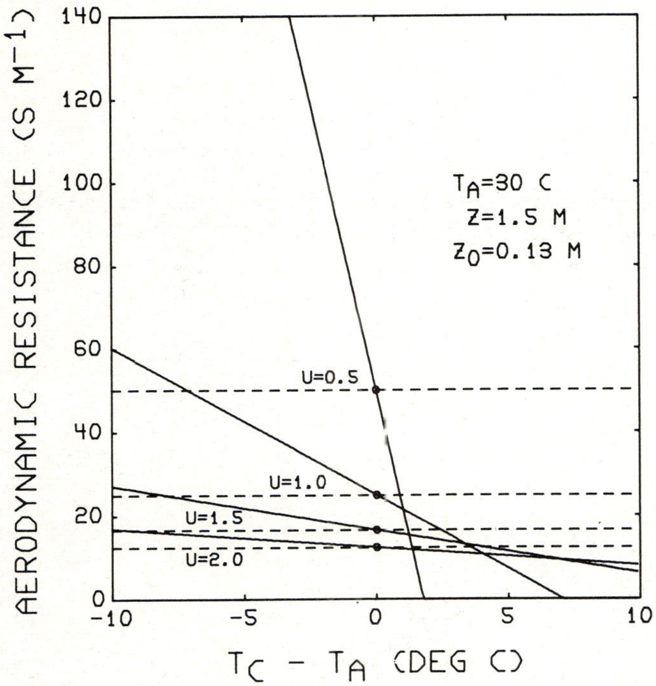

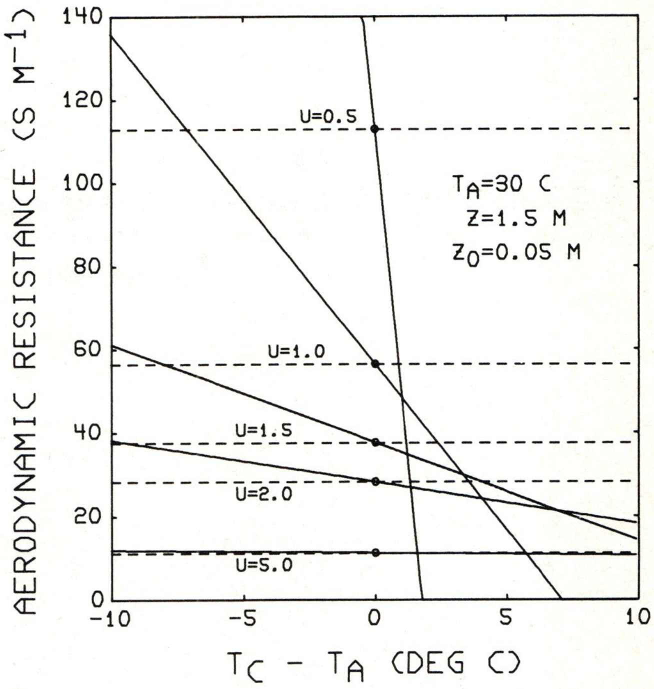

Fig. 1. Stability corrected (solid lines) and uncorrected (dashed lines) aerodynamic resistances for a surface roughness of 0.13 m at four values of windspeed, as a function of canopy-air temperature differences.

Fig. 2. Stability corrected (solid lines) and uncorrected (dashed lines) aerodynamic resistances for a surface roughness of 0.05 m at five values of windspeed, as a function of canopy-air temperature differences.

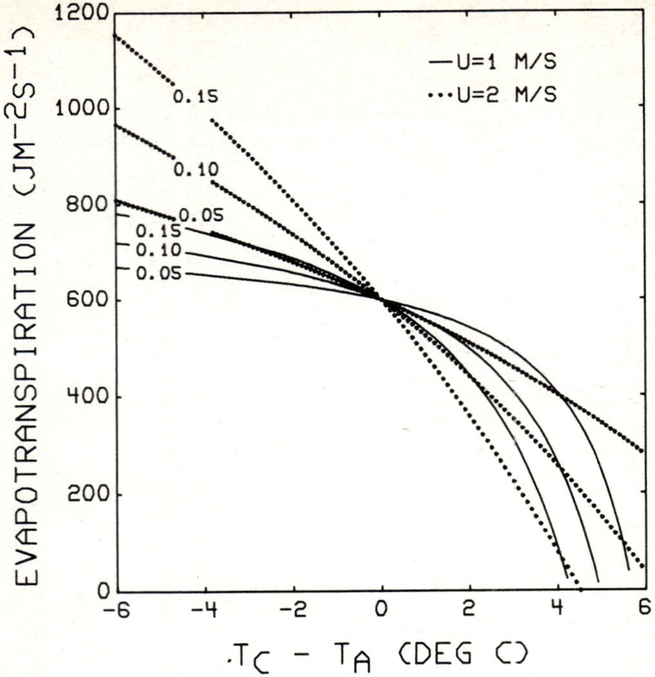

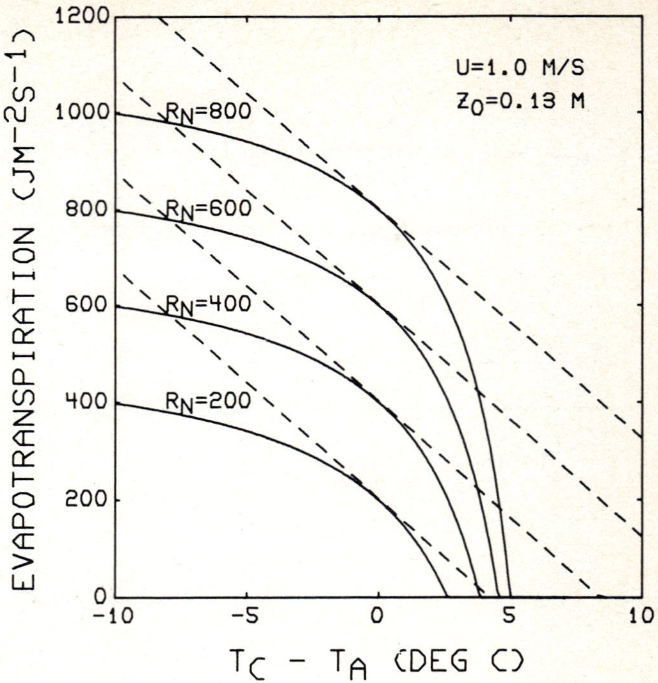

Fig. 3. Evapotranspiration calculated using Eq. (3) with stability corrected aerodynamic resistances for two windspeeds and three values of the surface roughness (lines labeled with roughness values), as a function of canopy-air temperature difference.

Fig. 4. Evapotranspiration calculated using stability corrected (solid lines) and uncorrected (dashed lines) aerodynamic resistances for four values of net radiation, as a function of the canopy-air temperature difference.

EXPERIMENTAL PROCEDURES

Data used to evaluate Eq. 3 were obtained at a number of locations during 1980 and 1982. In 1980, data were collected at Temple, TX; St. Paul, MN; Kimberley, ID; and Davis, CA. In 1982, the measurements were made entirely at Phoenix, AZ. The crops and dates of the measurements are given in Table I.

TABLE I

Measurement dates, locations, and crop parameters

Location	Dates	Crop	Height (cm)	Ground cover (%)
Temple, TX	13—17 June 1980	Sorghum	90	100
St. Paul, MN	02—03 July 1980	Alfalfa	80	100
Kimberly, ID	24—29 July 1980	Alfalfa	80	100
Davis, CA	04—05 August 1980	Tomato	60	80
Phoenix, AZ	1 March—15 May 1980	Wheat	60—85	100

At each location, net radiation, windspeed, and air temperature measurements were made. For row crops, the net radiometers were positioned 0.5 m above the plants and directly over the row. Windspeed and air temperature measurements were made at 1.3 m above the canopy (at Phoenix the instruments were 1.5 m above the soil surface). All instruments were calibrated before and after the experiments. Canopy temperatures were measured with an infrared thermometer having an 8° field of view and a 10.5 to 12.5 μm filter, between 13.30 and 14.00 h, local standard time.

Evapotranspiration data were obtained from weighing lysimeters for 30-min intervals at each location. All canopy temperature and meteorological measurements were made over the lysimeters. All crops grown in the lysimeters were well watered except at Phoenix, where differential irrigations were applied to three lysimeters.

Evapotranspiration rates were calculated using Eq. (3) with canopy temperature and meteorological inputs measured between 13.30 and 14.00 h. At Phoenix, the meteorological data were averaged over a 30-min interval. In the 1980 experiments, a 10-min average around the time of the canopy temperature measurement was used. In all cases the estimated evapotranspiration was compared to a 30-min average measured with lysimeters. Additional details of these experiments are given by Hatfield et al. (In press).

RESULTS AND DISCUSSION

Evapotranspiration rates were calculated using Eq. (3) for data from the locations listed in Table I. Calculations were made using both the neutral

(Eq. 4), and the stability corrected (Eq. 5) aerodynamic resistances. In all cases the surface roughness was taken as 0.05 times the crop height, except a value of 0.10 was used for tomatoes having 80% canopy cover. The soil heat flux (G) was assumed to be negligible. The error involved is thought to be small but this problem needs to be examined in detail.

These predicted values are compared with evapotranspiration values obtained from weighing lysimeters in Fig. 5. In the figure, the × symbols represent values predicted using the aerodynamic resistance for neutral conditions, and the ○ symbols represent predicted values using the stability corrected resistances. The 1 : 1 relation is indicated by the dashed line. The solid lines represent the best fit line obtained by linear regression for the two sets of data. The data calculated using the stability corrected resistances (○) fall close to the 1 : 1 line with the corresponding regression line being essentially the same as the 1 : 1 line. Almost all of the points calculated using the uncorrected resistances (×) are greater than the measured values, with the distance from the 1 : 1 line being greater for the higher values of evapotranspiration. It is evident that stability corrected resistance values should be used when calculating evapotranspiration using Eq. (3).

The data shown in Fig. 5 represent a range of canopy-air temperature differences from −4°C. Also, they were obtained at several locations having widely differing climates. These results support the use of surface temperature based models such as Eq. (3) for calculating evapotranspiration.

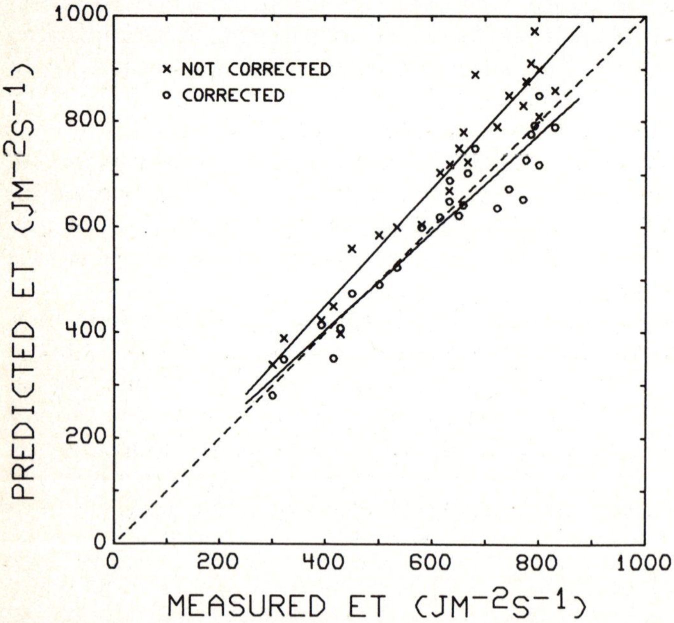

Fig. 5. Predicted values of evapotranspiration using stability corrected (○ symbols) and uncorrected (× symbols) aerodynamic resistances versus values measured using weighing lysimeters. The solid lines represent the regression lines for the two cases.

CONCLUSIONS

We conclude that remotely sensed surface temperatures, used in conjunction with net radiation, air temperature, windspeed, and surface roughness, can be used to estimate evapotranspiration. Surface temperatures can be obtained on a regional basis by use of air- of satellite-borne sensors. Some research has been directed toward estimating net radiation remotely. Questions that need to be answered include; is the spatial distribution of windspeed such that regional values can be extrapolated from point measurements; and, how can the surface roughness be estimated for each crop so that the proper stability correction can be made. It may be possible to design experiments to evaluate the surface roughness using surface temperature measurements in Eq. (3), for cases where the evapotranspiration can be obtained by independent means.

This paper describes how evapotranspiration can be estimated using remotely sensed surface temperatures. Remotely sensed data are obtained essentially instantaneously. To be most useful, evapotranspiration data should be representative of daily totals. In a companion contribution, Jackson et al. (1983, this issue) derived a coefficient which accounts for latitude, day-of-year, and time of day, that can be used to convert one-time-of-day measurements to daily totals, for clear days. Soer (1980), and Seguin and Itier (1983), also addressed this problem.

Although more effort will be required, the goal of estimating regional evapotranspiration by remote means appears attainable.

REFERENCES

Bartholic, J.F., Namken, L.N. and Wiegand, C.L., 1972. Aerial thermal scanner to determine temperatures of soils and of crop canopies differing in water stress. Agron. J., 64: 603—608.

Brown, K.W. and Rosenberg, N.J., 1973. A resistance model to predict evapotranspiration and its application to a sugar beet field. Agron. J., 68: 635—641.

Hatfield, J.L., Reginato, R.J. and Idso, S.B., In press. Evaluation of canopy temperature-evapotranspiration models over various surfaces. Agric. Meteorol.

Heilman, J.L., Kanemasu, E.T., Rosenberg, N.J. and Blad, B.L., 1976. Thermal scanner measurements of canopy temperatures to estimate evapotranspiration. Remote Sensing Environ., 5: 137—145.

Jackson, R.D., 1982. Canopy temperature and crop water stress. In: D. Hillel (Editor), Advances in Irrigation, Vol. 1. Academic Press, New York, NY, pp. 43—85.

Jackson, R.D., Reginato, R.J. and Idso, S.B., 1977. Wheat canopy temperature: a practical tool for evaluating water requirements. Water Resour. Res., 13: 651—656.

Jackson, R.D., Idso, S.B., Reginato, R.J. and Pinter, P.J., Jr., 1981. Canopy temperature as a crop water stress indicator. Water Resour. Res., 17: 1133—1138.

Jackson, R.D., Hatfield, J.L., Reginato, R.J., Idso, S.B. and Pinter, P.J., Jr., 1983. Estimation of evapotranspiration from one time-of-day measurements. Agric. Water Manage., 7: 351—362.

Monteith, J.L., 1963. Gas exchange in plant communities. In: L.T. Evans (Editor), Environmental Control of Plant Growth. Academic Press, New York, NY, pp. 95—112.

Monteith, J.L., 1973. Principles of Environmental Physics. Edward Arnold, London, 241 pp.
Monteith, J.L. and Szeicz, G., 1962. Radiative temperature in the heat balance of natural surfaces. Q.J.R. Meteorol. Soc., 88: 496—507.
Perrier, A., 1975a. Étude physique de l'evapotranspiration dans les conditions naturelles: I. Évaporation et bilan d'énergie des surfaces naturelles. Ann. Agron., 26: 1—18.
Perrier, A., 1975b. Étude physique de l'évapotranspiration dans les conditions naturelles: II. Expressions et paramètres donnant l'évapotranspiration réelle d'une surface "mince". Ann. Agron., 26: 105—123.
Perrier, A., 1975c. Étude physique de l'évapotranspiration dans les conditions naturelles: III. Évapotranspiration réelle et potentielle des couverts végétaux. Ann. Agron., 26: 229—243.
Seguin, B. and Itier, B., 1983. Using midday surface temperature to estimate daily evaporation from satellite thermal I.R. data. Int. J. Remote Sensing, 4 (In press).
Seguin, B. and Petit, V., 1980. Utilisation de la thermographie infrarouge pour l'estimation de l'évaporation. Analyse critique des résultats des mésures sur le site de la Crau. TELLUS Newsl. 16, Comm. Eur. Commun. ISPRA, Varese, Italy.
Soer, G.J.R., 1980. Estimation of region evapotranspiration and soil moisture conditions using remotely sensed crop surface temperatures. Remote Sensing Environ., 9: 27—45.
Stone, L.R. and Horton, M.L., 1974. Estimating evapotranspiration using canopy temperatures: field evaluation. Agron. J., 66: 450—454.
Sumayao, C.R., Kanemasu, E.T. and Brakke, T.W., 1980. Using leaf temperature to assess evapotranspiration and advection. Agric. Meteorol., 22: 153—166.
Verma, S.B., Rosenberg, N.J., Blad, B.L. and Baradas, M.W., 1976. Resistance-energy balance method for predicting evapotranspiration: Determination of boundary layer resistance and evaluation of error effects. Agron. J., 68: 776—782.

Agricultural Water Management, 7 (1983) 351—362
Elsevier Science Publishers B.V., Amsterdam — Printed in The Netherlands

ESTIMATION OF DAILY EVAPOTRANSPIRATION FROM ONE TIME-OF-DAY MEASUREMENTS

R.D. JACKSON, J.L. HATFIELD*, R.J. REGINATO, S.B. IDSO and P.J. PINTER, Jr.

U.S. Water Conservation Laboratory, Agricultural Research Service, U.S. Department of Agriculture, 4331 East Broadway Road, Phoenix, AZ 85040 (U.S.A.)

*Permanent address: Department of Land, Air, and Water Resources, University of California, Davis, CA 95616 (U.S.A.)

(Accepted 1 November 1983)

ABSTRACT

Jackson, R.D., Hatfield, J.L., Reginato, R.J., Idso, S.B. and Pinter, P.J., Jr., 1983. Estimation of daily evapotranspiration from one time-of-day measurements. *Agric. Water Manage.*, 7: 351—362.

The estimation of evapotranspiration (ET) on a regional basis requires remote sensing inputs. When obtained from air or space platforms, remotely sensed measurements are usually made at periodic intervals, and are essentially instantaneous in nature. A problem, then, is the estimation of daily values of ET from one time of day measurements. A technique is presented that allows the calculation of the coefficient necessary to convert one time of day measurements to daily totals. Input requirements are latitude, day of year, and time of day. This coefficient was applied to measured one time of day ET values and the results were compared to lysimetrically determined daily totals obtaind at five locations and for four crops. One time of day ET was also calculated using an ET model that requires remotely sensed surface temperatures. These values were converted to daily totals and compared with measured values. The results indicated that reliable estimates of daily total ET from one time of day measurements could be made for cloud free days. For cloudy days the results are less reliable, but they suggest that estimates may be improved by considering the amount and temporal distribution of cloud cover.

INTRODUCTION

The areal assessment of evapotranspiration (ET) by conventional methods requires numerous installations and considerable spatial interpolation, but may be repeatedly made. On the other hand, ET assessment by remote sensing methods can overcome the spatial problem, but is hampered by infrequent, essentially instantaneous, measurements that must be extrapolated to a longer time period such as a day to be meaningful.

During the past 10 years, several investigators have proposed that surface temperatures be used in the estimation of regional evapotranspiration (Bartholic et al., 1972; Brown and Rosenberg, 1973; Stone and Horton, 1974; Heilman et al., 1976; Jackson et al., 1977; Soer, 1980; Seguin and

0378-3774/83/$03.00

Itier, 1983). Most authors based their analyses on energy balance considerations, which lead to ET formulas of the type (Brown and Rosenberg, 1973):

$$\lambda E = R_n + G - \rho c_p (T_s - T_a)/r_a \qquad (1)$$

where λ is the heat of vaporization ($J\,g^{-1}$), R_n is the net radiation ($W\,m^{-2}$), G the soil heat flux ($W\,m^{-2}$), ρc_p the volumetric heat capacity of the air ($J\,kg^{-1}\,{}^\circ C^{-1}$), T_s the surface temperature (°C), T_a the air temperature (°C), and r_a the aerodynamic resistance ($s\,m^{-1}$).

Of the terms on the right hand side of Eq. (1), only T_s can be measured remotely. This can be accomplished with ground based radiometers or thermal scanners aboard air- or spacecraft. Net radiation can be estimated from remote measurements, but methods have not been developed to obtain R_n with the same resolution as T_s. The air temperature T_a must be measured on the ground. The error induced by using regional estimates of T_a has not been evaluated. Windspeed, implicit in the aerodynamic resistance r_a, must also be ascertained on the ground. A discussion of r_a in relation to ET estimation was given by Hatfield et al. (1983, this issue).

In addition to the problems involved with obtaining the various factors in Eq. (1) by remote measurements, the problem of estimating daily values of ET from essentially instantaneous measurements remains. Jackson et al. (1977) related daily totals of ET to $B(T_s - T_a)$, where the temperature difference was measured at one time of day. The factor B was found by regression. Seguin and Itier (1983) presented a detailed discussion concerning the theoretical basis for B, and showed that it could be expected to vary with windspeed and roughness. Soer (1980) used a model (Tergra) to calculate the energy balance and related the calculated daily value of ET to an instantaneous value for one time on a specific day. Converting one time of day measurements to daily totals appears feasible, at least for clear days. The conversion factor will be dependent on latitude, day of year, and time of day that the one time measurement was made.

In this report we assume that, for clear days, diurnal solar radiation and evapotranspiration can be adequately described by a sine function, and develop a method for calculating daily ET from one time of day values. The method accounts for latitude, day of year, and time of day. Solar radiation data are used in the development of the method, which is then applied to evapotranspiration data.

SOLAR IRRADIANCE

Monteith (1973) and Hirschmann (1974) showed that the instantaneous solar irradiance (S_i) that strikes the earth on a clear day closely conforms to the relation:

$$S_i = S_m \sin(\pi t/N) \qquad (2)$$

where S_m is the maximum irradiance at solar noon, t is the time (beginning at sunrise), and N is the time period between sunrise and sunset in units of t. At solar noon, $t = N/2$ and $S_i = S_m$. The diurnal trend of solar irradiance at Phoenix, AZ is shown in Fig. 1. The symbols represent values of irradiance accumulated several times each minute, averaged and recorded every 5 min throughout the day. The lines represent the best fit of a sine function ot the experimental data. The sine closely follows the data points except at sunrise and sunset. These data support the results of Monteith (1973) and Hirschmann (1974).

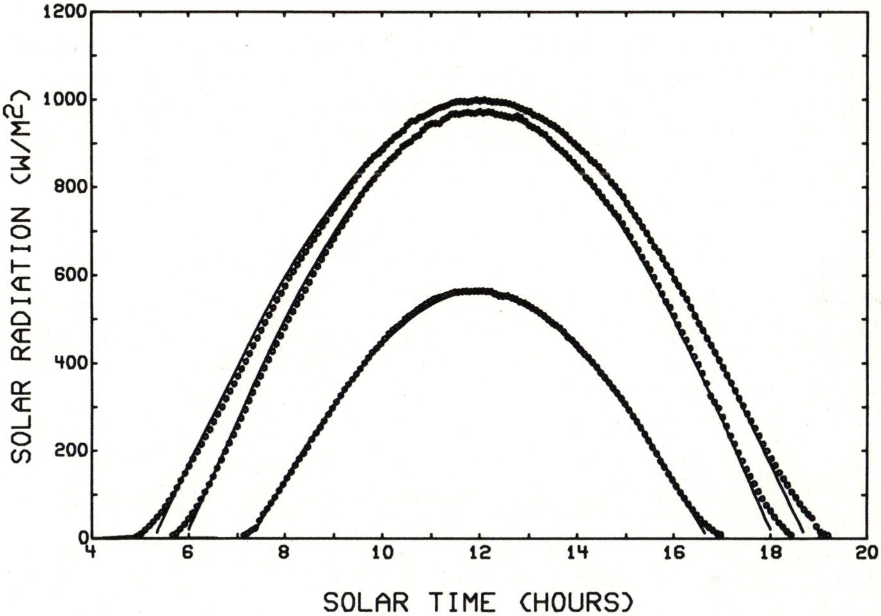

Fig. 1. Diurnal trend of solar radiation at Phoenix, AZ for three clear days: lower curve, 2 January 1981; middle curve, 10 April 1981; upper curve, 21 June 1981. Symbols represent experimental data. The lines were calculated from the best fit sine function.

Equation (2) can be integrated to give the total daily irradiance (S_d), i.e.:

$$S_d = \int_0^N S_m \sin(\pi t/N)dt = (2N/\pi)S_m \tag{3}$$

The ratio of total daily irradiance to the instantaneous irradiance at time t is:

$$S_d/S_i = 2N/[\pi \sin(\pi t/N)] \tag{4}$$

At solar noon $S_d/S_i = 2N/\pi$. The t and N terms in the argument of the sine must have the same units. The N in the numerator should have the same time units as S_i.

Equation (4) shows that the ratio S_d/S_i (thus, the ratio of total daily to instantaneous evapotranspiration, ET_d/ET_i), is both time of day and day of year dependent. We found that daylight period (N) could be expressed as:

$$N = a + b \sin^2 [\pi(D + 10)/365] \quad (5)$$

where a and b are latitude dependent constants, and D is the day of year. The term a is the length of the daylight period for the shortest day of the year and b is the amount that must be added to a to obtain the daylight period for the longest day of the year. A comparison of daylight hours for Phoenix published by the Nautical Almanac Office and those calculated using Eq. (5) are given in Table I for seven equally spaced days of the year. The maximum error occurred near days 50 and 300. Near these days the predicted values were about 7 min less than the actual. This error is acceptable within the context that Eq. (5) will be used here.

TABLE I

Daylight hours at Phoenix, AZ. Nautical Almanac office data, and approximate values calculated using Eq. (5)

Day of year	Almanac	Equation (5)	Difference (h)
50	11.13	11.01	-0.12
100	12.83	12.85	0.02
150	14.18	14.20	0.02
200	14.10	14.12	0.02
250	12.66	12.67	0.01
300	10.98	10.86	-0.12
350	9.95	9.94	-0.01

The constants a and b are readily evaluated for any location. The daylight periods for the winter and the summer solstice at each 10 degrees of latitude, taken from table 3.1 of Neiburger et al. (1973), were used to obtain expressions for a and b as a function of latitude. The regression equations are:

$$a = 12.0 - 5.69 \times 10^{-2} L - 2.02 \times 10^{-4} L^2 + 8.25 \times 10^{-6} L^3 - 3.15 \times 10^{-7} L^4 \quad (6a)$$

and

$$b = 0.123 L - 3.10 \times 10^{-4} L^2 + 8.00 \times 10^{-7} L^3 + 4.99 \times 10^{-7} L^4 \quad (6b)$$

where L is the latitude (minutes should be converted to decimal degrees). For Phoenix, AZ (33° 26′N), a = 9.93, and b = 4.43 from the Nautical Almanac data, and a = 9.79, b = 4.42 from Eqs. (6a,b). The difference between S_d/S_i calculated using actual values and those obtained from Eqs. (6a,b) was about 1%.

The use of N calculated from Eq. (5) in Eqs. (2), (3), and (4), will cause a small error. The sine function closely approximates the diurnal solar irradiance except near sunrise and sunset (Fig. 1). If the value of N in Eq. (2) is obtained from a best fit relation for numerous values of S_i throughout a diurnal cycle the resulting value of N would be less than the value obtained from an ephemeris. Using the values of N obtained for the lines shown in Fig. 1, and the actual daylight period values, we found the reduction to be about 0.945, which compares favorably with the value of 0.946 found by Hirschmann (1974). Using this factor, Eq. (5) becomes:

$$N = 0.945 \{a + b \sin^2 [\pi(D + 10)/365]\} \qquad (7)$$

where a and b are calculated from Eqs. (6a,b). Using N from Eq. (7) in Eq. (4) allows the calculation of the ratio S_d/S_i for any time of day and day of year, and for all latitudes between 60° S and 60° N.

Additional checks on the validity of this approach were made by graphically integrating solar irradiance curves for London (51° 30′N) from fig. 3.6 of Monteith (1973), and for Valparaiso (33° 02′S) from fig. 3 of Hirschmann (1974), obtaining a value for S_i from the curves, and forming the ratio S_d/S_i (units of MJ m^{-2} day^{-1}/(W m^{-2})). The results were: London, 0.0358 (0.0345), 0.0249 (0.0252), 0.0164 (0.0170), and Valparaiso, 0.0311 (0.0325), 0.0266 (0.0271), 0.0212 (0.0223). In all cases the predicted value is given followed by the measured value in parentheses. The three values for each location represent clear days near the summer solstice, near the spring equinox (fall in Chile), and the winter solstice. These data show that the method for predicting S_d/S_i is valid for a range of latitudes, in both the northern and southern hemispheres.

The ratio S_d/S_i obtained from Eq. (4) is a number used to multiply a measured value of S_i by to obtain S_d. For brevity, and to minimize confusion, we define:

$$J = S_d/S_i = ET_d/ET_i \qquad (8)$$

and will subsequently refer to the ratio as J. Furthermore, the time units (days and seconds) will be mixed in J so that a one time of day value of S_i or ET_i can be determined in W m^{-2}, and multiplied by J to give S_d or ET_d in units of MJ m^{-2} day^{-1}.

The ratio (J) of the total daily solar irradiance (MJ m^{-2} day^{-1}) to the instantaneous irradiance (W m^{-2}) at Phoenix is shown in Fig. 2. The symbols represent data obtained from July 1980 to July 1981. The line was calculated using Eqs. (4), (6a,b), and (7). The calculated value of J matched the measured values reasonably well.

It is important that the time of day be specified because the ratio increases with time for solar noon. This effect is shown in Fig. 3, where the ratio for solar noon, for 1 and 2 h from solar noon, and for a constant time of day (13.45 h MST) are plotted as a function of day of year. The line for 2 h from solar noon has a smaller yearly amplitude than the line for solar noon, indicating that time of day effects are a function of the day of year. The ratio for the constant time of day and 1 h from solar noon are nearly

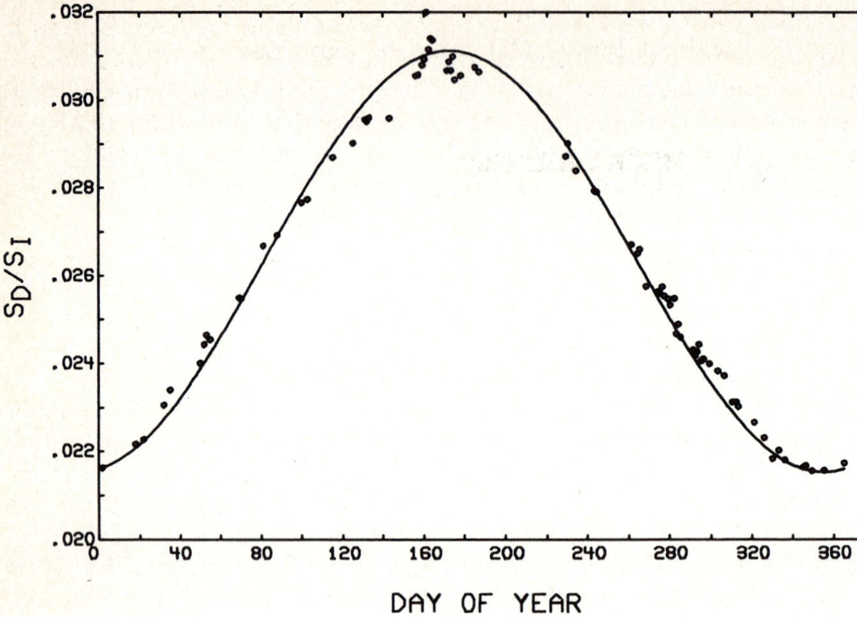

Fig. 2. Ratio of the total daily solar radiation (MJ m^{-2} day^{-1}) to the instantaneous irradiance at solar noon (W m^{-2}). The symbols represent experimental data taken at Phoenix, AZ during 1980—81. The line was calculated using Eqs. (4), (6), and (7).

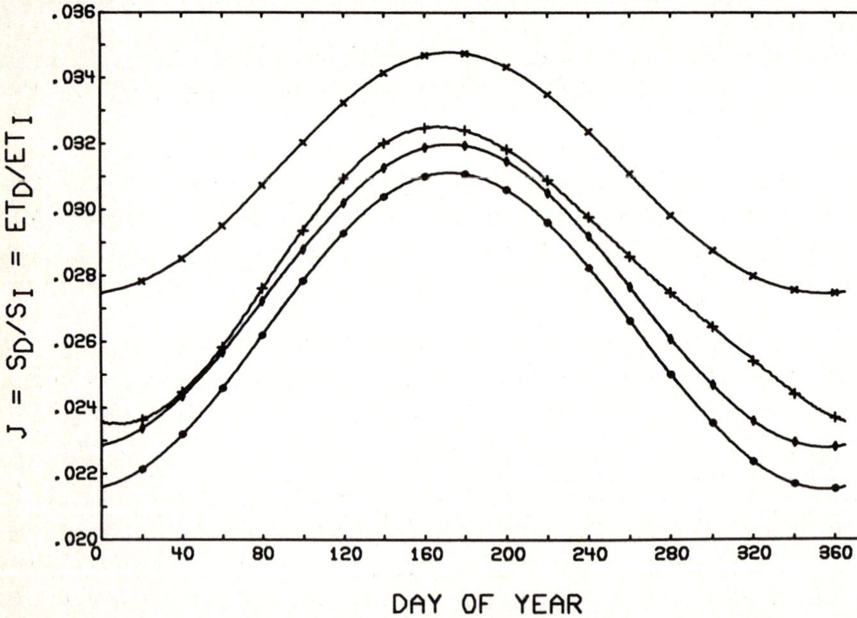

Fig. 3. Coefficient J to convert one time of day measurements to daily totals (MJ m^{-2} day^{-1}/(W m^{-2})). Symbols on the lines are for identification: circles identify the solar noon line; diamonds, 1 h from solar noon; x's, 2 h from solar noon; and the plus symbols identify a constant local time of day (13.45 h MST), all at 33° 26′N latitude.

the same in February but differ considerably in October. These results demonstrate the need for accounting for the time of day and day of year in calculating the ratio to convert instantaneous values to daily totals.

Assuming that ET can be described by a sine function, we can evaluate J for any location, day of year, and time of day and calculate daytime values of evapotranspiration from one time of day values. In arid areas where irrigation is practiced, nighttime ET may be appreciable. In such situations, an adjustment must be made to account for the nighttime contribution to the daily total.

MEASURED EVAPOTRANSPIRATION

Available data indicate that diurnal values of ET approximate a since function (for cloud free days), although it may be displaced from the solar sine function. In this section we will not be concerned with the sine function per se, but will use measured values of ET to determine how well daily totals of ET can be estimated from one time of day measurements, using the J coefficient. Most ET data have been reported as daily totals. Few data sets are available that include ET measurements throughout the day and allow the selection of one time of day (essentially instantaneous) values, as well as daytime and nighttime totals.

Evapotranspiration of wheat was measured at Phoenix, AZ using automatic weighing lysimeters (Van Bavel and Myers, 1962) that yielded a weight reading every 20 min. The difference between two readings multiplied by a calibration factor resulted in a value for ET representing the 20 min period. These values were summed to obtain a 24 h total, and also subtotaled for the daylight and the nighttime periods. Over the growing season, beginning in early December 1975, and ending in late May 1976, 122 days of data were obtained, of which 44 were cloud free. We used the ET for the 13.40 to 14.00 h time period for the midday value. A value of J was calculated for the appropriate day of year and time of day. The product of J and the midday ET yielded estimated values of daylight ET which we compared to the measured daylight totals. Results for the 44 cloud free days are shown in Fig. 4. These data represent bare soil and all stages of wheat growth from emergence to senescence. The highest ET value was about 25 MJ m^{-2} day^{-1} (about 10 mm of water). We can state that, for cloud free days, a reliable estimate of ET during the daytime hours can be made using one value obtained near midday.

A comparison of the predicted daylight values with 24 h totals indicated that nighttime ET in Phoenix was about 10% of the daytime values. This is only approximate since nighttime ET is dependent on the prevailing environmental conditions. We multiplied the predicted ET values by 1.1 for all 122 days of the 1975—76 wheat experiment and plotted them versus the measured 24-h totals, regardless of weather conditions. The results are shown in Fig. 5. Although more scatter is evident around the 1 : 1 line for these data than

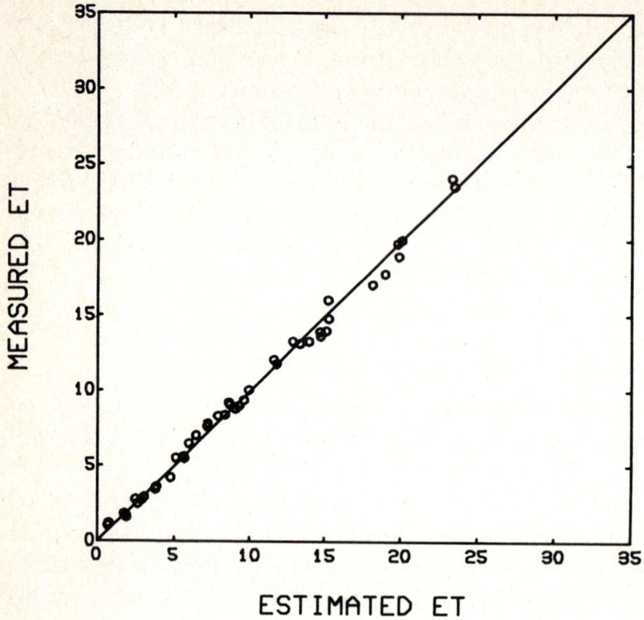

Fig. 4. Daily evapotranspiration (MJ m^{-2} day^{-1}) estimated from a one time of day (13.50 h) measurement versus measured evapotranspiration for the daylight period. The data points represent 44 clear days for wheat from emergence to senescence.

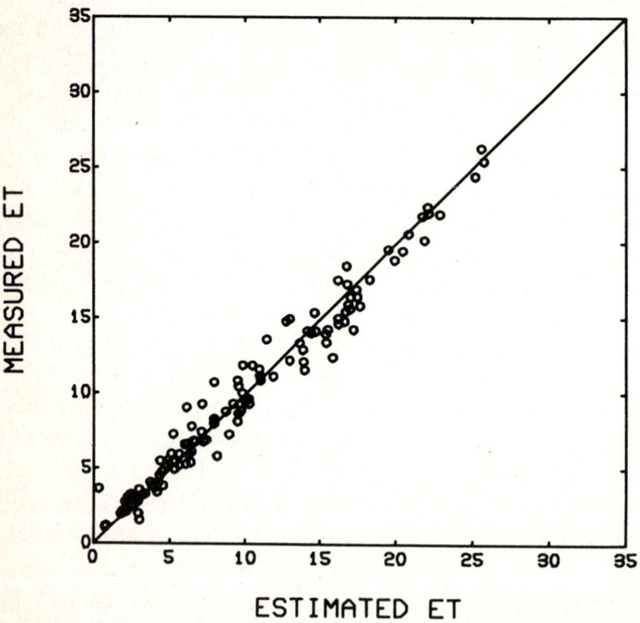

Fig. 5. Daily evapotranspiration (MJ m^{-2} day^{-1}) estimated from one time of day measurements increased by 10% to compensate for nighttime ET versus measured values of the 24 h total. The data represent all days for which data were available over a wheat growing season. Of the 122 days, 78 had various amounts of cloud cover.

for the cloud free days shown in Fig. 4, the degree of scatter is less than anticipated. It is possible to have a cloud free period during which the one time measurement is made, with clouds during the remainder of the day. Under these conditions the predicted value would overestimate the actual. A cloudy period at the time of the one time measurement would result in a underestimation of the actual, if sunshine prevailed at other times. If the cloud cover is relatively constant all day the predicted value should match the actual. The data in Fig. 5 suggest that ET estimates on cloudy days might be improved if the degree and temporal distribution of clouds were considered.

The data shown in Figs. 4 and 5 were from a wheat crop at Phoenix, AZ. A question may arise as to how well the method applies to different crops and to different locations. During 1980, an experiment was conducted at several locations in the U.S.A. that had lysimeters available for ET measurements (Hatfield et al., In press). Data for clear days were obtained from this experiment along with data taken some twenty years ago at Phoenix (from U.S. Water Conservation Laboratory Annual Reports, data of Drs. C.H.M. van Bavel and L.J. Fritschen). The data shown in Fig. 6 represent alfalfa, sudangrass, soybeans, and tomato, obtained at Phoenix, AZ; Davis, CA; Lincoln, NE; St. Paul, MN; and Kimberly, ID. The high value of 30 was

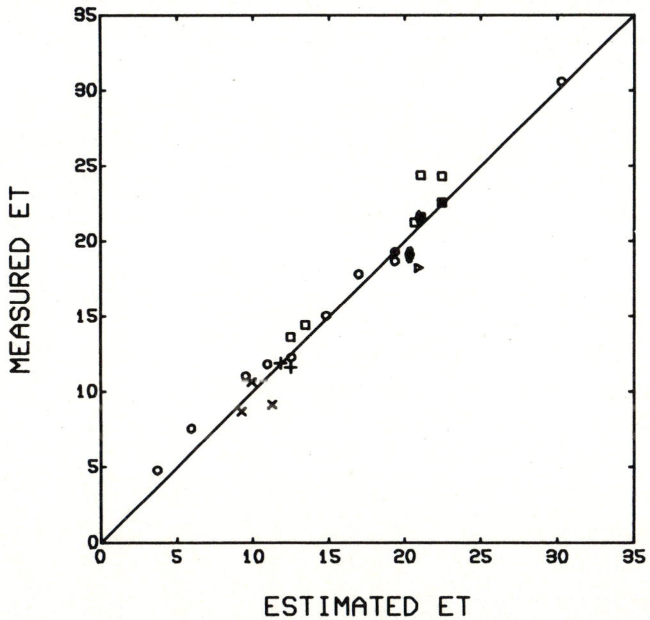

Fig. 6. Daily evapotranspiration (MJ m^{-2} day^{-1}) estimated from one time of day measurements versus measured values of the 24 h total. The circles represent data from alfalfa grown in Phoenix, AZ; the squares, sudangrass, Phoenix, AZ (solid squares were corrected to daylight values by subtracting measured nighttime ET); triangle, alfalfa, St. Paul, MN; diamonds, alfalfa, Kimberly, ID; x's, soybeans, Lincoln, NE; plus, tomato, Davis, CA.

noted as a very windy day at Phoenix. The two open squares near 25 appear distant from the 1 : 1 line. When these data were corrected for nighttime ET they fell near the line as indicated by the solid squares. No other adjustment for nighttime ET was made although the Phoenix data would obviously fall closer to the line if a 10% adjustment was made as it was in Fig. 5. Nighttime ET would be expected to be less in a cooler, more humid environment than exists in Phoenix. The necessity of including a factor to adjust for nighttime ET depends on local environmental conditions.

REMOTELY SENSED EVAPOTRANSPIRATION

An advantage of using remotely sensed parameters in the calculation of evapotranspiration is that relatively large areas can be evaluated. However, remotely sensed data can, in general, only be obtained at periodic times and thus requires a converison to daily values to be of maximum usefulness. Few data sets are available by which a one time of day remotely sensed value can be calculated, the daily total predicted, and the result compared with a measured value. We abstracted data for a few clear days from an experiment that was recently concluded and not as yet completely analyzed. In the experiment, surface temperatures were obtained using hand-held infrared radiometers. Air temperatures and windspeed were measured using standard techniques. ET was calculated using Eq. (1) as described by Hatfield et al., 1983.

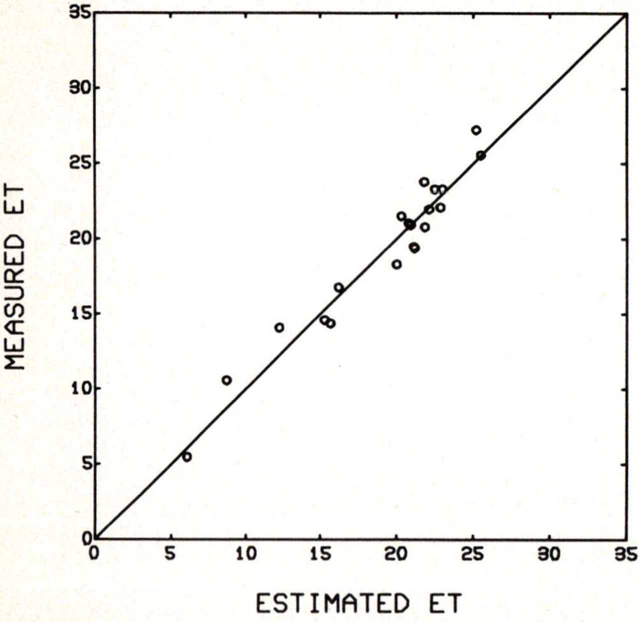

Fig. 7. Daily evapotranspiration ($MJ\,m^{-2}\,day^{-1}$) estimated from radiometrically measured surface temperatures and other energy balance components versus the measured 24 h totals. Data are for wheat at Phoenix, AZ, 1982.

The one time ET was converted to a daily total by use of the appropriate *J* factor and compared with measured values obtained from lysimeter data. The results for twenty measurements are shown in Fig. 7. The data fall reasonably close to the 1 : 1 line. These results indicate that remotely sensed, one time of day data can be used to estimate daily values of ET.

CONCLUDING REMARKS

The conversion of one time of day evapotranspiration values into daily totals can be made using a factor readily calculated from day of year and time of day for latitudes between 60° S and 60° N. The one time of day measurements should be taken within about 2 h of solar noon. We have presented data obtained from weighing lysimeters to demonstrate the validity of the method. However, any accurate means of measuring ET could be used.

The fact that clouds posed less of a problem than anticipated (Fig. 5) may be fortuitous. However, the data suggest that it may be possible to develop a means of using an estimate of cloud cover to improve the daily ET estimates made from one time-of-day measurements.

This demonstration that one time of day values of ET can reliably estimate daily totals is an added step towards achieving the goal of regional ET assessment, which can only be obtained using remote means. The rapid development of techniques to assess evapotranspiration that requires a minimum of ground based data are warranted.

REFERENCES

Bartholic, J.F., Namken, L.N. and Wiegand, C.L., 1972. Aerial thermal scanner to determine temperatues of soils and of crop canopies differing in water stress. Agron. J., 64: 603—608.

Brown, K.W. and Rosenberg, N.J., 1973. A resistance model to predict evapotranspiration and its application to a sugar beet field. Agron. J., 68: 635—641.

Hatfield, J.L., Perrier, A. and Jackson, R.D., 1983. Estimation of areal evapotranspiration based on remotely sensed surface temperature. Agric. Water Manage., 7: 341—350.

Hatfield, J.L., Reginato, R.J. and Idso, S.B., In press. Evaluation of canopy temperature-evapotranspiration models over various surfaces. Agric. Meteorol.

Heilman, J.L., Kanemasu, E.T., Rosenberg, N.J. and Blad, B.L., 1976. Thermal scanner measurements of canopy temperatures to estimate evapotranspiration. Remote Sensing Environ., 5: 137—145.

Hirschmann, J.R., 1974. The cosine function as a mathematical expression for the process of solar energy. Sol. Energy, 16: 117—124.

Jackson, R.D., Reginato, R.J. and Idso, S.B., 1977. Wheat canopy temperature: a practical tool for evaluating water requirements. Water Resour. Res., 13: 651—656.

Monteith, J.L., 1973. Principles of Environmental Physics. Edward Arnold, London, 241 pp.

Neiburger, M., Edinger, J.G. and Bonner, W.D., 1973. Understanding our Atmospheric Environment. Freeman, San Francisco, CA, 293 pp.

Seguin, B. and Itier, B., 1983. Using midday surface temperature to estimate daily evapotranspiration from satellite thermal I.R. data. Int. J. Remote Sensing, 4 (In press).

Soer, G.J.R., 1980. Estimation of regional evapotranspiration and soil moisture conditions using remotely sensed crop surface temperatures. Remote Sensing Environ., 9: 27—45.

Stone, L.R. and Horton, M.L., 1974. Estimating evapotranspiration using canopy temperatures: field evaluation. Agron. J., 66: 450—454.

Van Bavel, C.H.M. and Myers, L.E., 1962. An automatic weighing lysimeter. Agric. Eng., 43: 580—583.

Agricultural Water Management, 7 (1983) 363—378
Elsevier Science Publishers B.V., Amsterdam — Printed in The Netherlands

INFRARED REMOTE SENSING FOR MONITORING RAINFALL

D.G. MOORE[1], J.C. HARLAN[2], J.L. HEILMAN[3], D.O. OHLEN[4] and W.D. ROSENTHAL[5]

[1] *EROS Data Center, U.S. Geological Survey, Sioux Falls, SD 57198 (U.S.A.)*
[2] *Remote Sensing Institute, South Dakota State University, Brookings, SD 57006 (U.S.A.)*
[3] *Remote Sensing Center and Department of Soil and Crop Sciences, Texas A&M University, College Station, TX 77843 (U.S.A.)*
[4] *Technicolor Government Services, Inc., EROS Data Center, Sioux Falls, SD 57198 (U.S.A.)*
[5] *Texas Agricultural Experiment Station, Blacklands Research Center, Temple, TX 76503 (U.S.A.)*

(Accepted 12 November 1982)

ABSTRACT

Moore, D.G., Harlan, J.C., Heilman, J.L., Ohlen, D.O. and Rosenthal, W.D., 1983. Infrared remote sensing for monitoring rainfall. *Agric. Water Manage.*, 7: 363—378.

Evaluations of thermal infrared satellite data from TIROS-N and the Heat Capacity Mapping Mission (HCMM) showed that rainfall distribution patterns could be reliably detected on images acquired up to at least three days after the event. The temperature relationship decreased eight days after the event when soil variations influenced the signal. A time-series analysis reduced thermal variability normally observed over diverse landscapes and increased the sensitivity of the procedures. The method of repetitive low-resolution thermal observations could be operationally employed over large geographic regions with currently available satellite systems. The results would augment the existing rain gauge stations by increasing the spatial sensitivity and the reliability of detection and mapping individual rainfall events.

INTRODUCTION

Soil moisture is a dynamic variable crucial to processes affecting man and earth. The distribution and spatial variability of soil moisture provides many difficulties for developing a monitoring program. Point measurements are of limited use for extrapolation. Heilman et al. (1978) summarized a workshop at which remote-sensing-oriented discipline scientists defined needs for synoptic, temporal soil moisture data in programs such as agricultural yield estimates, assessments of trafficability, monitoring of desertification, modeling runoff, and predicting outbreaks of crop diseases and pests. They concluded that no existing remote sensing capability could provide an adequate

0378-3774/83/$03.00

measure of soil moisture for the diverse uses and geographic area of need.

Applications range from concern with the moisture at or near the soil surface to concern with moisture at profile depths of many meters. The complexity and scientific thrust for monitoring this variable is well documented by the frequency of publications describing the variety of approaches being tested. The purpose of this paper is to briefly document existing remote sensing capabilities as a direct measure of soil moisture and to present a technique of time-series analyses for assessing rainfall distribution.

Remote sensing can provide repetitive synoptic measurements. However, a relationship between moisture content and the spectral data recorded by the sensor must exist before a direct measure is available. Moore et al. (1975) demonstrated that spectral data as measured from the Skylab S-192 scanner could discriminate between wet and dry surfaces of bare fields. They found the longer ($> 1\ \mu m$) reflected infrared and the thermal infrared wavelengths were capable of this discrimination. Idso et al. (1975a) demonstrated that stages of drying of a bare soil were related to radiance in the reflected solar region. However, the moisture estimates were only for the top few millimeters of the soil profile. Reginato et al. (1976) reported that thermal infrared measurements related to near-surface soil moisture after the surface was visually dry. Idso et al. (1975b) continued the evaluation and demonstrated that surface temperature was sensitive to moisture changes in the top 4 to 5 cm of the soil profile.

Where vegetation cover is incomplete, reflected and emitted signals are emanating from both the vegetation and the soil surfaces. A plant canopy that completely covers the soil surface permits the use of the canopy temperature to monitor the soil-water-plant regime (Tanner, 1963). However, investigators have shown that even with full canopy cover thermal exitance from the soil surface can affect near-nadir remote temperature measurements of the crop canopy (Blad and Rosenberg, 1976). Jackson (1982) recently concluded that it was not feasible to relate profile soil moisture to a one-time plant temperature measurement. He suggests additional research using temporal, multispectral approaches to monitor changes in plant biomass, the onset of stress, and other factors that may provide clues to soil moisture.

Heilman et al. (1981) developed and tested models to separate canopy from background soil temperatures when using a nadir-viewing remote sensor for incomplete canopy covers. They concluded canopy temperatures could be estimated if appropriate considerations were given to soil background radiance, emissivity variations, and sky irradiance. Heilman and Moore (1980), who tested the approach throughout a full growing season of barley, found an excellent temperature relationship throughout the growing season between 0 to 4 cm depth volumetric soil moisture and corrected radiometric temperatures. However, considerable ancillary information, only part of which is available on a synoptic basis, must be available to employ the approach.

Microwave wavelengths have increased soil and cloud penetration ability and, therefore, have potential for relating to profile soil moisture. However, both Schmugge (1978) and Newton et al. (1982) concluded that practical depths of measurement sensitivity to moisture are typically less than 10 cm. The effects of canopies, surface roughness, and uncertain depth of penetration, among others, limit the understanding and practical use of microwave wavelengths.

Most studies have been site specific and were designed to reduced interactive effects. Such effects dominate synoptic soil moisture information applications. A concerted effort will be required to advance the techniques. Even under limited landscape variability, no sensor or combination of sensors has yet proven effective for soil moisture measurement except for the very near surface.

In dryland agriculture, the soil moisture input is predominantly from precipitation. The National Oceanic and Atmospheric Administration (NOAA) has established and maintained an extensive rain gauge network for point measurements of precipitation occurrences and amounts. These provide the average condition for a particular time and location based upon observations for many years. Because many of the recording stations are up to 50 km apart, any single event which is not wide spread has a low probability of being recorded. Mapping the spatial distribution can only be accomplished to within tens of kilometers for any single rainfall event using the existing gauging network. An investigation is underway to determine if presently available satellite data can provide information to spectrally and spatially record rainfall events as a supplement to the existing NOAA gauging network.

MATERIALS AND METHODS

Site description

Two study sites were selected: the entire state of South Dakota and the Washita River Watershed of central Oklahoma. The Oklahoma study has been reported by Rosenthal et al. (1982).

South Dakota. A network of 1042 rain gauges was established by the South Dakota Department of Water and Natural Resources with daily recording of the previous 24 h of rainfall for the months of June through September of 1979. Land cover in South Dakota varied from a large proportion of row crops (corn and sorghum) interspersed with alfalfa, small grain, and pasture in the southeast to a large percentage of rangeland with a minor portion of small grain (principally wheat) in the northwest. Corresponding average annual precipitation in rainfall equivalents ranges from 61 cm in the southeast to 35 cm in the northwest. The precipitation is typically frontal in the fall, winter, and spring and convective during the summer. Soils vary from warm, moist prairie (Udic Ustolls) in the southeast to cool, very dry plain (Aridic Borolls) in the northwest.

Washita Watershed, OK. Rain gauges established by the U.S. Department of Agriculture were located on a 5 km grid throughout the watershed. Pasture, dryland winter wheat, and sorghum are the predominant agricultural crops. The extent of the pasture varied across the watershed from 100% coverage in the east to less than 10% in the west. Wheat was interspersed throughout the watershed, varying from 70% in the west to 50% coverage in the central portion. Sorghum was growing in 10% of the watershed, primarily in the central area. A wide range of soil types was present, ranging from clay (Udic Ustochrepts) in the eastern half to a sandy loam (Pachic Argiustolls) in the western quarter of the watershed.

Data sources

Both the Heat Capacity Mapping Mission (HCMM) and TIROS-N satellites were used for the investigations. The HCMM satellite data were from a two-channel scanning radiometer (0.55—1.1 and 10.5—12.5 μm). The satellite was in a sun-synchronous orbit at an altitude of 620 km. The corresponding ground coverage of the instantaneous field of view (IFOV) was 0.5 km $\times$ 0.5 km at nadir for the visible channel and 0.6 km $\times$ 0.6 km at nadir for the thermal channel. The noise equivalent temperature difference (ne$\triangle$t) of the thermal channel was 0.4 K at 280 K. The HCMM collected data at approximately 02.30 and 13.30 Local Standard Time (LST) with repeat cycles of 5 or 16 days depending on latitude.

The TIROS-N near-polar, sun-synchronus orbiting satellite included an Advanced Very High Resolution Radiometer (AVHRR) which had spectral responses of 0.55—0.90, 0.725—1.10, 3.55—3.93, and 11.5—12.5 μm. The orbit provided coverage at approximately 04.00 and 15.00 LST. The IFOV at the nominal altitude of 833 km was approximately 1.1 km on the ground. The 10 bit data had a ne$\triangle$t of approximately 0.2 K at 280 K. Daily repeat coverage was available.

Data processing and map registration

The daytime AVHRR data for South Dakota were acquired for two consecutive dates, 21 and 22 July 1979. The data were entered and processed using the Interactive Digital Image Manipulation System (IDIMS) at the U.S. Geological Survey's (USGS) Earth Resources Observation Systems (EROS) Data Center. Subscenes of South Dakota were extracted from each original scene. Each subscene was registered to a 1 km Alber's Equal-Area Conic map projection. The second order transformation equation had residual errors no larger than one pixel for registration of the control points to the map base. Digital Line Graph (DLG) data containing state and county boundaries were acquired from the USGS National Mapping Division (NMD) and were registered to the same 1 km grid. A binary mask was created to use the DLG image to mask all areas except those within South Dakota in the TIROS-N data. Attempts were made to remove clouds and cloud shadows by training and executing a maximum likelihood classifier and creating a cloud mask image for each of the two dates. The two masks were combined with the image data to remove image regions where clouds were present in either of the two images. Rain gauge data were registered into the data base.

The HCMM thermal infrared data of Oklahoma for three dates: 13, 24 and 29 July 1978, were prepared into grey-level maps using a computer system at Texas A&M University. The grey-level maps were overlain to a 1:250 000 scale topographic map where the rain gauge locations had been plotted. Apparent surface temperature averages were calculated from the HCMM data for a 3-by-3 matrix of pixels coincident with each rain gauge. Multidate data were normalized by subtracting air temperature differences between the dates from the corresponding dates. Rain gauges located under clouds or having less than 60% pasture in the surrounding 5 km^2 area were removed from the data set. For those gauge sites that remained, day to day surface apparent temperature differences between 24 and 13 July, and between 29 and 13 July were calculated, where 13 July was a date with dry soil conditions.

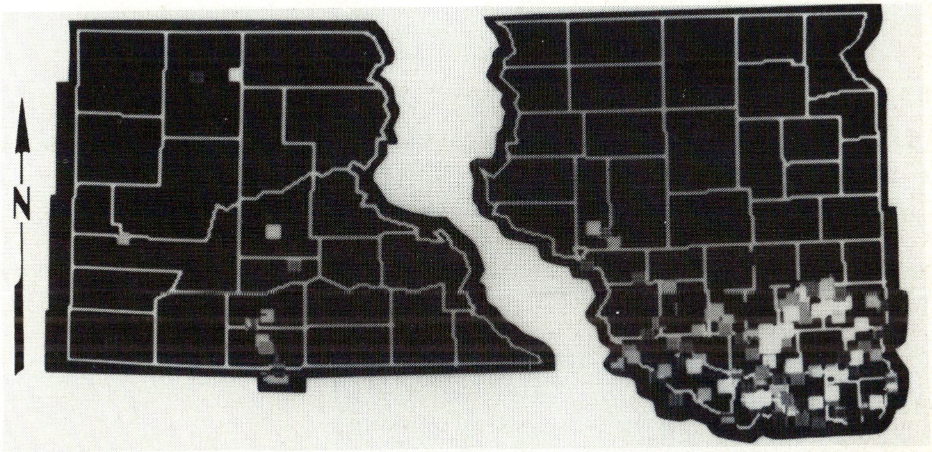

Fig. 1. Spatial display of South Dakota precipitation for the 24-h period preceding the 22 July AVHRR overpass (original in color). The lightest toned squares represent rainfall amounts of greater than 4.4 cm. Progressively darker tones represent 0.6 cm increments to the darkest tone being a trace to 0.6 cm. The original color display contained eight definite levels. State and county boundaries identify geographic locations. The east and west portions of the state were displayed and photographed individually using a curved television screen and will, therefore, not register together. The approximate length of South Dakota is 650 km.

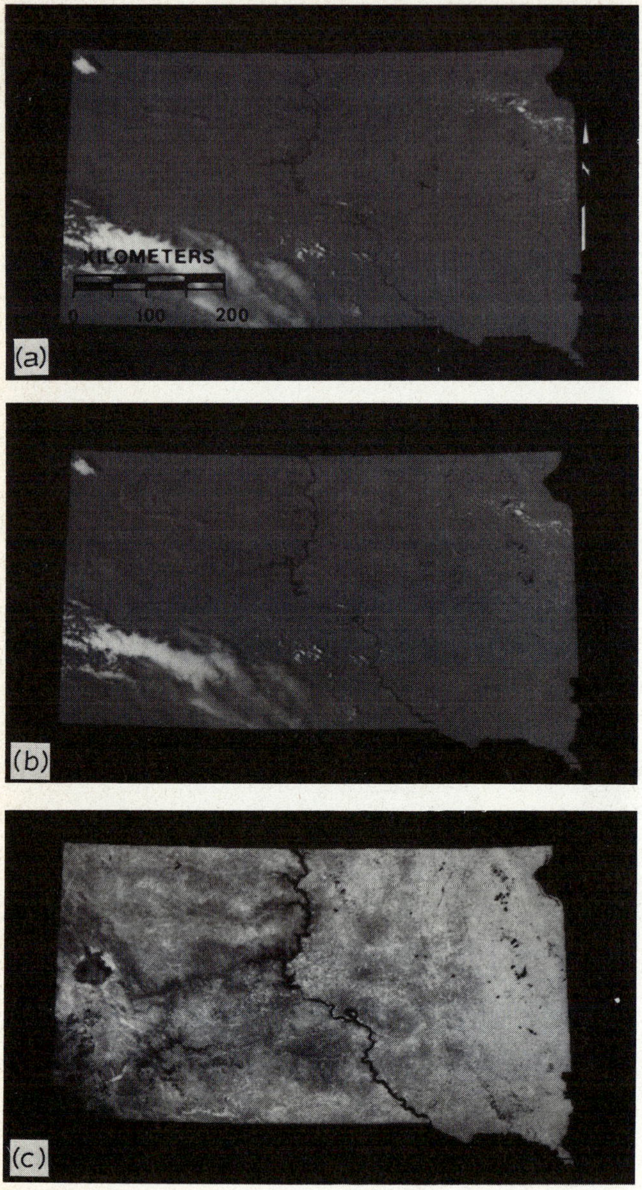
KILOMETERS
0
100
200
(a)
(b)
(c)

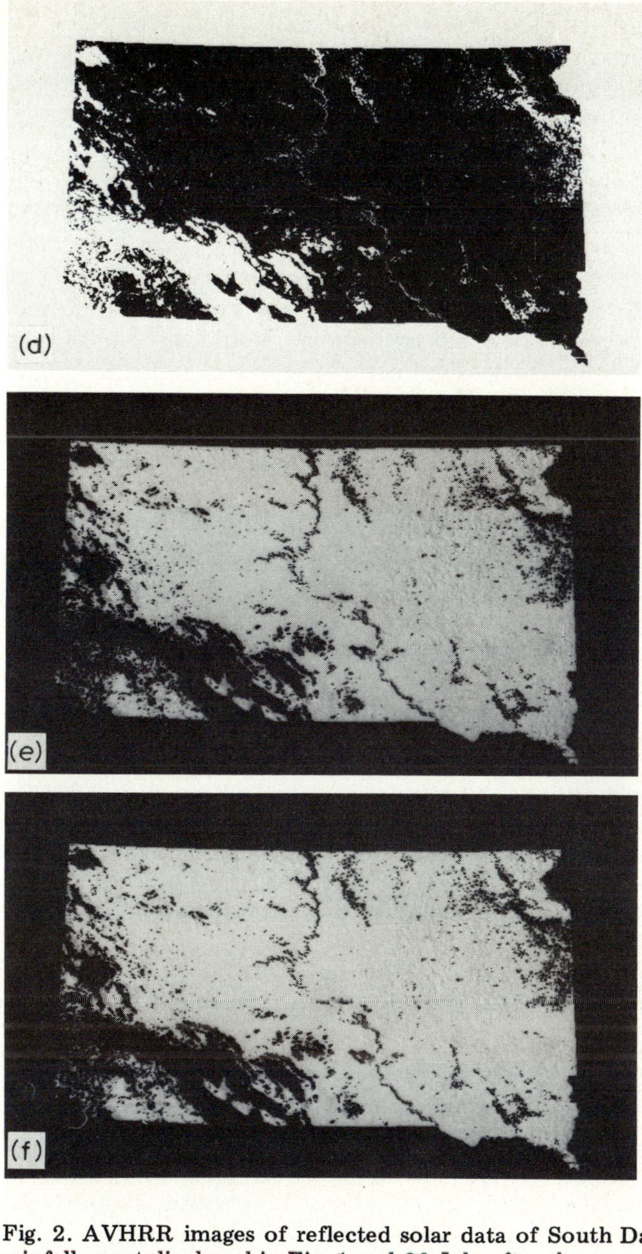

Fig. 2. AVHRR images of reflected solar data of South Dakota. 21 July was prior to the rainfall event displayed in Fig. 1 and 22 July after the event. The cloud mask in (d) shows cloud locations as light tones. Clouds are masked as dark in (e) and (f). (a) 21 July (0.55—0.09 μm). (b) 21 July (0.725—1.10 μm). (c) 22 July: Difference image (b) - (a). (d) Cloud mask. (e) Difference image with mask, 22 July minus 21 July (0.55—0.90 μm). (f) Difference image with mask, 22 July minus 21 July (0.725—1.10 μm).

RESULTS AND DISCUSSION

South Dakota. Rainfall occurred at 243 of the 1042 gauge sites in South Dakota during the 24 h period prior to the 22 July satellite overpass. The maximum recording was 8.9 cm. To establish the spatial distribution of rainfall, the data were displayed with the state and county boundaries (Fig. 1). The light squares represent higher levels of precipitation (>4.4 cm).

Figs. 2 and 3 include AVHRR data of South Dakota before and after the rainfall. Spectral variations in the 0.55—0.90 μm image (Fig. 2a) do not respond to changing land cover patterns; whereas, variations in the 0.725—1.10 μm image (Fig. 2b) have an increased response to the general cropping pattern. It is expected that the 0.725—1.10 μm region would be sensitive to variations in land cover since the reflective infrared spectral response generally increases as vegetation increases. The 0.55—0.90 μm spectral band includes both the chlorophyll absorption spectral region where vegetation reflectance is low and the infrared spectral region where vegetation reflectance is high. Pixel by pixel subtraction of the spectral values of the 0.55—0.90 μm band (Fig. 2a) from the 0.725—1.1 μm band (Fig. 2b) created the difference image in Fig. 2c. The effects of spectral overlap (0.725—0.90 μm) are reduced, and the resulting image reveals more of the vegetation pattern detail. Eastern South Dakota generally appears lighter on the image than does the western part of the state, which is predominantly rangeland. Note the close correlation between the cool patterns (light tones) in the thermal infrared image of Fig. 3a and the vegetation patterns (light tones) in the difference image of Fig. 2c.

The pre-rainfall image was subtracted from the post-rainfall image as presented in Figs. 2e,f and 3c. This method assumes that spectral change associated with an event will be greater than a spectral change associated with the other temporal variables (changes in vegetation, sun angle, etc.). This would especially be true for the reflected solar regions but would not be as consistent for the thermal region. Regional variations in air temperature, humidity, wind speed and direction, or other factors affecting the energy budget could have a significant influence on the thermal signal over this large a geographic area.

The difference images in Figs. 2e,f and 3c were prepared using both cloud and geographic masks. The masked regions are dark tones, areas of no change are medium tones; and a reflectance increase or cooler temperature in the second date as compared to the first date is light tones. The cloud mask in Fig. 2d was created by multispectral classification from both dates. Note that the cloud mask obscures regions that are cloud covered on either one of the dates. This could be avoided by using a cloud free image acquired reasonably close to the date to be analyzed, subtracting the 0.55—0.90 μm region of one date from another, and creating a cloud mask which included clouded areas only from the image being analyzed. Since the 0.55—0.90 μm channel is relatively invariant with land cover changes, an image separated by many days from the date in question would probably not vary significantly.

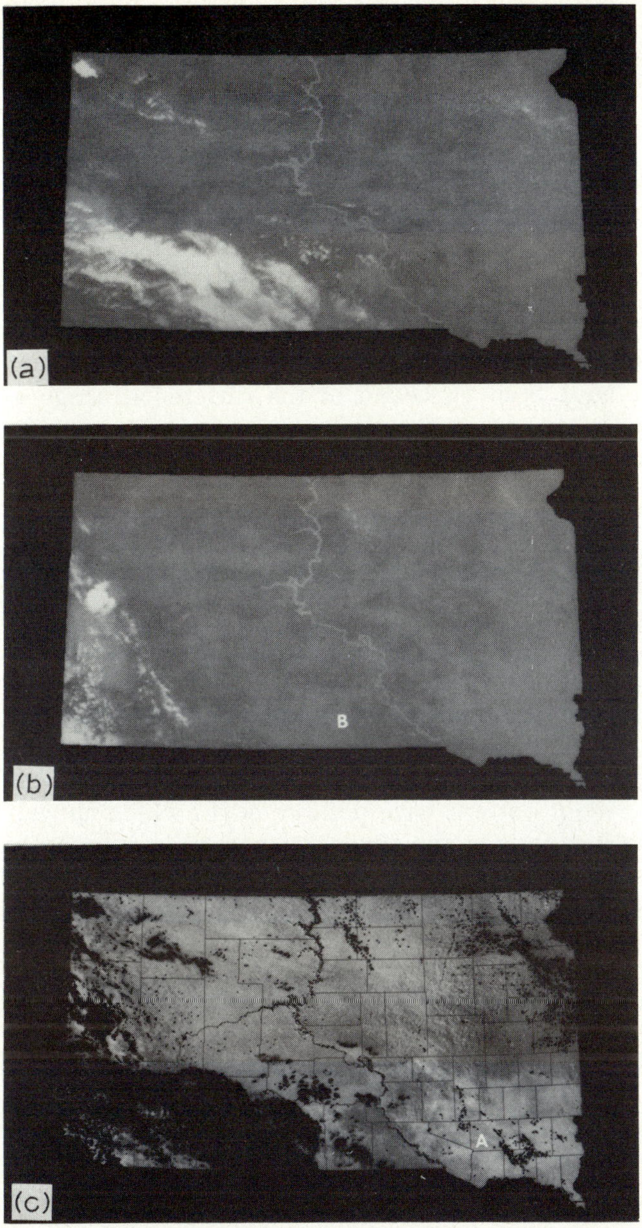

Fig. 3. AVHRR images of thermal infrared (11.5—12.5 μm) data of South Dakota. Points A in (c) and B in (b) denote rainfall locations as defined in Fig. 1. Light tones are cool. The difference image in (c) contains county boundaries for comparison with Fig. 1. (a) 21 July. (b) 22 July. (c) Difference image 22 July minus 21 July with cloud mask as dark.

Figs. 2e and 2f indicate that the rainfall event documented in Fig. 1 did not create a spectral change in the reflective portions of the spectrum. However, the thermal change in Fig. 3c followed the distribution of rainfall in Fig. 1. Latent energy of evaporation was a large factor in the energy budget, although changes in soil thermal properties and transpiration could have also contributed to the observed cooling. The statistical correlation between the amount of rainfall for the 202 gauges receiving rainfall which were not cloud covered on either date and the temperature difference between the two dates (Fig. 3c) was r=0.163* (P>0.05). This low coefficient of determination (r^2=0.02) indicated that the temperature change was not associated with changes in thermal inertia or transpiration since both would vary as the quantity of absorbed moisture varied.

The thermal patterns in Figs. 3a and 3b which were associated with vegetation patterns in Fig. 2c, have been cancelled in the difference image of Fig. 3c. Using time-series data to spectrally observe the same location for event detection decreases the complexity of interpretation by eliminating the need to model all the thermal variations for a single date to determine their actual physical causes. The only knowledge required to interpret the difference image is that wet surfaces are cooler than dry surfaces in midday. However, thermal data can be misleading because of any factor that changes the energy budget can cause differences in apparent temperatures. Note the subtle cool pattern of the north central region which appears lighter in Fig. 3c. This could be associated with regional air temperature or humidity variations. However, this pattern does not have the magnitude of temperature change that the rainfall pattern at A has, nor does it have the shape of a typical rainfall pattern. Location B in Fig. 3b also correlates with a rainfall occurrence (Fig. 1). It is a cool pattern with definite edges which appears light in this single date image. However, the cloud cover from 21 July masks the region on the difference images.

Washita Watershed, OK. On 21 July 1978, a small storm passed over the central and northeast parts of the watershed (Fig. 4). The rainfall pattern was approximately 18 km wide. Precipitation amounts across the watershed ranged from 0.0 to 7.5 cm. The soil was dry before the storm. The storm passed over areas of intermixed sorghum and pasture, and all of the soil types represented in the watershed.

Fig. 5 shows a 24 July HCMM apparent temperature map of the watershed. The dark areas in the south and northwest correspond to clouds (locations C and D). The dark band extending southwest to northeast across the center of the watershed had the same general shape and width as the rainfall contour map for the 21 July storm (Fig. 4). No clouds were apparent within this region. Because 3 days elapsed between the rainfall and thermal image, the vegetation and soil surfaces were visually dry. Any thermal indicator would require that a change in thermal inertia and/or transpiration be large enough to be detected and spatially correlated with the rainfall distribution.

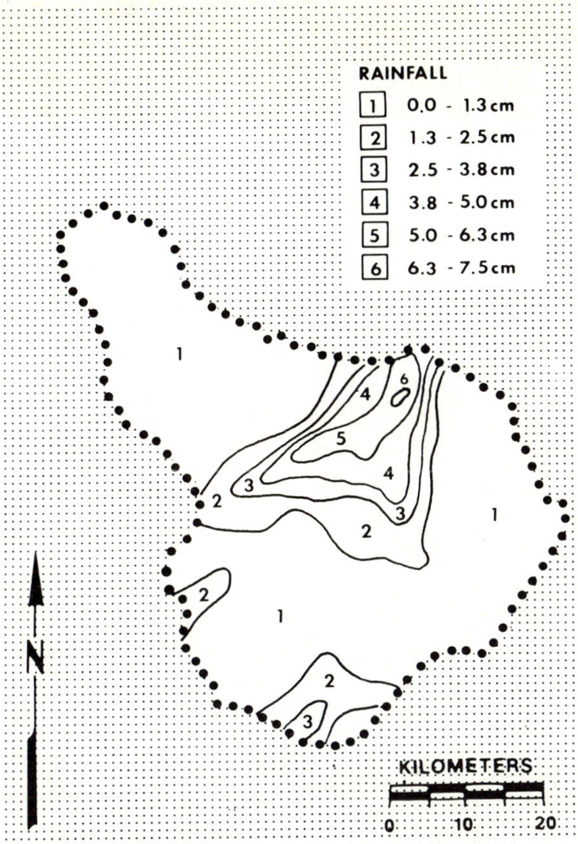

Fig. 4. Rainfall contour map for the 21 July, 1978, event for the Washita River Watershed (after Rosenthal et al. 1982). Chickasha, OK is located approximately in the center of the watershed.

The Antecedent Precipitation Index (API) is a method proposed by Kohler and Linsley (1951) for use as a soil moisture depletion model:

$$API_i = P_i + (API_{i-1} * K_i) \qquad (1)$$

where API_i is the antecedent precipitation index on a given day i, K is the depletion factor which is a function of time, soil depth, and the inverse of mean daily temperature, and P_i is the effective precipitation or infiltration. Effective precipitation was evaluated for the Washita River Watershed area of Oklahoma as being $P_i = PP^{0.891}$ where PP is daily precipitation (cm) and K for the 0 to 15 cm surface soil layer was determined to range from 0.84 in July to 0.99 in December (B.J. Blanchard, NASA Goddard Space Flight Center, Greenbelt, MD, personal communication, 1978). Blanchard et al. (1981) related API to surface soil moisture content for their central Oklahoma study area using these coefficients.

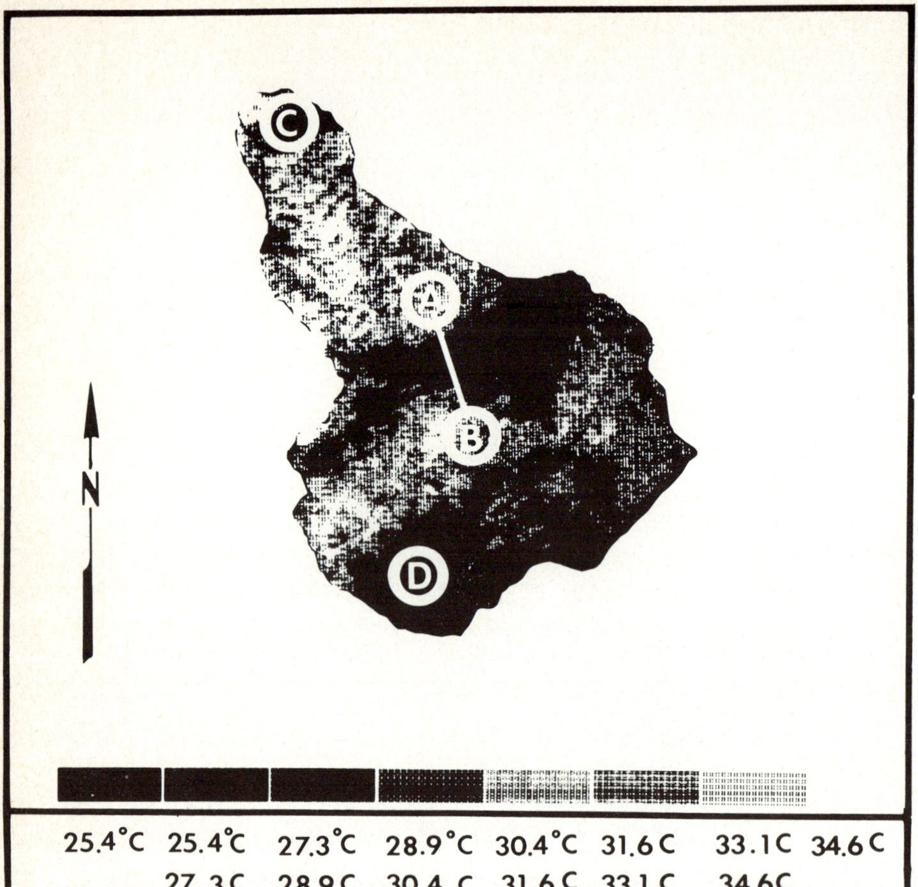

Fig. 5. Thermal infrared apparent temperature map from the HCMM satellite scene at 14.00 LST on 24 July 1978. Equivalent blackbody temperatures from colder than 25.4°C (black) to warmer than 34.6°C in 1.5°C increments. Data for the transect from A to B are plotted in Fig. 6. Locations C and D are clouds (after Rosenthal et al. 1982).

A transect between A and B in Fig. 5 was used to further illustrate in Fig. 6 the relationship between the thermal data and API. Cooler areas corresponded with heavier precipitation, despite the variations in land use, terrain, and soils along the transect. Apparent temperature differences between 24 July and 13 July, and 24 July API's for all non-cloud covered rain gauges in the watershed were highly correlated (R^2=0.58**, P<0.01). The relationship appeared to be independent of soil type for the transect observations (Fig. 7a).

The relationship between apparent temperature differences from 29 July to 13 July and 29 July API had a lower coefficient of determination but was statistically highly significant (R^2=0.25**, P<0.01). The selected transect in

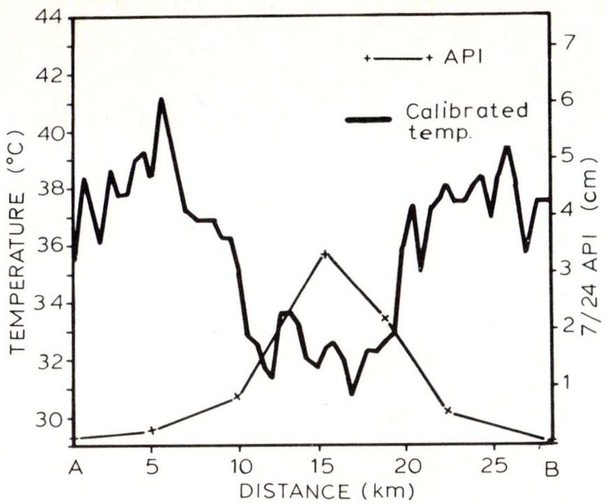

Fig. 6. 24 July 1978, radiometric temperatures and antecedent precipitation index (API) variations plotted along transect A to B in Fig. 5 (after Rosenthal et al. 1982).

Fig. 7b, shows a trend that the temperature response was dependent on soil type. Sandy soils had a smaller thermal response to moisture differences than did loamy soils. The API values above 1 cm tended to follow the same regression line shown in Fig. 7a for 24 July.

CONCLUSIONS

The use of frequent multidate thermal infrared remote sensing can improve both the spatial resolution of data and the probability of rainfall event detection as compared to using only the existing NOAA network of rain gauges. Surface wetting by rainfall, in an otherwise dry region, changes the energy budget which is thermally detectable with existing remote sensing systems at least one day after the event. After the surface dries, thermal inertia and evapotranspiration remain sufficiently altered so that the increased moisture is detectable for an additional several days without interference of land use, terrain, and soil factors. Eight days after a rainfall up to 6.5 cm, the relationship was not as clearly defined and the effects of soil patterns on the thermal signal complicated the interpretation. Therefore, a satellite with daily midday repeat coverage in regions with midday clear sky conditions at least once every 4 or 5 days would provide an effective system for rainfall monitoring. In regions where cloud cover would not permit that frequent a number of successful observations, soil moisture is normally not as limiting a factor in plant growth.

Difference images from multidate satellite data reduce spectral variability resulting from landscape and vegetation patterns since these change little over a few days. Regional air temperature, humidity, and/or wind patterns must be considered when interpreting thermal data. Therefore, normalization procedures such as air temperature subtraction may be required.

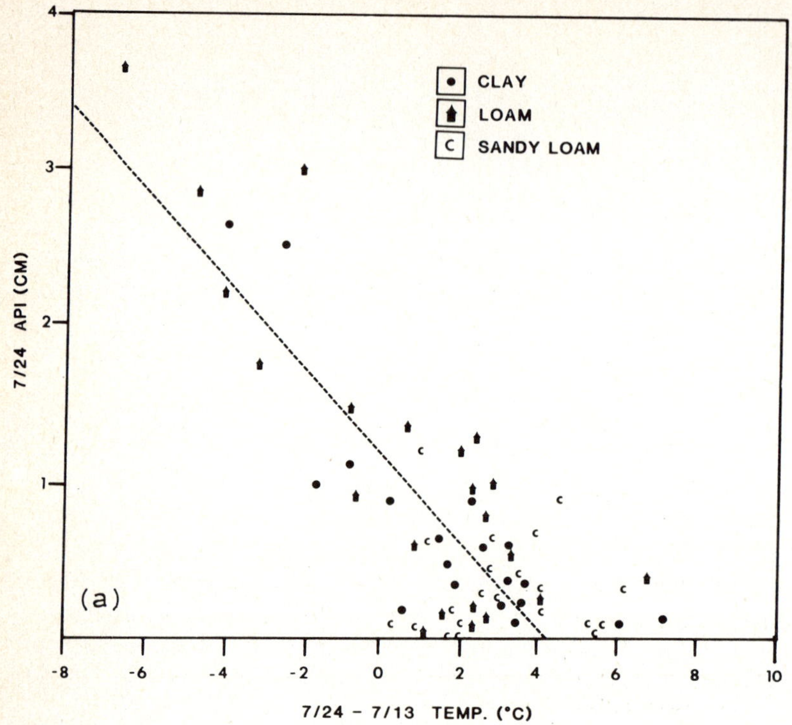

The 1 km resolution of present polar orbiting satellite systems is sufficient for regional applications. Data from NOAA 7 are currently available and could be processed and analyzed at a reasonable cost for regions as large or larger than the Great Plains. Additional information, such as green biomass estimates, could also be made using the reflected solar spectral intervals of this satellite.

The time-series measurements should be more fully evaluated as an operational tool for rainfall event detection under varying rainfall amounts, soil and terrain differences, antecedent soil moisture, and differences in land cover. The approach may also be an appropriate tool for obtaining other synoptic resource information available through multispectral analysis of the low-resolution data.

ACKNOWLEDGEMENTS

The work was performed under the U.S. Geological Survey Contract Number 14-08-0001-20129 and National Oceanic and Atmospheric Administration Grant NA 80 AA-D-00032.

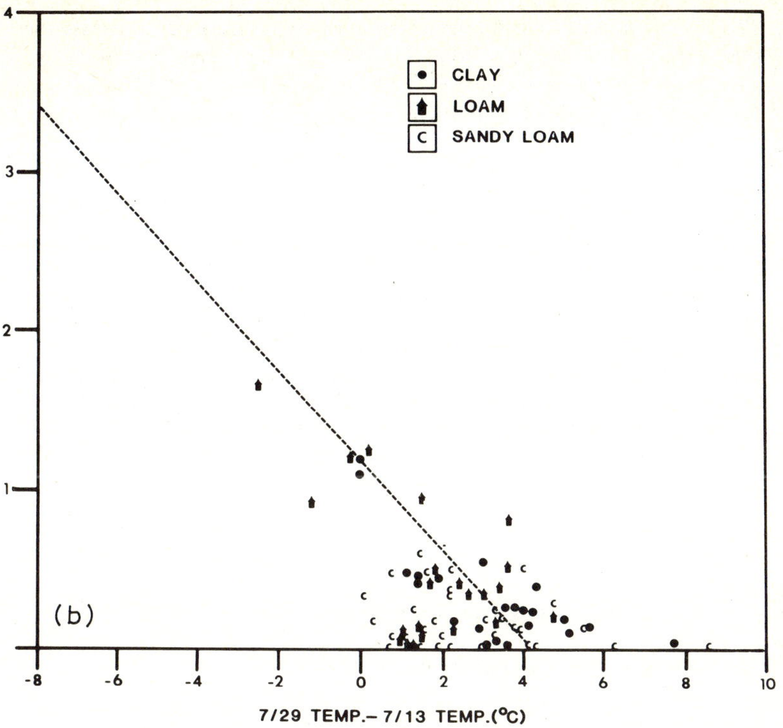

Fig. 7. Plots of antecedent precipitation indices versus normalized surface temperature differences at rain gauge sites. The dashed line in (a) is a best fit line. The same line is repeated in (b) which is not a statistical best fit line through the data values in (b) (after Rosenthal, et al. 1982). (a) 24 July. (b) 29 July.

REFERENCES

Blad, B.L. and Rosenberg, N.J., 1976. Measurement of crop temperatures by leaf thermocouple, infrared thermometry and remotely sensed thermal imagery. Agron. J., 68: 635–641.

Blanchard, B.J., McFarland, M.J., Schmugge, T.J. and Rhoades, E., 1981. Estimation of soil moisture with API algorithms and microwave emission. Water Resour. Bull., 17: 767–774.

Heilman, J.L. and Moore, D.G., 1980. Thermography for estimating near-surface soil moisture under developing crop canopies. J. Appl. Meteorol., 19(3): 324–328.

Heilman, J.L., Myers, V.I., Moore, D.G., Schmugge, T.J. and Friedman, D.B., 1978. Proc. Soil Moisture Workshop. NASA Conference Publication 2073, 17–19 January 1978, Scientific and Technical Info. Office, NASA, Washington, DC, 220 pp.

Heilman, J.L., Heilman, W.E. and Moore, D.G., 1981. Remote sensing of canopy temperature at incomplete cover. Agron. J., 73: 403–406.

Idso, S.B., Jackson, R.D., Reginato, R.J., Kimball, B.A. and Nakayama, F.S., 1975a. The dependence of bare soil albedo on soil water content. J. Appl. Meteorol., 14: 109–113.

Idso, S.B., Schmugge, T.J., Jackson, R.D. and Reginato, R.J., 1975b. The utility of surface temperature measurements for the remote sensing of surface soil water status. J. Geophys. Res., 80: 3044–3049.

Sensors that operate in the microwave region of the electromagnetic spectrum can be active or passive devices. Active devices (radar) transmit energy to the scene and measure the energy that is scattered back from the scene and returned to the receiving antenna. Passive devices (radiometers) measure microwave radiation that is naturally emitted from the scene. Both active and passive devices can be used to measure near-surface soil moisture. In this paper, we will discuss only passive microwave measurement.

The intensity of emission at microwave wavelengths is proportional to the product of the temperature and emissivity of the surface (Rayleigh-Jeans approximation). This product is referred to as the brightness temperature. Figure 1 shows that, regardless of polarization or view angle, there is a wide response of the brightness temperature as the soil moisture content changes (Newton, 1977, Newton and Rouse, 1980).

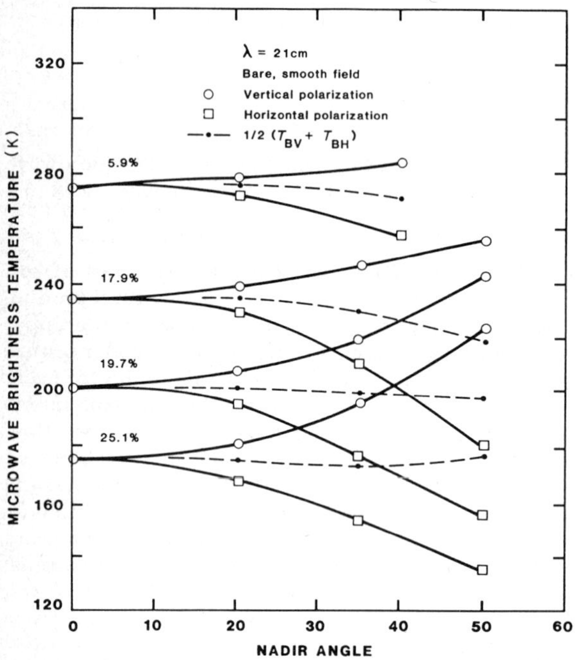

Fig. 1. Microwave brightness temperatures (21 cm wavelength) as a function of incident angle and gravimetric soil moisture content. Measurements were made for a smooth, bare soil. T_{BV} and T_{BH} are brightness temperature at vertical and horizontal polarizations, respectively.

Brightness temperature also depends upon the surface roughness (Newton, 1977), as shown in Fig. 2, and, in addition, at a given moisture content, is affected by soil texture. However, Schmugge (1980) suggested that this last effect could be minimized by normalizing the soil moisture to the field capacity. Newton (1977) proposed that brightness temperature was related to the soil water potential, independently of soil texture. Since the depen-

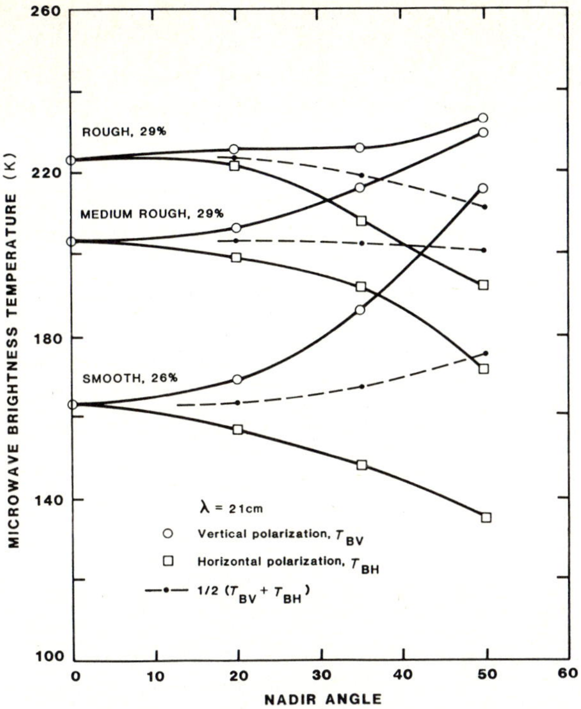

Fig. 2. Microwave brightness temperatures (21 cm wavelength) as a function of incident angle for three bare fields of the same soil moisture content but different surface roughness. T_{BV} and T_{BH} are brightness temperatures at vertical and horizontal polarizations, respectively.

dence on the physical temperature can be minimized by normalizing brightness temperature to the soil temperature as shown in Fig. 3 (Newton, 1977), a more general relation could be developed between soil moisture availability and brightness temperature as measured by a passive microwave sensor, provided the surface temperature and the relation between water content and water potential can be measured or estimated.

A final complication may arise from the fact that vegetation acts as an absorbing layer, decreasing the measured brightness temperature. This effect is illustrated in Fig. 4 which summarizes results of Basharinov and Shutko (1978). In Fig. 4, the vegetation factor is defined as one minus the effective transmissivity of the vegetation. The effects of surface roughness and of vegetation factors will decrease as the wavelength increases.

A more general and significant problem in the way of practical application is that the microwave measurement directly reflects only the soil moisture condition in the upper few centimeters of the soil profile (Wilheit, 1978; Black and Newton, 1981). The microwave emissivity of soil is determined by the dielectric contrast at the air-soil interface. The magnitude of this contrast is determined by the soil moisture in a layer down to a depth equal to a few

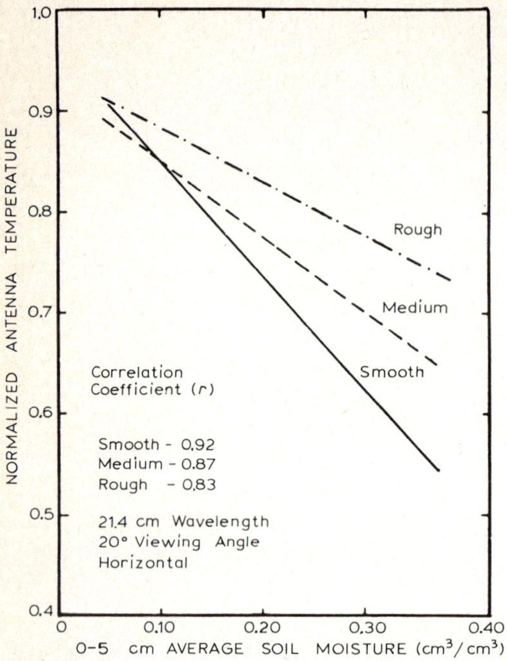

Fig. 3. Brightness temperatures normalized by the physical soil temperature as a function of the 0—5 cm average volumetric soil moisture content for three surface roughness conditions.

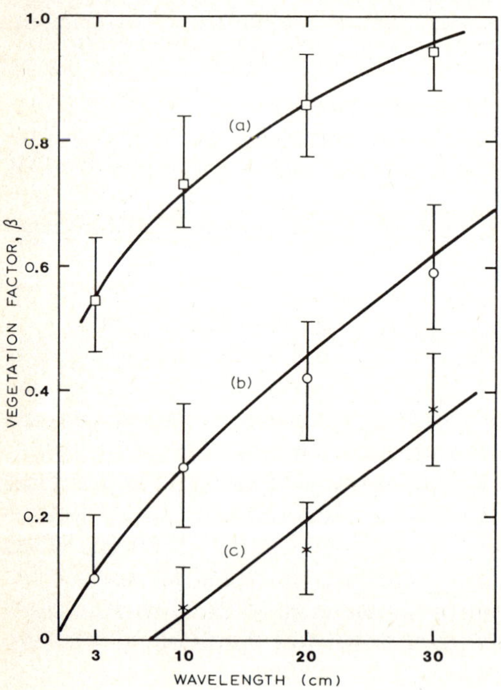

tenths of the microwave wavelength. At a wavelength of 21 cm, this depth is on the order of 2 — 5 cm, and is termed the reflectivity sampling depth. While the emissivity is controlled by this relatively shallow layer, the total microwave energy generated by the entire soil volume is influenced also by the physical soil temperature to a depth of the order of a wavelength, as shown by Newton et al. (1982). This depth is termed the thermal sampling depth.

A significant correlation between brightness temperature and near-surface soil moisture has been established for a wide range of conditions (Schmugge, 1980). However, for the microwave measurement to be useful in agricultural applications, the depth sensitivity of the measurement must be further investigated, and in addition, techniques must be developed for using near-surface measurements in estimating root zone soil moisture.

We report here experimental and theoretical investigations, conducted to verify the depth sensitivity of the microwave measurement and to assess some of the problems associated with using near-surface measurements to estimate root zone soil moisture.

EXPERIMENTAL PROCEDURES

During the summer of 1980 a field experiment was conducted at Texas A&M University, on four bare fields, ranging in surface roughness from smooth to very rough (Makanvand and Newton, 1982, Newton et al., 1982). During the experiment, microwave brightness temperatures were measured frequently at wavelengths of 21 cm, 6 cm, and 2.8 cm over an 18-day period. Soil moisture profiles were measured gravimetrically at ten locations within each field, with a gamma ray attenuation technique at four locations within each field, and with a neutron probe at six locations within each field. Soil temperature profiles were measured with thermocouples at two locations within each field, and surface soil temperature was measured using a thermal radiometer, boresighted with the microwave antennas. Surface roughness was measured manually several times during the experiment, at two locations within each field.

On the 4th day of the observation period the plots were irrigated for 18 hours using a sprinkler system that supplied 19 mm of water per hour. A complete set of measurements was acquired twice prior to irrigation, and then at irregular intervals depending on the drying rate, until the soil profile had dried to a condition in which there was relatively minor additional change.

The microwave measurements made before the irrigation (dry soil) and immediately after irrigation (saturated soil) were correlated to the gravimetric

---

Fig. 4. Dependence of the vegetation factor, defined as one minus the effective transmissivity of the vegetation, upon wavelength for three types of vegetation: (a) small grains; (b) broad leaf cultures; and (c) mixed forest. From Basharinov and Shutko (1978) after Schmugge et al. (1980).

soil moisture measurement in the top two centimeters to establish a linear regression between brightness temperature and soil moisture. This relationship was then used to convert all subsequent brightness temperatures to soil moisture content.

Meteorological measurements, required as inputs for a soil moisture — heat flow model (Van Bavel and Hillel, 1975; Lascano and Van Bavel, 1979) were made at 30-min intervals throughout the experiment. A continuous simulation of the soil moisture and soil temperature profile, using the data, was made by the method of Lascano and Van Bavel (1982). This simulation produced soil moisture and soil temperature profiles every hour for the entire observation period of 18 days to a depth of 1.8 m. The simulated moisture and temperature profiles were used as inputs to a microwave radiation model (Black and Newton, 1981) to calculate brightness temperatures for the smooth field.

RESULTS

Fig. 5 shows a plot of the gravimetrically measured soil moisture content for three surface layers of increasing thicknesses. By comparing the rate at which the microwave brightness temperature changes as a function of time with the data in Fig. 5, it can be decided empirically which surface layer corresponds most closely to the microwave measurements. Fig. 6 shows estimates of near-surface soil moisture made from the actual brightness temperature measurements at three wavelengths and using the calibration for the upper 2 cm, as explained earlier. The measurements were made at approximately the same time of day to minimize diurnal effects. A comparison of

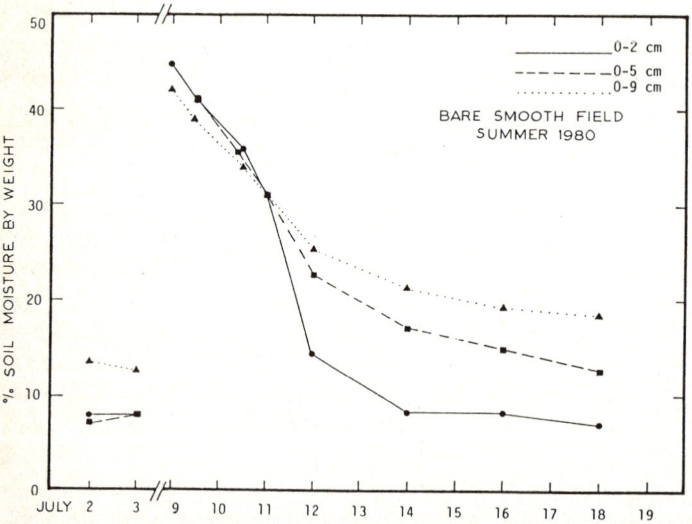

Fig. 5. Gravimetrically-measured soil moisture content, for three surface layer thicknesses, as a function of date, for a smooth bare soil.

Figs. 5 and 6 suggest that, at the longest wavelength (21 cm), the microwave data pattern comes close to the recorded pattern for the top 2 cm. This agreement is further demonstrated by the regression of the microwave-based soil moisture estimates on actual measured values as shown in Fig. 7.

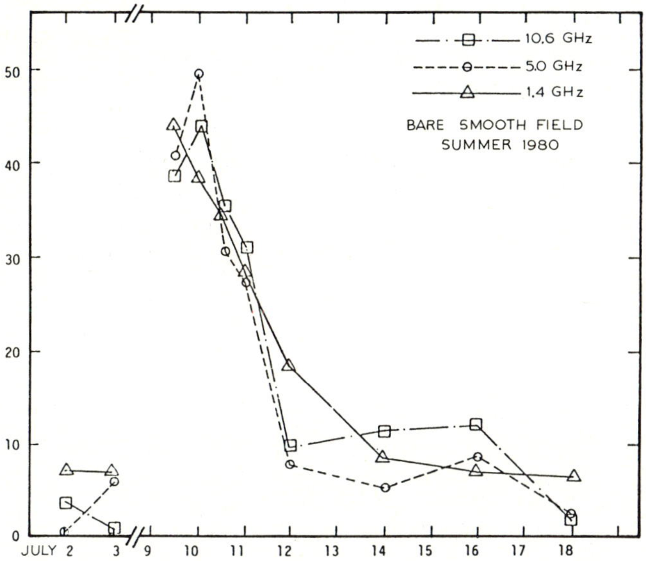

Fig. 6. Near surface soil moisture content estimates from brightness temperature measurements at three frequencies, as a function of date, for a smooth, bare soil.

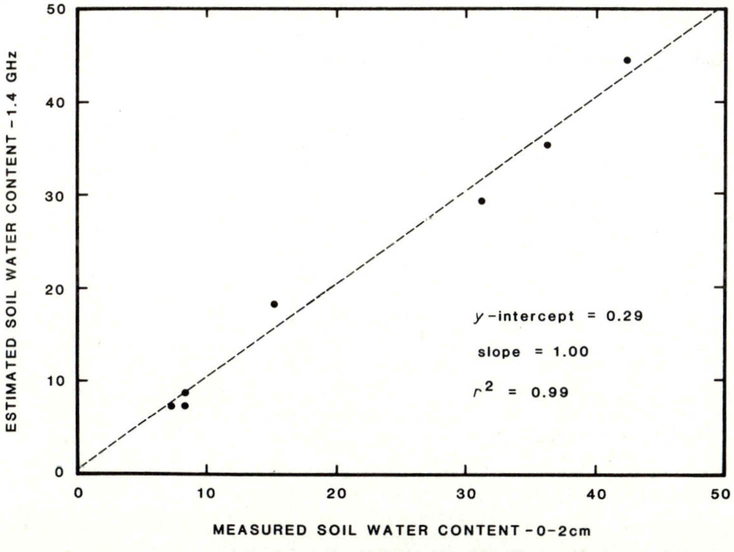

Fig. 7. Microwave estimates (1.4 GHz) of soil water content compared with measured gravimetric soil water content in the 0—2 cm layer.

However, the events in the top 5 and top 7 cm are not reflected in the microwave-based estimates. For the shorter wavelengths (6.0 and 2.8 cm) the agreement is even less close. This preliminary result is consistent with theoretical results previously discussed by Wilheit (1978) and Black and Newton (1981).

Since the brightness temperature is a function of both soil moisture and soil temperature, diurnal variations of moisture and temperature can have a significant impact on the microwave measurements. Figs. 8 and 9 show, respectively, the simulated soil temperature and soil moisture values, calcu-

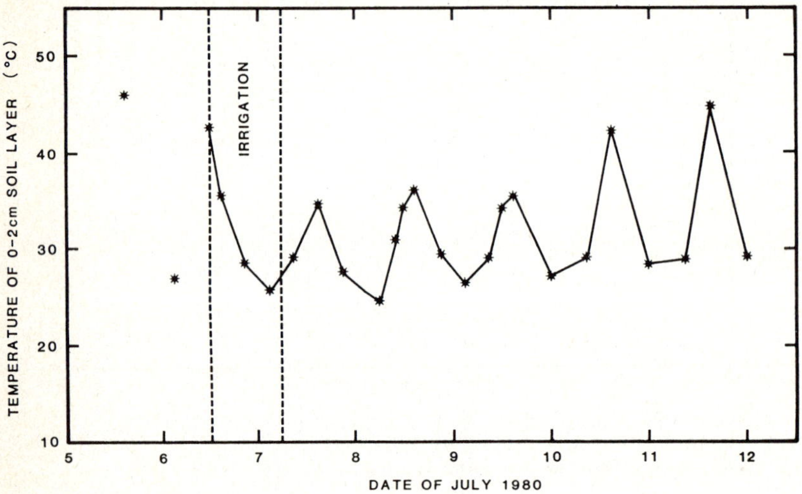

Fig. 8. Simulated soil temperature for the upper 2 cm of a smooth, bare soil.

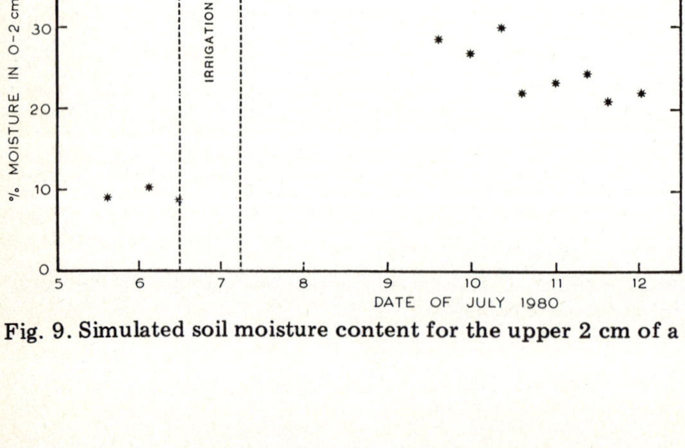

Fig. 9. Simulated soil moisture content for the upper 2 cm of a smooth bare soil.

lated with the model, for the upper 2 cm of a smooth, bare soil. Diurnal variations of soil temperature were on the order of 10°C for wet conditions, and 15°C for dry conditions. Diurnal soil moisture variations were also present, but were less symmetric due to the fact that the average moisture in the layer is changing rapidly over the period of time presented.

Fig. 10 shows that diurnal variations of the calculated brightness temperature exhibit the same periodicity as the diurnal temperature variations. However, the magnitude of the diurnal brightness temperature variation is moderated by changes in soil moisture.

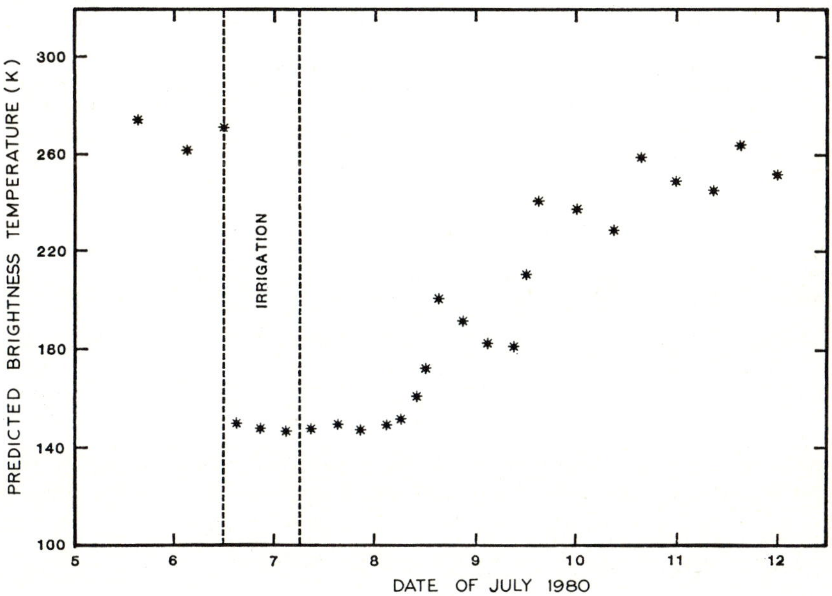

Fig. 10. Simulated 21 cm brightness temperature for a smooth, bare soil.

Converting the brightness temperature to an effective emissivity by normalizing the brightness temperature to the measured or estimated soil temperature minimizes the impact of the variation in soil temperature. The calculated effective emissivity, shown in Fig. 11, parallels the soil moisture values shown in Fig. 9. This is further demonstrated by Fig. 12 which compares calculated effective emissivity with calculated soil moisture. These results suggest that, because of the diurnal variations, it may be necessary to obtain microwave measurements at the same time for each measurement date.

The use of near-surface soil moisture to estimate root zone soil moisture is difficult and remains to be demonstrated. Jackson (1980) adapted a soil moisture profile model proposed by Hillel et al. (1975) to relate estimates of surface layer moisture to the average moisture to a depth of one meter. From these results, it would appear that a general method for interpreting the surface soil moisture estimate can be made only by using a dynamic rather than a static approach, and by utilizing generally available antecedent hydrological and meteorological data, as well as a knowledge of the nature of the soil profile for which the interpretation is made. Work in this direction is in progress and deals with vegetated soils as well as bare surfaces.

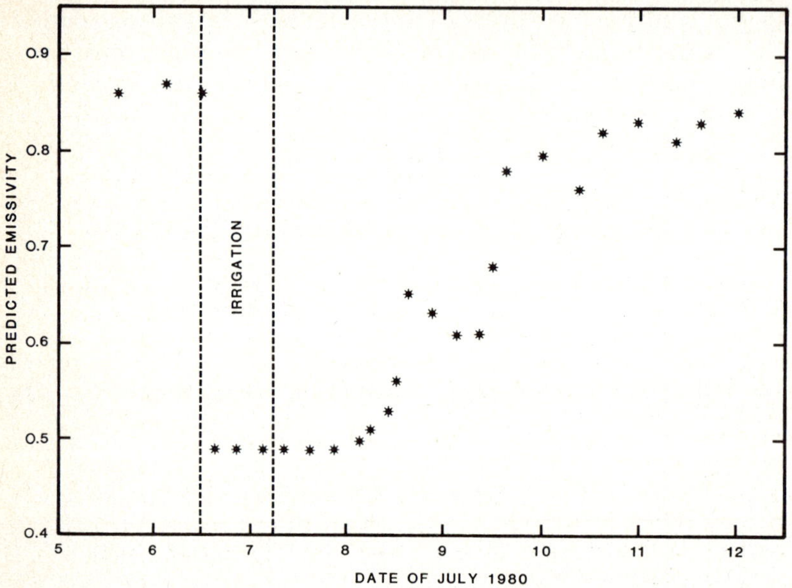

Fig. 11. Simulated effective emissivity for a smooth, bare soil. Effective emissivity was computed by normalizing the simulated brightness temperature by the 0—2 soil temperature.

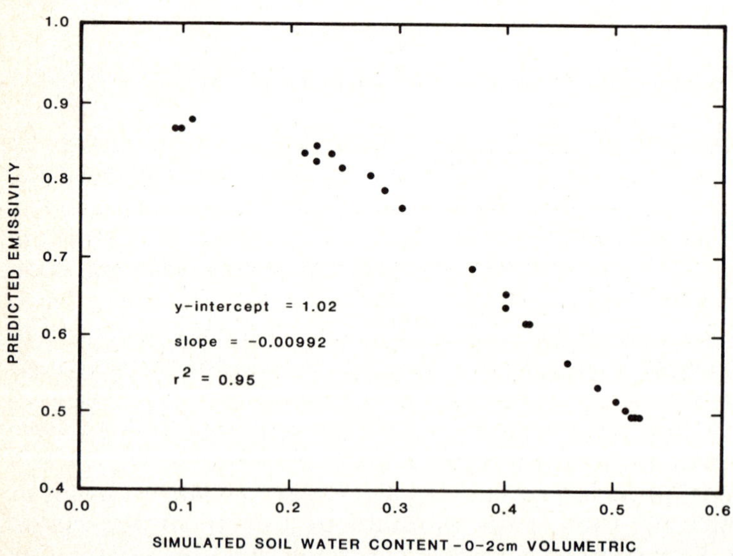

Fig. 12. Predicted microwave emissivity compared with simulated volumetric soil water content in the 0—2 cm layer.

ACKNOWLEDGMENTS

The authors would like to thank R.J. Lascano and M.R. Smith, Departments of Soil and Crop Sciences and Remote Sensing Center, respectively, Texas A&M University, for their assistance in field measurements and the simulation studies.

REFERENCES

Basharinov, A. and Shutko, H.M., 1978. Determination of the moisture content of the earth's cover by superhigh frequency (microwave) radiometric methods: a review. Radiotek Electron., 23: 1778—1791. (English Translation, Radio Eng. Electron. Phys., 23: 1—12.)

Black, Q.R. and Newton, R.W., 1981. Airborne microwave remote sensing of soil moisture. AgRISTARS Rep. SM-T1-40457, Tech. Rep. RSC-108, Remote Sensing Center, Texas A&M University, College Station, TX, 171 pp.

Hillel, D., Van Bavel, C.H.M. and Talpaz, H., 1975. Dynamic simulation of water storage in fallow soil as affected by mulch of hydrophobic aggregates. Soil Sci. Soc. Am. Proc., 39: 826—833.

Jackson, T.J., 1980. Profile soil moisture from surface measurements. J. Irr. Drain. Div. ASCE, 106 (IR2): 81—92.

Lascano, R.J. and Van Bavel, C.H.M., 1979. CONSERVB: a numerical method to compute soil water content and temperature profiles under a bare surface. Progr. Rep. RSC-3058-1. Remote Sensing Center, Texas A&M University, College Station, TX, 75 pp.

Lascano, R.J. and Van Bavel, C.H.M., 1982. Experimental verification of a model to predict soil moisture and temperature profiles. Soil Sci. Soc. Am. J. (In press).

Makanvand, S., and Newton, R.W., 1982. Ground truth report: 1980 Texas A&M University soil moisture experiment. Tech. Rep. RSC-122, Remote Sensing Center, Texas A&M University, College Station, TX, 200 pp.

Newton, R.W., 1977. Microwave remote sensing and its application to soil moisture detection. Tech. Rep. RSC-81. Remote Sensing Center, Texas A&M University, College Station, TX, 500 pp.

Newton, R.W. and Rouse, J.W., Jr., 1980. Microwave radiometer measurements of soil moisture content. IEEE Trans. Antennas Propag., AP-28: 680—686.

Newton, R.W., Black, Q.R., Makanvand, S., Blanchard, A.J. and Jean, B.R., 1982. Soil moisture information and thermal microwave emission. IEEE Trans. Geosci. Remote Sens. (special issue IGARSS), GE-20(3): 275—281.

Schmugge, T.J., 1980. Effect of soil texture on the microwave emission from soils. Tech. Mem. 80632, NASA Goddard Space Flight Center, Greenbelt, MD.

Van Bavel, C.H.M. and Hillel, D.I., 1975. A simulation study of soil heat and moisture dynamics as affected by a dry mulch. In: Proc. 1975 Summer Computer Simulation Conference, San Francisco, CA. Simulation Councils, Inc., La Jolla, CA, pp. 815—821.

Wilheit, T.T., 1978. Radiative transfer in a plane stratified dielectric. IEEE Trans. Geosci. Elect., GE-16: 138—143.